The Mathcad 2001i H

The Mathcad 2001i Handbook

D. Kiryanov

CHARLES RIVER MEDIA, INC.
Hingham, Massachusetts

Acquisitions Editor: Brian Sawyer
Production: A-List Publishing
Cover Design: The Printed Image

CHARLES RIVER MEDIA, INC.
20 Downer Avenue, Suite 3
Hingham, Massachusetts 02043
781-740-0400
781-740-8816 (FAX)
info@charlesriver.com
www.charlesriver.com

This book is printed on acid-free paper.

D. Kiryanov. *The Mathcad 2001i Handbook.*
ISBN: 1-58450-265-7

Printed in the United States of America
02 7 6 5 4 3 2 First Edition

CHARLES RIVER MEDIA titles are available for site license or bulk purchase by institutions, user groups, corporations, etc. For additional information, please contact the Special Sales Department at 781-740-0400.

Table of Contents

Introduction

The subject of this book is the popular program Mathcad 2001i, by MathSoft. Mathcad 2001i helps solve various mathematical tasks and make high-level professional calculating reports.

While writing this book, I have tried to keep two goals in mind. First, I wanted to cover basic calculating and outline the user interface. I also wanted to do all of this while using a step-by-step method, going from simple to complicated, to give readers the chance to study Mathcad independently. Even if you have never solved mathematical tasks with this power computing system, you can learn its key features and tools from the information provided in this book. Secondly, I know from personal experience that a problem-oriented method can, at times, be very useful. So, if there is a particular task that you need solved, you can just go to the appropriate chapter and use this book as a reference to guide you through it. Though a short outline of math concepts and terms is given at the beginning of each section that deals with the solution to a certain problem, I also assume that readers have the basic mathematical knowledge.

Part I presents the basics of Mathcad and describes how to work in its math editor. Parts II and III look at some issues relating to practical mathematical tasks and provide examples with various listings. Part IV includes information on how to create professional reports in Mathcad and how to develop effective methods of working for advanced users.

All listings are autonomous and work without any additional modules or plug-ins. Listings just contain calculations without comments (in order to make them more readable), and all explanations can be found in the text. Almost all graphics that are the result of the given listings are presented. Sections designated by an asterisk contain information on special numerical algorithms, useful tips, or my own algorithms. Beginners may skip these sections.

What is Mathcad? First of all, you should know that Mathcad includes several integrated components:

- A power text processor that lets you input, edit, and format a text or mathematical expression
- A calculating processor that makes calculations using input formulas and built-in calculating methods
- A symbolic processor, which is, in fact, an artificial intelligence system
- A huge data repository containing both engineering and mathematical information presented as an interactive e-book

General Information

1 ┊ Getting Started

In this chapter, we'll discuss the purpose of the Mathcad 2001i software and, in order to get acquainted with its basic functional capabilities, consider the basic techniques of its usage *(see Sections 1.1 and 1.2)*. Those of you who have previous experience with earlier releases of this software, starting with Mathcad 7, and have a good working knowledge of its editor, can skip this chapter. The main focus will be on the Mathcad 2001i user interface components, which are intuitive and rather similar to those of other Windows applications *(see Section 1.3)*, and the efficient usage of the Mathcad Help system *(see Section 1.4)*.

1.1. MATHCAD'S PURPOSE

Mathcad is a mathematical editor, enabling the user to perform various scientific and engineering calculations, from elementary arithmetic to complex implementations of numeric methods. Mathcad's intended audience includes students, scientists, engineers, and technical specialists. Thanks to the fact that it is easy to use, and to its visualization of mathematical operations, large library of built-in functions and numerical methods, as well as its admirable toolkit for representing the results (such as various types of graphs and powerful tools for preparing printed documents and Web pages), Mathcad has become one of the most popular mathematical applications in use today.

In contrast to most other modern mathematical software, Mathcad 2001i was developed according to the WYSIWYG principle (WYSIWYG stands for "What You See Is What You Get"). As a result, it is very simple to use, particularly on account of the fact that you don't need to first develop a program that implements the required mathematical calculations and then execute it. Instead, it is sufficient to simply type in mathematical expressions using the built-in formula editor in a format, which is very close to the commonly accepted one, and get the result immediately. In addition, it is possible to produce a printed copy of the document or to create a Web page representing the document as it looks on the PC screen when you work with Mathcad.

Developers of the Mathcad software have made all possible efforts to let the user without a programming background (such users make up a majority of scientists and engineers) take full advantage of modern computer science and technologies. Basic skills of PC usage are sufficient for efficient work with the Mathcad editor. On the other hand, professional programmers, such as the author of this book, can get much more from Mathcad's functionality by creating various programmatic solutions that significantly extend Mathcad's built-in capabilities.

According to real-world requirements, mathematicians usually need to perform one or more tasks from the list provided below:

- Type in various mathematical expressions (for future use in calculations or in order to create documents, presentations, or Web pages).
- Perform mathematical calculations.
- Represent calculation results in graph form.
- Perform input of initial data and output the results to text files or database files in different data formats.
- Prepare reports in the form of printed documents.
- Prepare Web pages for publishing the results on the Internet.
- Obtain reference information in the field of mathematics.

Mathcad is ideally suited for solving all these (and even some additional) problems:

- Mathematical expressions and text are entered using a built-in formula editor, which provides a large set of functional capabilities and is as straightforward as the formula editor built into Microsoft Word.

■ Mathematical calculations are performed immediately, according to the entered formulae.

■ Graphs of various types (according to the user's choice) provide a rich set of formatting capabilities and are inserted directly into the documents.

■ Data input and output can be performed to and from files in different formats.

■ Documents can be printed directly from Mathcad, exactly as they are seen by the user on the PC screen, or saved in RTF format for further editing using more specialized text editors, such as Microsoft Word.

■ Documents can be saved as Web pages, and pictures inserted into the documents are saved automatically.

■ Symbolic calculations enable users to reference mathematical information immediately. The Help system, Resource Center, and built-in electronic books allow you to quickly find the required reference or example.

Thus, Mathcad combines various closely integrated components: these include a powerful text editor for input and editing of both text and formulae; a calculation processor for performing calculations according to the formulae typed in by the user; and a symbolic processor, which, actually, is an Artificial Intelligence (AI) system. The combination of these components provides a convenient environment for various mathematical calculations, providing a rich functionality for documenting the results of your work.

1.2. FIRST ACQUAINTANCE WITH MATHCAD

In this section, we'll look somewhat ahead and show you how to get a jump start with Mathcad, enter formulae of mathematical expressions, and get the first calculation results. Later we'll provide a more detailed discussion of the materials provided in this section. For the moment, however, to get acquainted with the Mathcad environment, we'll only demonstrate some of its capabilities. If you encounter some problems when you attempt to reproduce some of the actions described here on your PC (for example, expression entry or creating graphs), it is recommended that you look for appropriate information in some of the subsequent sections of this chapter.

After you install Mathcad 2001i on your computer and start it, the main application window, shown in Figure 1.1, will appear. The general appearance of this window is similar to that of a typical Windows application. The title bar and menu bar toolbars (standard toolbar and formatting toolbar) are at the top of the window, and below you'll see the application's worksheet. When you start

Mathcad, a new document is automatically created. The status bar is at the bottom of the application window. If you have at least some experience with standard Windows applications such as text editors, you'll intuitively understand the function of most toolbar buttons.

When you start Mathcad for the first time, the **Tip of the Day** dialog will appear in the foreground. To close this window, click the **Close** button; the cursor points to this in Figure 1.1. To disable the option that displays the **Tip of the Day** window every time you start the Mathcad application, clear the **Show tips on startup** checkbox. To view the next tip, click the **Next Tip** button. When Mathcad starts, it also displays another window — Resource Center, which actually represents a separate program intended to simplify the navigation of the Mathcad capabilities by providing a large number of examples illustrating solutions of the most common mathematical, physical, and engineering problems. When working with Mathcad, you can either use this window as a good complement to the built-in Help system or pay no attention to it.

*The Resource Center contains a good tutorial on Mathcad 2001i intended for beginners. To view it, follow the **Overview and Tutorials** link on the home page of the Resource Center, then select the **Getting Started Tutorial** from the **Contents**. Standard capabilities of viewing material are provided by the navigation toolbar and hyperlinks system, which is similar to that of a normal browser. If you encounter any problems, view the materials provided in Chapter 17 of this book.*

After you close the **Tip of the Day** dialog, you can start entering expressions into the new, empty document. The easiest (but, probably, not the best for the beginner) method of data input is typing the formulae using the keyboard. To perform the first easiest calculation by the formulae, proceed as follows:

- Mark the place within the document where you need to enter a new expression by clicking it with the mouse.
- Enter the left part of the expression.
- Enter the equal sign <=>.

For the moment, let us reserve the discussion of more reliable entry techniques for future consideration. Now we are going to provide an example of the simplest calculation. For example, to calculate the sine of some number, it is sufficient to enter an expression such as sin(1/4) = from the keyboard. After you press the equal sign key, the result will be immediately displayed in the right part of the expression (Listing 1.1).

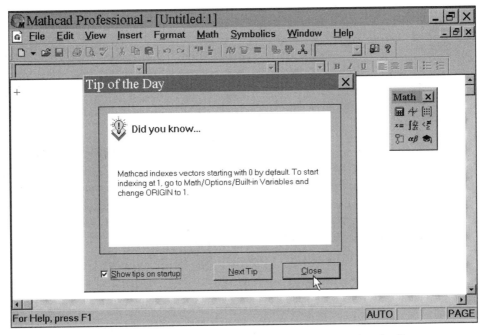

FIGURE 1.1. The main window of the Mathcad 2001i application after the first startup

LISTING 1.1. Simplest Calculation Example

$$\sin\left(\frac{1}{4}\right) = 0.247$$

Here and later on, the whole contents of the Mathcad document worksheet along with the calculation results are presented in listings.

A similar technique can be used for performing more complex calculations using the whole range of Mathcad's built-in special functions. The simplest way of entering function names is to type them from the keyboard, like we did in the example provided earlier with the sine calculation. However, there is a better method, allowing one to avoid possible errors when you are typing function names. To use this method when entering built-in function names into new expressions, do the following:

1. Determine the position within an expression where you need to insert the function.
2. Click the toolbar button labeled *f(x)* on the standard toolbar (in Figure 1.2 this button is pointed at by the cursor).
3. The **Insert Function** dialog will appear. From the **Function Category** list, select a function category (in our case this will be the **Trigonometric** category).
4. From the **Function Name** list select the name of the required built-in function (in our example this will be **sin**). If you have difficulties with this selection, view the help displayed in the text field at the bottom of the **Insert Function** window when you select a function.
5. Click **OK** to insert the selected function into your document.
6. Fill in the missing arguments of the newly inserted function (in our case this will be 1/4).

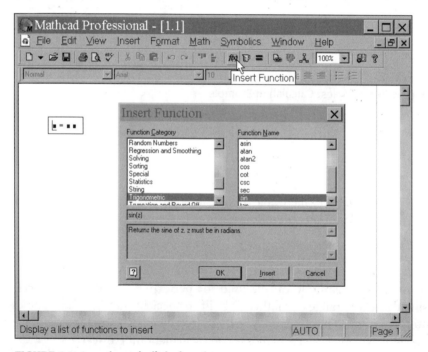

FIGURE 1.2. Inserting a built-in function

As a result, the expression shown in Listing 1.1 will be entered. To obtain the result, you need only enter the equal sign.

Most numeric methods provided by Mathcad are implemented as built-in functions. For a brief overview of the special functions and numeric methods that can be used in your calculations, scroll down the lists of these functions in the **Insert Function** *dialog. These functions and built-in numeric algorithms will be discussed in detail in Part III, and a complete list of the built-in functions is provided in Appendix A.*

Certainly, not all symbols can be entered directly from the keyboard. For example, it is not evident how to insert symbols such as integration or differentiation operators into the document. For this purpose, Mathcad provides special toolbars very similar to the ones provided by the Microsoft Word's built-in formula editor for the same purpose. One of these toolbars — the **Math** toolbar — is shown in Figure 1.1. It provides built-in tools for inserting typical mathematical objects into your documents, such as operators, graphs, program elements, etc. Figure 1.3 represents a more detailed example of this toolbar along with the document being edited.

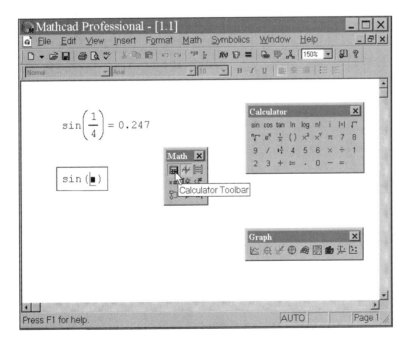

FIGURE 1.3. Using the **Math** toolbar

This toolbar contains nine buttons. Clicking each of these buttons, in turn, opens just another toolbar. Using these additional toolbars, you can insert various objects into Mathcad documents. As shown in Figure 1.3, the two leftmost buttons of the **Math** toolbar need to be clicked. Therefore, two additional toolbars are present on the screen — **Calculator** and **Graph**. The names of these toolbars easily allow you to guess the types of objects that you can insert using them.

More detailed descriptions of these and other input toolbars will be provided later in this chapter (see Section 1.3).

For example, the expression from Listing 1.1 can be entered using the buttons exclusively from the **Calculator** toolbar. To do so, simply click the leftmost button of the top row of this toolbar (sin). The result of this action is shown in Figure 1.3 (the expression enclosed by rectangular frame). Now one needs to enter the 1/4 expression within the brackets (in the placeholder designated by the black rectangle). To do so, click the 1, /, and 4 buttons of the **Calculator** toolbar sequentially, and then the = button on the same toolbar to get the result (which will definitely be the same as the one shown in Listing 1.1).

As you can see, similar to most other Windows applications, Mathcad provides various methods of inserting mathematical symbols into your documents. Depending on one's previous experience with Mathcad and individual preferences, the user can select any of these methods.

If you are a beginner starting to use the Mathcad editor, we strongly recommend that you use the toolbars and the preceding method of inserting the functions using the Insert Function dialog. This will allow you to avoid many typing errors.

The procedures discussed previously demonstrate the usage of Mathcad as a normal calculator with an extended set of functions. From the mathematician's point of view, the capabilities of specifying variables and operations that include user-defined functions are of greater interest and importance. These and many other operations are implemented in Mathcad according to the general rule that can be formulated as follows: "User input must follow the rules generally accepted in mathematics." Therefore, let us provide appropriate examples (Listings 1.2

and 1.3) without spending time on comments (however, if you have difficulty understanding these listings, see additional explanations in the appropriate sections of this chapter). Pay special attention to the usage of the assignment operator, which is used for assigning values to the variables in the first line of Listing 1.2. The Assignment operator, like other symbols, can be entered using the **Calculator** toolbar. Assignment operation is designated by a symbol other than the equal sign in order to emphasize its difference from the calculation operation. Using the equal sign means that the expression value must be calculated from left to right, while using the ":=" symbol means that the value of the expression in the right part must be assigned to the variable in the left part.

LISTING 1.2. Using Variables in Calculations

$$x := 1.2 \quad y := 55 \quad z := 4$$

$$\frac{\left(x^2 \cdot 250\right)}{\sqrt[5]{y}} \cdot \ln(z \cdot \pi) = 408.814$$

LISTING 1.3. Creating a User-Defined Function and Calculating Its Value for x=1

$$a := 2$$

$$f(x) := x^a - \frac{2}{|x - 5|}$$

$$f(1) = 0.5$$

In Listing 1.3 the $f(x)$ function is defined. Its graph is shown in Figure 1.4. To create this graph, click the **Graph** toolbar button corresponding to the graph type that you need (in Figure 1.4 the mouse cursor points to the toolbar button that corresponds to the graph type provided in this illustration). Then, as the graph placeholder appears, specify the values that will be plotted by the axes. In our case, it was required to enter x into the placeholder along the abscissas axis and $f(x)$ — for the ordinates (Y).

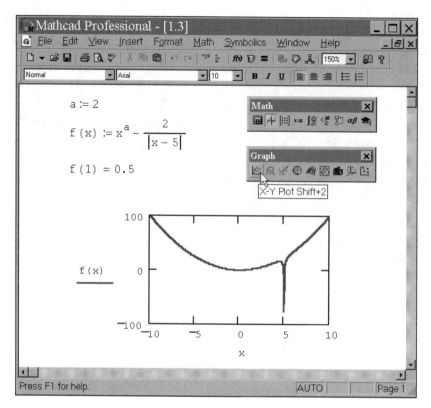

FIGURE 1.4. Plotting the graph of the function (see Listing 1.3)

NOTE

Compare the contents of Listing 1.3 and Figure 1.4. This style of data representation will be used in the whole book. Listings are fragments of the document worksheet that can be used independently, without any additional code (except for the cases when code is specially mentioned). You can enter the contents of any listing into a new (blank) document, and it will work exactly the same way as shown in this book. For the sake of simplicity and to make the examples more illustrative, graphs are represented by separate illustrations. In contrast to the contents of Figure 1.4, further illustrations will not duplicate the code of listings. If the descriptive text below the illustration refers to the listing, this means that the graph presented in this illustration can be inserted into the document directly below the listing being mentioned.

One of the most impressive capabilities of Mathcad is represented by so-called symbolic calculations, which enable you to provide analytical solutions for most tasks. According to the author's personal opinion, Mathcad "knows" mathematics at least at the level of a qualified scientist. Appropriate and skillful usage of the Mathcad's AI will enable you to avoid a large amount of routine calculations — for example, when using integrals and derivatives, as shown in Listing 1.4. In this listing, notice that the expressions are represented in the most common and traditional form. The only specific feature is that it's necessary to use the symbolic calculation character ($\rightarrow$) instead of the equal sign (=). By the way, Mathcad's editor allows you to enter this character either using the **Evaluation** toolbar or the **Symbolic** toolbar. Integration and differentiation symbols can be entered using the **Calculus** toolbar.

LISTING 1.4. Symbolic Calculations

$$\int \frac{\ln(a \cdot x)}{x^b} dx \rightarrow \left[\frac{-(b \cdot \ln(a) - \ln(a) + 1)}{\left(-2 \cdot b + b^2 + 1\right)} \cdot x - \frac{1}{(b-1)} \cdot x \cdot \ln(x) \right] \cdot x^{-b}$$

$$\frac{d}{dx} \frac{\left(x^2 \cdot 250\right)}{\sqrt[5]{y}} \cdot \ln\left(z \cdot \pi\right) \rightarrow 500 \cdot \frac{x}{y^{\frac{1}{5}}} \cdot \ln\left(z \cdot \pi\right)$$

In this section, we have covered only a small part of the calculation functionality provided by the Mathcad system. However, the examples presented here produce an adequate impression of its aims and purposes. It is even possible that by producing this impression of simplicity, the author has "lost" part of his audience, since the most impatient users might close the book and proceed with solutions of their tasks on their own. If you are among such users, we'd like to advise you to use the second and third parts of this book as a reference, while the fourth part can be recommended for those who need a better representation of the obtained results. Later on, in this and subsequent chapters of this section, we'll concentrate on Mathcad's basics and discuss these topics in more detail.

1.3. USER INTERFACE

The Mathcad user interface is intuitive and similar to that of other typical Windows applications. Its most important controls are listed below:

- Main menu or menu bar
- **Standard** and **Formatting** toolbars
- The **Math** toolbar and additional toolbars, which provide various add-on mathematical toolbars
- Worksheet
- Status bar
- Pop-up menus, also known as context menus or right-click menus
- Dialogs

Most commands are available both via menus (application main menu or right-click menu) and toolbars or the keyboard.

1.3.1. The Menu Bar

The menu bar is located at the top of the Mathcad main application window. It contains the following nine menu items:

- File — Commands that are used to create a new document, open or save the existing one, send it via e-mail, or print it
- Edit — Commands that are related to various text-editing operations, such as copying, pasting, deleting text fragments, etc.
- View — Commands that manage the document layout in the Mathcad editor window and commands used to create animation files
- Insert — Commands that allow you to insert various objects into documents
- Format — Commands that provide formatting capabilities for text, formulae, and graphs
- Math — Commands that manage the calculation process
- Symbolics — Commands for symbolic calculations
- Window — Commands that manage windows with different documents
- Help — Commands that provide access to the context-dependent Help system, **Resource Center**, **Tip of the Day** dialog and information on the Mathcad version

To select the required command, click an appropriate menu item with the mouse and select the menu command that you need. Some commands are available via submenus or nested menus, as shown in Figure 1.5. To run such a command (for example, to display the Symbolic toolbar), select the **Toolbars** command from the **View** menu, and then select the **Symbolic** option from the submenu.

Now and later on, when describing specific actions performed via the sequence of menu commands, we'll use the brief notation by separation menu command names by slashes. Using this notation, the command mentioned above can be described as follows: View | Toolbars | Symbolic.

Notice that menu commands containing submenus are designated by arrows (like the **Toolbars** command in Figure 1.5). Furthermore, some menu commands have status checkmarks that visually show if the respective menu command is currently selected (or not selected). For example, the screenshot provided in Figure 1.5 shows that the **Status Bar** command from the **View** menu and the names of the three toolbars are marked, which means that the Mathcad window displays the status bar and the three toolbars. On the other hand, menu items such as **Ruler**, **Regions**, and names of mathematical toolbars are not selected, which means that their respective options are currently off.

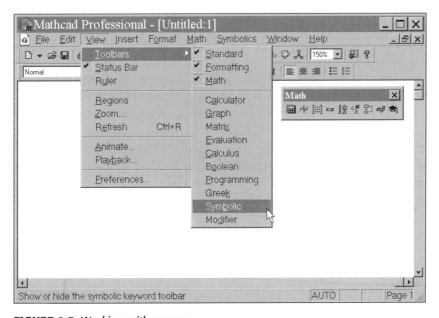

FIGURE 1.5. Working with menus

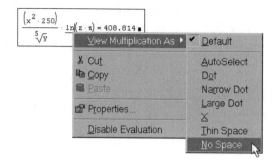

FIGURE 1.6. Right-click menu

The left part of the status bar (which is located at the bottom of Mathcad window) displays a brief help string describing the currently selected menu command. For example, on the screenshot shown in Figure 1.5 the mouse cursor points to the Symbolic command, and therefore the help string displayed in the status bar is as follows: "Show or hide the symbolic keyword toolbar."

Similar to most typical Windows applications, besides the main menu, Mathcad provides the capability of performing the same actions using pop-up menus displayed when the user clicks the right mouse button anywhere within a document (Figure 1.6). Because the commands available via right-click menus depend on the position within an object (the so-called context), these menus are also known as context menus. Depending on the context, Mathcad determines which operations might be currently needed and, based on this information includes appropriate commands on the right-click menu. Because of this, right-click menus are often easier to use than the application main menu, since in this case, you don't need to remember exactly how to access the required command. Like the application main menu, the right-click menu can also have submenus (for example, Figure 1.6 illustrates the operation of changing the display of the multiplication sign for the selected formula using submenus of the right-click menu; it is particularly remarkable that the only way of performing this operation in Mathcad is by using the right-click menu).

1.3.2. Toolbars

Toolbars are used for quick access to the most frequently used commands (actually, with only a click of the mouse). All actions available via toolbars are also available in the application main menu. Figure 1.7 displays the Mathcad main window with the three main toolbars located directly below the menu bar. Toolbar buttons within toolbars are grouped according to the command types:

- Standard — Contains buttons providing access to most standard commands, such as file operations, file editing, insertion of objects, and access to Help systems
- Formatting — Contains buttons providing quick access to formatting commands (such as changing font type and size, or modifying text or formula alignment options)
- Math — Provides commands for inserting mathematical symbols and operators into documents

Button groups within a toolbar are separated by vertical delimiters. When you point to any of the toolbar buttons with the cursor, the pop-up tooltip appears near that button containing a brief help text explaining the aim of that button (see, for example, Figures 1.3 and 1.4). Besides the pop-up tooltip, more detailed Help information on the operation provided by the currently selected button is displayed in the status bar.

The Math toolbar contains icons that, when selected, display an additional nine toolbars (Figure 1.8), which, actually, are the ones used for inserting mathematical operations into your documents. In previous releases of Mathcad these mathematical toolbars were also known as palettes. To display any of these toolbars, click the appropriate button on the Math toolbar (see Figure 1.3). Brief descriptions of these mathematical toolbars are supplied below:

- Calculator — Used for inserting main mathematical operations. This toolbar got its name because of the similarity between the set of buttons it provides to those provided by a typical calculator.
- Graph — Used for inserting graphs.
- Matrix — Used for inserting matrices and matrix operators.
- Evaluation — This toolbar is used for inserting calculation management and control operators.
- Calculus — Used for inserting integration, differentiation, and summation operators.
- Boolean — Contains buttons used for inserting logical (Boolean) operators.
- Programming — This toolbar is used for programming using the Mathcad programming capabilities.
- Greek — The buttons of this toolbar are used for inserting the symbols of the Greek alphabet.
- Symbolic — The buttons of this toolbar are used for inserting symbolic operators.

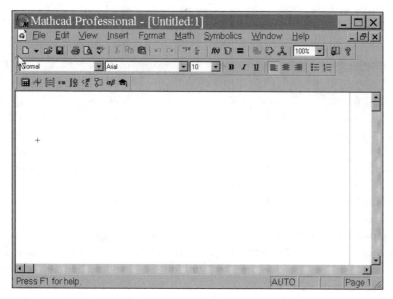

FIGURE 1.7. Main toolbars

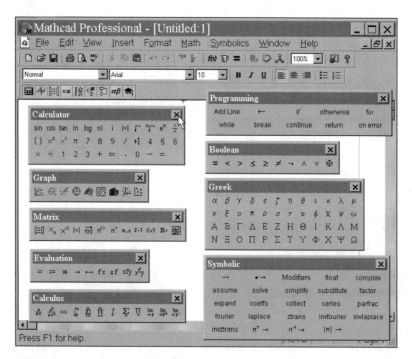

FIGURE 1.8. Mathematical toolbars

When you point to mathematical toolbar buttons with the cursor, in most cases, the pop-up tooltip also displays the hotkey combination you can use to access this toolbar. Using this hotkey combination produces the same result as clicking that toolbar button. In most cases, using keyboard shortcuts is more convenient than clicking toolbar buttons. However, it requires that the user have a significant amount of experience.

1.3.3. Customizing Toolbars

Much like most typical Windows applications, Mathcad enables the user to customize the toolbars according to his individual requirements and preferences.

For example, you can do the following:

- Display or hide toolbars.
- Move any toolbar to the preferred position within the screen and change their form.
- Switch from fixed to floating toolbars and vice versa.
- Customize main toolbars by determining the set of available buttons.

Displaying or Hiding Toolbars

You can display or hide any toolbar by selecting the name of the appropriate toolbar from the View | Toolbars menu (see Figure 1.5). You can also hide any toolbar by using the context menu, which will be displayed after you right-click any point within a toolbar (for example, any toolbar button). To hide a specific toolbar button, right-click it and select the Hide command from the pop-up menu. Furthermore, if the toolbar is *floating* (i.e., its position within the main window is not fixed) like all toolbars shown in Figure 1.8, you can hide it by clicking the Close button (in Figure 1.8 the cursor is pointing to the Close button of the Calculator toolbar).

Mathematical toolbars, in contrast to the main ones, can be displayed or closed by clicking an appropriate button on the Math toolbar. The display mode of mathematical toolbars is reflected by the status of the appropriate button on the toolbar (pressed or released) as shown in Figure 1.3, 1.4, or 1.8.

Creating Floating Toolbars

To switch any toolbar to the floating mode, proceed as follows:

1. Point the mouse cursor to the first (see Figure 1.7) or the last delimiter of the toolbar (the first delimiter usually has a typical 3D-type look, while the last one is usually normal).

2. Click and hold the left mouse button — after doing so, you'll see the characteristic outline of the toolbar.
3. Drag the toolbar to the desired position on the screen (you can evaluate the new position of the toolbar by the position of the toolbar outline).
4. Release the mouse button. The toolbar will be switched to the floating mode and move to the position previously taken by its outline.

The result of dragging the main toolbars is shown in Figure 1.9. Notice that floating toolbars now have a title bar displaying the toolbar name. To switch the floating toolbar to the fixed mode, point the mouse cursor to the title bar and drag it with the mouse to the top of the window. When the toolbar approaches the upper boundary of the window, it starts to be "attracted" to it. When you notice this effect, release the mouse button, and the toolbar will be switched to the fixed mode. Notice, that you can fix toolbars not only to the menu bar at the top of the window, but also to any window border.

Most mathematical toolbars only appear in the floating mode.

NOTE

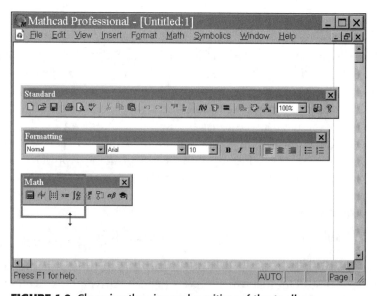

FIGURE 1.9. Changing the size and position of the toolbars

Mathcad provides an alternative method for switching toolbars from a to a floating mode, which at the same time is the easiest one. To use this method, simply double-click the toolbar's first (or the last) delimiter. To switch the toolbar to the fixed mode, simply double-click its title bar.

Moving Toolbars within a Screen

To move toolbars to any desirable screen position, first switch them to the floating mode. Floating toolbars can be moved to any position within the screen by simply dragging them with the mouse.

You can move toolbars to any position within the desktop, even if the main Mathcad window is not maximized.

Fixed toolbars can be moved along the window boundaries. To do so, point the mouse cursor to the starting or ending delimiters and drag them with the mouse along the window boundary. By doing this, you will find that it is easy to position several toolbars into one vertical or horizontal line.

To change the form of the floating toolbar (i.e., to reshape its outline in order to specify another number of button rows/columns), point the mouse cursor to the panel edge and, when the characteristic toolbar outline appears (Figure 1.9), click and hold the left mouse button and drag with the mouse until the toolbar outline takes the required shape. After you release the button, the toolbar shape will change.

Customizing the Toolbar Buttons

This customization allows you to change the number and selection of the buttons available on any of the three main toolbars (**Standard**, **Formatting** and **Math**). This might be useful, for example, if you need to remove rarely used buttons in order to make the screen less cluttered (especially if you are forced to use low screen resolution). To change the set of available toolbar buttons, right-click the toolbar with the mouse (anywhere except for the title bar) and select the **Customize** command from the pop-up menu. The **Customize Toolbar** dialog will appear (Figure 1.10). This dialog contains two lists, the left of which lists the buttons that can be included in the list of available toolbar buttons while the right list contains the names of the buttons that are currently available for the toolbar.

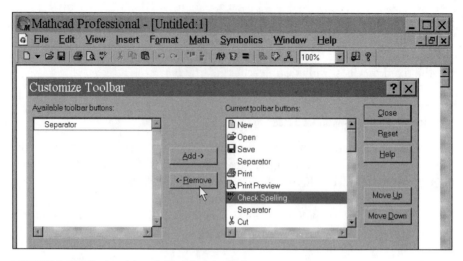

FIGURE 1.10. Customizing the toolbar buttons

To remove the button (or button delimiter) from the toolbar, select its name in the right list and then click the **Remove** button. To add a new button to the list, select the name of the required button in the right list and then click **Add**. To change the position of the specific button, select its name from the right list and move it to the desired position by clicking the **Move Up** or **Move Down** buttons.

To confirm your customizations, click the **Close** button or close the dialog by clicking the **Close** button at the top-right corner. To reset the default set of available toolbar buttons, click the **Reset** button.

1.3.4. The Worksheet

The main part of the Mathcad window is taken up by the worksheet, where the user enters mathematical expressions, text fields, and programming elements. In order to master the document navigation it is important to know how to properly customize the worksheet.

Data Entry Cursor

When carefully viewing some pictures presented in this chapter (see Figure 1.7, for example), you'll notice the data entry cursor displayed as a small crosshair cursor (displayed in red on the screen). This cursor marks the document position where you can currently enter formulae or the text. To move this cursor to the desired position, simply click the required position with the mouse or press the arrow keys. If you click a formula or start entering data at a free position, the cursor will be replaced by the so-called editing lines, which mark the position within a formula or text that you are currently editing (see Figure 1.3).

The usage of the data entry cursor and consideration of the document-editing techniques will be covered in detail in Chapter 2.

Document Layout

The Mathcad document is structured according to the principle of positioning both formulae and text within the worksheet area, which initially is similar to the blank sheet of paper. To show or hide the positioning of regions containing mathematical expressions, text, or graphs, you are provided with the option of enabling or disabling the region boundary display. To enable or disable this option, use the View I **Regions** commands from the main menu. If this option is enabled, the document will look like the one shown in Figure 1.11 (for comparison, see the document shown in Figure 1.4).

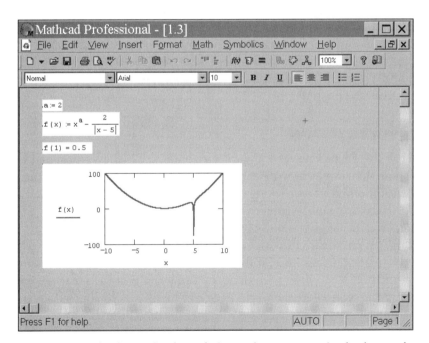

FIGURE 1.11. Viewing region boundaries against a contrasting background

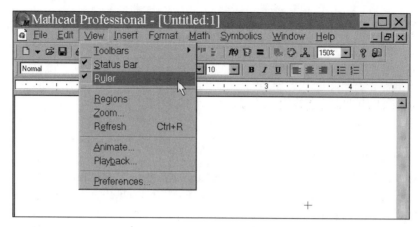

FIGURE 1.12. Visualizing the ruler

If you carefully view Figure 1.11 and some other illustrations provided in this chapter, you'll notice the vertical page separation line in the rightmost position of the worksheet. If your document is large, then somewhere within it there will also be a dashed horizontal page separation line. These lines specify how the document will be paginated when you send it to printer. To change the page layout, select the File / Page Setup commands from the main menu.

The horizontal ruler simplifies the process of positioning objects within a document page. This ruler is displayed below the toolbars at the top of Mathcad window (Figure 1.12). To display or hide the ruler, select the View I Ruler commands from the main menu.

Navigating within a Document

To scroll the document vertically or horizontally, use the vertical and horizontal scroll bars, by simply moving their sliders (in which case the document will be scrolled smoothly) or by clicking one of the slider edges with the mouse (in which case the scrolling will be uneven). To move the cursor within a document, you can also use the <PgUp> and <PgDn> keys. Notice, that in all above-mentioned cases the cursor position doesn't change, rather, we are simply viewing the document contents. Furthermore, if the document is large, the most convenient navigation method is viewing its contents by using the Edit I Go to Page menu commands. When you select the Go to Page command, the dialog will open, allowing one to quickly navigate to the page with the specified number.

To move the cursor up or down or from left to right within a document, press the arrow keys on the keyboard. When the cursor fits within regions containing

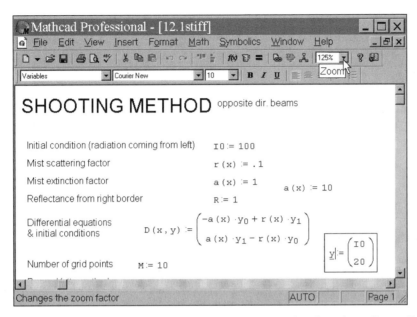

FIGURE 1.13. Re-scaling the document display using the drop-down list available in the toolbar

text of formulae, it turns to the two blue editing lines — horizontal and vertical. When the cursor moves further within a region, editing lines are moved one position in the appropriate direction. When the cursor moves outside the region boundaries, it once again turns to the red crosshair cursor.

Of course, it is also possible to move the cursor by clicks of the mouse. If you click a blank area within a document, the data entry cursor (red crosshair) will appear at that point; if you click the mouse within a region, editing lines will appear.

Scaling the Document

Re-scaling the document has no influence on its content, but, rather, simply defines the size of characters and graphics displayed on the screen.

To re-scale the screen display, go to the appropriate field on the **Standard** toolbar, as shown in Figure 1.13. Clicking this field with the mouse opens a drop-down list, from which you can select the display scale (available options range from 25 to 200%). The value of 100% corresponds to the document page size that will be obtained when you print the document. Compare the illustrations provided in Figures 1.13 and 1.14, where the same document is shown using different scale options.

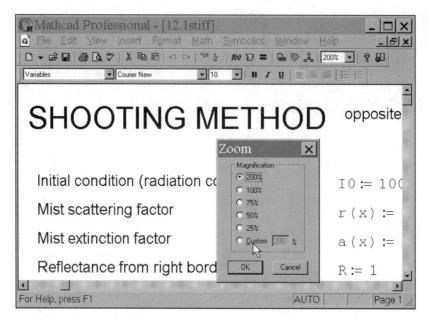

FIGURE 1.14. The **Zoom** dialog

To select a document scale value different from one of the available toolbar settings, use the View I Zoom commands from the main menu. After you select these options from the menu, the Zoom dialog will appear (Figure 1.14), where you can select one of the radio buttons corresponding to the desired magnification values. To select the scale value manually, select the **Custom** option and then enter the desired value (in percentage of the real scale of the page) in the text field next to the option. Click OK to confirm your selection.

Editing Multiple Documents

All the previous illustrations represented examples of editing a single document maximized to consume the whole Mathcad application window. However, it is also possible to edit several documents simultaneously, each in a separate window. These windows can be positioned on the screen in any order. To set the preferred option, click the **Window** menu and then select one of the following commands: **Cascade**, **Tile Horizontal**, or **Tile Vertical**. As a result of this operation, all windows will either cascade within the main application window or tile (vertically or horizontally) in such a way as to cover the whole main window area one above the other or side by side, as shown in Figures 1.15–1.17.

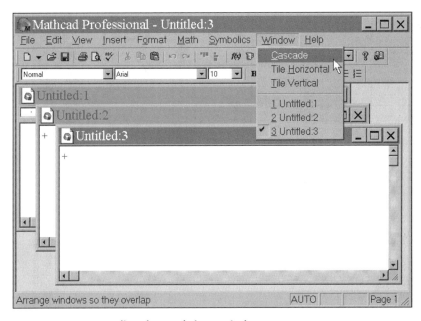

FIGURE 1.15. Cascading the worksheet windows

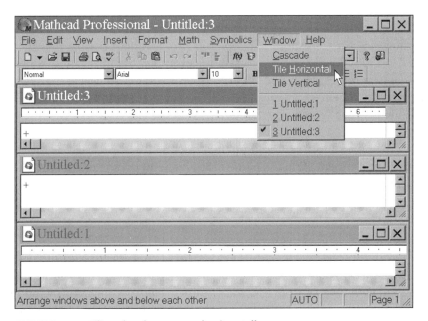

FIGURE 1.16. Tiling the documents horizontally

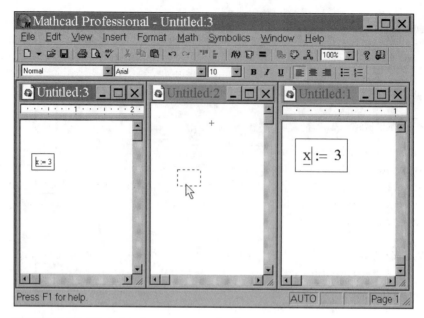

FIGURE 1.17. Tiling the documents vertically

Notice that each document window has its own cursor (either the crosshair cursor or editing lines, depending on the position within the document). Furthermore, for each document, it is possible to enable or disable the ruler display, specify an individual scale value (see, for example, the illustration shown in Figure 1.17), or set the region boundary display mode. Also notice that you can edit only one document at a time. The title bar of the window containing the active document is highlighted. The document window can be activated either by clicking the mouse anywhere within its area or by selecting its name from the Window drop-down menu. Names of all opened documents are located at the lower part of the Window menu, and the name of the currently active document is checked.

Having positioned several documents within the main application window area, you can resize them or change the position of each window by dragging with the mouse.

The multiple-document mode is convenient for copying objects from document to document using the Drag-and-Drop technology. To do so, click the required object with the mouse and drag it to the opened window of another document without releasing the left mouse button. Position the object as required and then release the mouse button (see Figure 1.17).

When using the multiple-document mode, you can maximize any document to the full-screen area. This is done by using the window control at the top-right corner of the window. When the document window is expanded to the full-screen mode, its window controls move to the application main menu bar. To manage the documents when working in this mode, use the commands available in the **Window** menu.

1.3.5. The Status Bar

The status bar is located at the bottom of the Mathcad window, directly below the horizontal scroll bar, and it is shown in most illustrations provided in this chapter. The status bar displays the most important information on the editing mode — the fields being separated by delimiters (Figure 1.18). The information fields display the following data (from left to right):

- Context-dependent help string displaying information on the currently selected menu command
- Calculation mode: automatic (AUTO) or manually selected (Calc F9)
- Current keyboard mode (CAP)
- Current mode of the numeric keyboard mode (NUM)
- Current page number

To display or hide the status bar, select the **View | Status Bar** options from the main menu.

FIGURE 1.18. The status bar

1.4. THE REFERENCE INFORMATION

Mathcad supplies several sources of reference information that can be accessed via the Help menu (shown in Figure 1.19).

- Reference information concerning various aspects of Mathcad usage, including the following:
    - Mathcad Help — The help system containing technical support information

- Developer's Reference — Additional help chapters intended to support the developers of stand-alone applications created using Mathcad's programming language
- Author's Reference — Additional help chapters intended for authors developing their own electronic books

■ Electronic books — A collection of calculations supplied with hyperlinks and interactive examples of Mathcad programs:

- Resource Center — Stand-alone application designed as an electronic book providing ready-to-use solutions demonstrating the usage of Mathcad 2001i functionality
- Open Book — The menu item that allows you to open an existing e-book, which might be created by the user or obtained from another author

The technique of creating e-books and the capabilities provided by the Resource Center are covered in detail in the last part of this book (see Chapter 17). You'll also find additional methods of obtaining the reference information, including Internet resources, the Mathcad user community, and Mathcad developers.

Besides the items listed above, the **Help** menu contains the following items:

■ Tip of the Day — Opens the **Tip of the Day** dialog, which is typical for most Windows applications. This window displays useful tips in random order (see Figure 1.1).
■ Mathcad Update — Enables one to check the official MathSoft site to determine if any Mathcad 2001i updates are available.
■ About Mathcad — Displays information on the current Mathcad version and developers of this software.

If you need to access reference information when working with Mathcad, select the Help | Mathcad Help commands from the main menu, press the <F1> key or click the question mark (Figure 1.20) on the standard toolbar.

The Mathcad help system is context-dependent, which means that its content is defined by the position of the cursor within the document when it was called. For example, the illustration provided in Figure 1.19 shows the cursor (editing lines) pointing at the matrix transposition operator. Therefore, if you open the Help system at this point, the **Mathcad Help** window will be opened at a point describing the matrix transposition operation (Figure 1.21).

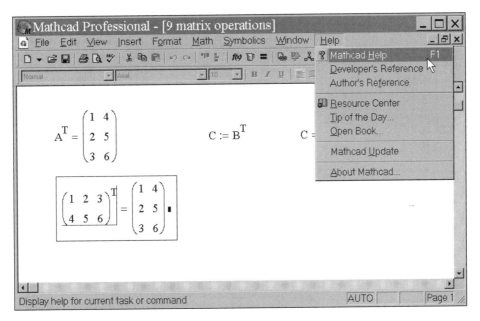

FIGURE 1.19. The **Help** menu

FIGURE 1.20. The **Help** and **Resource Center** buttons on the Standard toolbar provide access to Mathcad reference information

The Mathcad Help system window is formatted using the style characteristic of the Windows operating system. It comprises two panes, the left displays the Help topics list (the **Contents** tab), while the right pane displays the text of the selected help topic. You can temporarily hide the left pane by clicking the **Hide** button on the toolbar at the top of the window. To restore the display of the hidden left pane, click the **Show** toolbar button, which replaces the **Hide** button after you hide the content.

The Help system is based on the hyperlinks mechanism, which enables navigation between the help topics. The text of the selected Help topic is loaded into the right pane of the Help window. Navigation buttons such as **Back**, **Forward**, and **Home** provide an easy way of returning to the Help topics that were previously viewed.

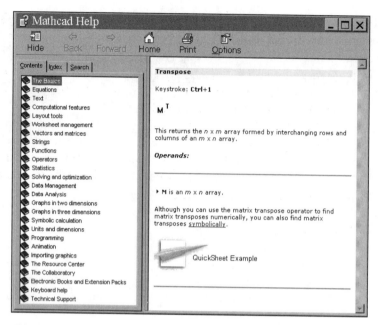

FIGURE 1.21. The **Mathcad Help** system window

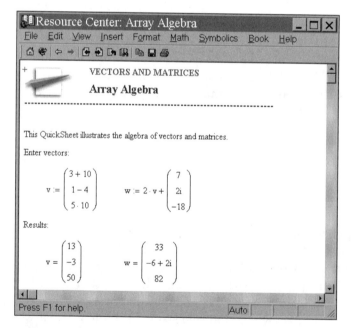

FIGURE 1.22. Displaying an example from the Resource Center

Quite often, the Help topic texts include the so-called QuickSheet Example buttons. When you click such buttons, appropriate examples from the Resource Center are called, which are related to the contents of the currently opened Help topic. For example, clicking such a button in the window shown in Figure 1.21 will open the **Resource Center** window shown in Figure 1.22, with the example illustrating the usage of vector and matrix operations.

The **Contents** tab in the left pane of the Help window initially displays only the main chapters of the Help system, which are labeled by the icons displaying the closed book (Figure 1.21). You can open subsections by double-clicking the title of the required help topic. Note that the book icon will change to display the opened book, while the titles of specific topics are marked with the question mark icons (Figure 1.23). Clicking specific subtitles displays the text of the selected Help topic in the right pane of the Help window.

Notice that the on-line Help system contains many more topics than subtitles listed on the **Contents** tab. To display most topics on the screen, it is necessary to follow one or more links. Also notice that the Help system includes a powerful searching tool that lets you quickly find other topics related to the one currently displayed. To access this tool, click the **Related Topics** button, the example of which is shown in Figure 1.23. Clicking this button opens the **Topics Found** dialog listing the Help topics related to the one currently displayed (Figure 1.24). To navigate to one of these topics, select an appropriate item from the list and click the **Display** button, or simply double-click the required list item.

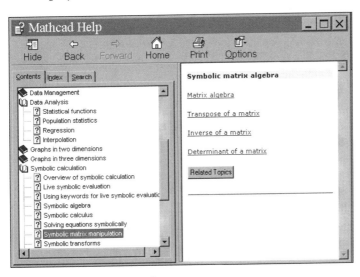

FIGURE 1.23. Using the Help system

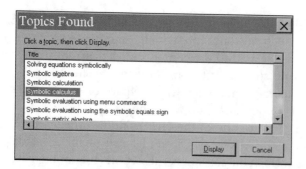

FIGURE 1.24. The **Topics Found** window displaying the list of Help topics related to the one currently displayed

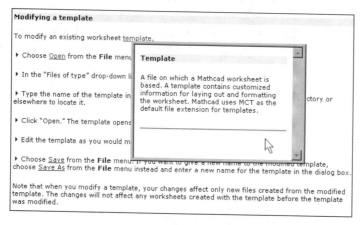

FIGURE 1.25. The pop-up window containing the term definition

Some hyperlinks don't open another page of the on-line Help system. Rather, they open a window containing a specific term and its definitions (Figure 1.25). To close this window after viewing the information, simply click the left mouse button anywhere within that window.

The left pane of the Mathcad Help window displays the contents of any one of the following three tabs:

- Contents — Displays the titles of the Help topics sorted by chapters and subtitles
- Index — Lists of Help topics sorted alphabetically (Figure 1.26)
- Search — Provides the capabilities of searching the on-line Help system by keywords or phrases (Figure 1.27)

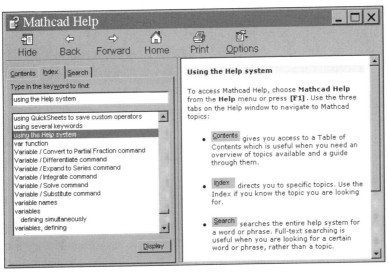

FIGURE 1.26. The index of the available Help topics

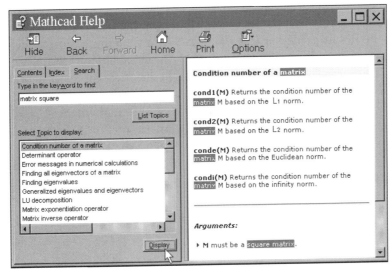

FIGURE 1.27. Searching the on-line Help system by keywords or pharses

On the Index tab, you can either select the required Help topic from the Select Topic to display list or enter two or more starting characters in the Type in the keyword to find text field. To navigate the Help contents, double-click the required list item, after which either the selected Help topic will be displayed in the right pane of the Help window or the **Topics Found** dialog will open, allowing you to select the topic from another list (Figure 1.24).

If you know the exact title of the required Help topic, go to the **Index** tab. If you need to search for the Help topic by the keywords, proceed as follows:

- In the left pane of the **Mathcad Help** window click the **Search** tab (Figure 1.27).
- Enter the keywords in the uppermost text field (if there is more than one keyword, separate them with the space character). For example, if you want to find information on the square matrices, enter "matrix square" text into this field.
- Click the **List Topics** button to display the list of the available topics.
- The list of available topics containing the keywords that you have entered will appear in the large field below, labeled **Select Topic to display**. To view a specific topic from this list, select its title and then click the **Display** button.
- The topic text will be displayed in the right pane of the **Mathcad Help** window, and the keywords used for the search procedure will be highlighted.

To conclude this chapter, let us note that both the Mathcad on-line Help system and the Resource Center (*see Chapter 17*) are something more than just texts and examples providing a description of Mathcad capabilities. Both can be considered stand-alone self-sufficient tutorials in the field of several branches of higher mathematics (the Resource Center is even considered an interactive one). Both reference systems cover main definitions, the mathematical basis of most operations, and algorithms implementing numeric methods. From the author's point of view, some topics are explained better in the on-line Help system and Resource Center than anywhere else. Therefore, the Mathcad on-line Help system represents a valuable source of information.

2 Editing Documents

This chapter concentrates on the main techniques of editing Mathcad documents. The first section is dedicated to the basic principles used to create new documents and covers saving calculation results in files *(see Section 2.1)*. The next three sections consider methods of editing formulae *(see Section 2.2)*, text *(see Section 2.3)*, and fragments of Mathcad documents *(see Section 2.4)*. Finally, to conclude this chapter, we'll provide the most important considerations on the principles of printing documents and sending them via e-mail *(see Sections 2.5 and 2.6)*.

2.1. WORKING WITH DOCUMENTS

All Mathcad calculations are organized within worksheets, which are initially blank. The user can add formulae and text to the worksheet. Mathcad worksheets are also known as Mathcad documents. This term, of course, doesn't reflect the exact sense of the English term "worksheet," but, on the other hand, its meaning is much more in line with Windows terminology. Each document is a stand-alone collection of mathematical calculations stored within a separate file. The document represents both the listing of the Mathcad program and the results of its execution. At the same time, it also represents a report, suitable for printing or publishing on the Web.

2.1.1. Creating a Blank Document

When you start Mathcad from the Windows Start menu (for example, by clicking the Start button and selecting Programs | MathSoft Apps | Mathcad 2001i Professional from the Start menu), the Mathcad main window will appear with a new blank document, which by default is named "Untitled:1".

If you want to create a blank new document when Mathcad is already running, do one of the following:

- Press the keyboard combination <Ctrl>+<N>.
- Click the New button on the toolbar.
- Select File | New from the main menu.

In Mathcad 2001i the New button on the standard toolbar looks as shown in Figure 2.1. When you click the blank page icon or press the keyboard shortcut <Ctrl>+<N>, the system creates a new blank document. If you click a small helper button next to the blank page icon (it is labeled with an arrow), you'll open a drop-down list of templates that can be used to create a new document. To create a blank document, select the **Blank Worksheet** item from this list (in Figure 2.1 this is a single item available in the list). A similar list will be displayed when you are using the third method of creating a new document (via the menu), however, in this case, the list will be displayed in the New dialog. To create a blank document, select the **Blank Worksheet** option from the **Worksheet Templates** list and click OK (Figure 2.2).

FIGURE 2.1. The **New** button

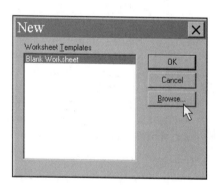

FIGURE 2.2. The **New** dialog

As a result of any of these actions, a new blank document will appear in the Mathcad window, named Untitled:2, Untitled:3, etc., depending on the number of new blank documents that you have already created. (The procedure of saving documents will be covered in Section 2.1.3.)

2.1.2. Creating Documents Based on Templates

After you work with Mathcad for some time and get some experience, you'll most likely come up with the idea of creating new documents based on previously developed ones rather than from scratch. There are two methods of performing this task:

- Open an existing document and save it under another name.
- Use predefined templates.

Document template is a dummy document containing formulae, graphs, and text, including layout, formatting parameters, default selecting of the calculation mode, etc. In the previous section, we considered the procedure of creating a new document based on the blank template. To select another template (either supplied with Mathcad or user-defined), click the **Browse** button in the New window (Figure 2.2). The **Browse** dialog will appear, enabling you to specify the location of the required Mathcad document template file (Figure 2.3). By default, template files have the .mct filename extension, which stands for "Mathcad Template." Select the desired template from the list of available ones and click **Open**.

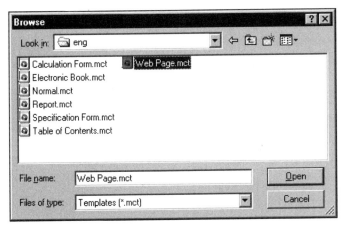

FIGURE 2.3. Opening a template file

As a result of these actions, the system will create a new document containing predefined formatting and layout settings (more detailed information on the document settings will be provided in Chapter 16). For example, if you select the Web Page.mct template for creating a new document (Figure 2.3), the document created based on this template will have the Web page design suggested by Mathcad developers (Figure 2.4). Naturally, such a template must be used for publishing the calculation results on the Internet.

To create your own user-defined template, proceed as follows:

1. Edit the document by including formulae, text, and graphs; format it and specify other document settings.
2. Select File | Save As commands from the main menu.
3. Browse the file system to find the folder where Mathcad templates are stored (for example, \Program Files \MathSoft\Mathcad 2001i Professional \Template) and open that folder (Figure 2.5).
4. From the Save as type list select the Mathcad Template (*.mct) option.
5. Specify a name for the new template (for example, "user1").
6. Click Save.

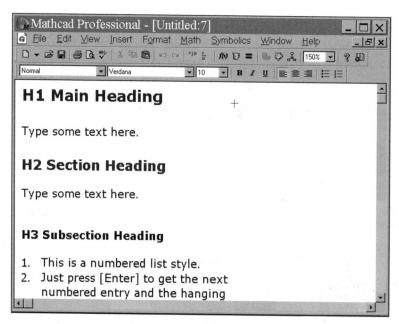

FIGURE 2.4. New document based on the default Web page template

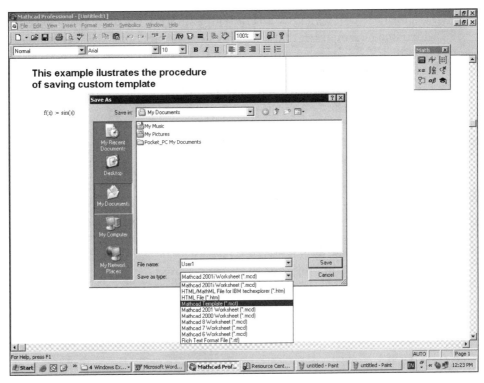

FIGURE 2.5. Saving a document template

A new template will be added to the list of existing ones, and, when you go to create a new document, it will be available both via the File | New menu command and via the New button on the toolbar (Figure 2.6). The saved template will be automatically added to the list of templates displayed when the user creates a new document.

FIGURE 2.6. The saved template appears in the list of available templates

You can edit saved document templates by opening them as normal Mathcad documents (remember that template files have .mct filename extension).

You can save Mathcad templates in any directory on your hard drive. Also notice that normal Mathcad documents (in MCD format) can be used as templates. However, in this case you'll need to remember where you saved the Mathcad document template or browse the file system to find it.

2.1.3. Saving Documents

To save your document in Mathcad format, select the File I Save commands from the main menu, press the <Ctrl>+<S> keys on the keyboard, or click the Save button on the standard toolbar. If you are saving the document for the first time, the system will display the Save dialog, where you'll be prompted to specify the name for the new file (Figure 2.7).

To save the existing document under another name, use the File I Save As commands from the main menu. The previous version of you document in this case will be saved.

To protect the results of your work, create backup copies of your files on a regular basis.

Mathcad 2001i allows you to save documents in various formats (Figure 2.7):

- Mathcad 2001i Worksheet (*.mcd) — the newest and the most powerful format, which is used by default.
- Mathcad 6–2001 Worksheet (*.mcd) — formats used in previous releases of Mathcad. This capability is useful, particularly if you are working with several Mathcad versions (for example, consider a situation when you are participating in a complex project and are exchanging your files with other developers who use different Mathcad versions). If this is the case, save your files in a format that can be understood by the earliest version of Mathcad that you are dealing with. Notice, however, that capabilities provided by earlier versions (for example, the sets of built-in functions) are limited in comparison to the later Mathcad versions. As a result, such functions will not be available if you save your documents in earlier formats.
- Mathcad Template (*.mct) — as was already mentioned before, Mathcad Template format is intended for saving custom templates (see Figure 2.7).

FIGURE 2.7. Mathcad 2001i allows the user to save documents in different formats

- Rich Text Format File (*.rtf) — save your files in this format only for further editing using text editors (this might be useful for creating reports). For example, if you save your document as an RTF-file, you'll be able to edit this file using Microsoft Word or any other text processor, since most of such applications support this format.
- HTML File (*.htm) — Web page format. Documents saved as HTML-files can be published on the Internet or immediately viewed using any browser.
- HTML/MathML File for IBM techexplorer (*.htm) — Web page format intended specifically for techexplorer Hypermedia Browser, a Web browser with extended math capabilities developed by IBM. The capability of saving documents in this format was first introduced with Mathcad 2001i. This format

is particularly advantageous, since the user can view files saved in this format with techexplorer Hypermedia Browser when browsing the Internet, and, on the other hand, open such files in Mathcad as normal MCD files. In contrast, files saved using the two previous options can't be opened for subsequent editing in Mathcad, but, rather, are needed for creating reports that can be opened using other applications.

When you save files in the latter two formats (*.htm), Mathcad, besides the file itself, creates an additional folder where it saves all other files containing illustrations in JPEG format. These files are necessary for browsers to display HTML documents correctly. The name of this additional folder is composed from the file name specified by the users and "_data" suffix.

2.1.4. Opening Existing Documents

To open an existing document for editing, select the File | Open commands from the main menu, press <Ctrl>+<O> keys on the keyboard, or click the Open button on the standard toolbar. When the Open dialog appears, select the file that you need to edit and click OK.

Furthermore, you can open MCD files by double-clicking them in Windows Explorer.

To open a Mathcad document located somewhere on the Internet rather than in your LAN, proceed as follows:

1. Start the Resource Center (**Help | Resource Center**).
2. When the Resource Center window opens (Figure 2.8), click the button labeled with the globe icon.
3. In the Address field, specify the URL of the Web page containing the required Mathcad document, for example: http://www.mathsoft.com.
4. Press <Enter>.

Internet resources contain lots of Mathcad files created by developers who are ready to share their experience. To solve a complex task, it is advisable that you get acquainted with colleagues' work or even borrow some useful fragments from their programs. Mathcad developers have included browser functionality into the design of the Resource Center application chiefly for such purposes, enabling you to view Mathcad documents published on the Internet.

FIGURE 2.8. Specifying the required URL in the **Address** field

2.1.5. Closing Documents

To close an active document, use one of the following techniques:

1. Click the Close button in the top-right corner of the document window (Figure 2.9).
2. Select the File I Close commands from the main menu.
3. Press the <Ctrl>+<W> keyboard combination.
4. Close the Mathcad editing session (by selecting the File I Exit commands from the main menu, by closing the Mathcad main window, or from the Windows taskbar). All opened Mathcad documents will be closed when closing Mathcad session, both active and inactive.

Figure 2.9. Window management controls

If the changes that you introduced during Mathcad editing sessions were not saved, Mathcad will prompt you to save your work by displaying an appropriate dialog. You can either save the file or discard the changes. If you click the Cancel button, you'll return to the editing session.

2.2. ENTERING AND EDITING FORMULAE

The formula editor built into Mathcad enables you to enter and edit mathematical expressions quickly and efficiently. Still, some aspects of its usage can't be considered intuitive. This is because it is necessary to avoid errors when performing calculations according to these formulae. Therefore, it is strongly recommended that you spend some time getting acquainted with the specific features of the formula editor. This will save you much more time when it comes to the real work.

2.2.1. Interface Components

Once again, let us return to the Mathcad user interface and its components (some of them were covered in Chapter 1):

- Mouse pointer — Plays its typical role, as in other Windows applications, namely, follows the mouse movements
- Cursor — Must reside within a document. Can take one of the following three forms:
  - Red crosshair cursor — Marks the blank position within a document where the user can enter text of formulae
  - Editing lines — Blue horizontal (underline) and vertical (insertion line)lines that highlight specific parts in text or formulae
  - Text insertion point — Vertical blue line, similar to the editing lines but intended for text input
- Placeholders — appear within incomplete formulae in positions, which must include symbol or operator:
  - Symbol placeholder — Solid black rectangle
  - Operator placeholder — Black rectangular outline

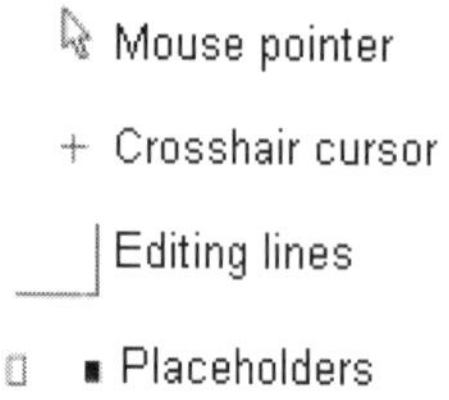

   ⬚ Mouse pointer

   + Crosshair cursor

   | Editing lines

  ⬚ ▪ Placeholders

FIGURE 2.10. Document-editing interface components

Cursors and placeholders related to the formula-editing procedures are shown in Figure 2.10.

2.2.2. Inserting Formulae

You can enter a mathematical expression into any blank position within a Mathcad document. To do so, place the crosshair cursor into the desired position by simply clicking with the mouse, and start typing the formula. As you do so, Mathcad creates a *math region* intended for storing formulae interpreted by the Mathcad processor. Let us demonstrate the sequence of actions with the example of entering the following expression x^{5+x} (Figure 2.11):

1. Click the mouse anywhere within a document.
2. Press the <x> key. A formula region will appear at that point, containing a single character — x. This character will be highlighted by the editing lines.
3. Enter the exponentiation operator by pressing the <^> key or by clicking the exponentiation button on the **Calculator** toolbar. The placeholder for entering the value of the power will appear in the formula, highlighted by the editing lines.
4. Enter the remaining symbols, in the following sequence: <5>, <+>, <x>.

Thus, to insert a formula into the document, simply start typing in symbols, numbers, or operators, such as + or /. In all such cases, a math region is created at the data entry point. This region contains both a formula and editing lines. In the latter cases, if a user starts entering a formula with the operator, placeholders appear automatically depending on the operator type. If you don't fill in the data in the placeholder positions, the Mathcad processor will be unable to interpret the formula (Figure 2.12).

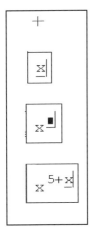

FIGURE 2.11. An example illustrating the process of entering the formula

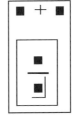

FIGURE 2.12. Examples illustrating the operator entry procedures

2.2.3. Moving Editing Lines within Formulae

To edit the formula, click it with the mouse. The editing lines will appear within the selected formula. Go to the position within that formula from which you want to start editing. Use one of the following techniques to move editing lines within a formula:

- Point the mouse cursor to the required position and click the left mouse button.
- Press arrow keys, the space key, and the <Ins> key on the keyboard:
  - Arrow keys are intended to move the editing lines up, down, left, or right.
  - The <Ins> key moves the data entry point (vertical editing line) to the opposite end of the horizontal editing line.
  - The space character is needed to separate various parts of the formula.

If you press the space key sequentially when entering the formula provided in the example presented above (see Figure 2.11), the editing lines will change their position cyclically, as shown in Figure 2.13. Take a look at Figure 2.14. If you press the left arrow key ($\leftarrow$) when the editing lines take the position shown at the uppermost illustration, they will move left. If you press the space key now, entry lines will alternately highlight each of the two parts of the formula.

TIP

By getting accustomed to using the space character for moving the editing lines within a formula, you can significantly simplify the process of working with Mathcad.

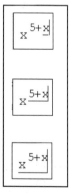

FIGURE 2.13. Changing the position of the editing lines using the space key

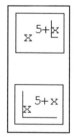

FIGURE 2.14. Changing the position of the editing lines using the left arrow key ($\leftarrow$)

Thus, the combination of arrow keys and space key enables you to easily move the editing lines to any desired position within a formula. You'll get accustomed to this technique with little difficulty when you gain some experience with it. Sometimes placing the edit line into the required position within a formula using the mouse is not a simple task. Consequently, using the keyboard for this purpose is preferable in Mathcad.

2.2.4. Editing Formulae

When editing formulae in Mathcad, proceed according to your intuition and previous experience of working with other text editors. Most formula-editing operations are implemented in the most natural way. Some of these operations, however, are somewhat different from the commonly accepted ones due to the particular features of Mathcad as a calculating system. Let us consider the most basic methods of editing formulae.

Inserting Operators

There are unary operators (operators with a single operand, such as matrix transpose operator or negation operator) and binary operators (operators with two operands, such as addition (+) or division (/)). When the user inserts a new operator into the document, Mathcad determines the number of operands that it requires. If one or both operands are lacking at the insertion point, Mathcad automatically places one or two placeholders near the operator.

When you are inserting an operator, that expression within a formula that the editing lines highlight becomes the first operand or the newly inserted operator.

To insert an operator into the formula, proceed as follows:

1. Place the editing lines in the part of the formula that you want to become the first operand.
2. Insert the required operator by clicking an appropriate toolbar button or by pressing a keyboard combination.

To insert an operator before the expression selected by the editing lines rather than after it, press the <Ins> key before typing an operator. This will move the vertical editing line (insertion point) backward, which is important, for example, when inserting operators such as negation.

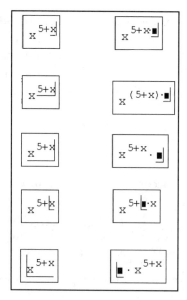

FIGURE 2.15. Examples illustrating insertion of an operator into various parts of a formula

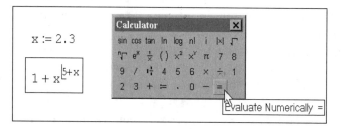

FIGURE 2.16. Insertion of the numerical evaluation operator

Figure 2.15 provides several examples illustrating the insertion of the summation operator into various parts of the formula covered in the example shown in Figure 2.11. The left column shows different positions of the editing lines within a formula, while the right column shows the result of insertion of the summation operator (by pressing the <+> key). As is evident from this example, Mathcad will automatically include parentheses as necessary, to make the selected part of the formula the first term.

Some Mathcad operators will be inserted correctly regardless of the position of the editing lines. The **Evaluate Numerically** operator (=), which, according to its definition, must produce the numeric value of the whole formula, is one such operator. The example shown in Figure 2.16 illustrates the insertion of this operator into the formula using the **Calculator** toolbar. The results of this action are shown in Listing 2.1.

LISTING 2.1. Calculating a Simple Expression

```
x := 2.3
```

$$1 + x^{5+x} = 438.133$$

Selecting Part of a Formula

To select part of a formula within a specific math region (Figure 2.17), proceed as follows:

1. Put the expression that you want to select between the editing lines. Use the arrow keys and space key as necessary.
2. Point the mouse cursor to the insertion line, click and hold on the left mouse button.
3. Drag the mouse cursor along the underline. The selected fragment of the formula will be highlighted by an inverted color.
4. Having selected the required fragment, release the mouse button.

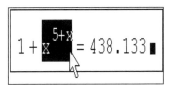

FIGURE 2.17. Selecting a fragment of formula

You can also select fragments of your formulae without using the mouse. To do so, press and hold the <Shift> key, and click the arrow keys to move the highlight. Instead of moving the editing lines, this selects an appropriate fragment of the formula. Most users consider this method more convenient than the previous one.

Deleting Fragments of a Formula

To delete the specific fragment of your formula, proceed as follows:

1. Select the fragment that you want to delete.

2. Press the <Del> key.
3. The formula fragment placed before the vertical (insertion) line can be deleted by pressing the <BackSpace> key. In some cases, for example, when working with complex formulae, you might need to press the <BackSpace> key several times to achieve the desired effect.

There is another method of deleting the formula fragment: select it and then press the <Ctrl>+<X> keyboard combination. By doing so, you'll cut the fragment from the document and insert it into the clipboard. This technique is convenient if the deleted fragment is needed for future use.

Cutting, Copying, and Inserting Formula Fragments

To edit a fragment of your formula, proceed as follows:

1. Select the required fragment or simply place it between the editing lines. This can be done using the mouse or by pressing the arrow keys and space bar.
2. For editing the selected fragment, use the **Edit** command from the main menu (Figure 2.18), right-click menu (Figure 2.19), or appropriate toolbar buttons. In addition, you can also use appropriate keyboard shortcuts:
   - Cut or <Ctrl>+<X> — to cut the selected fragment to the clipboard
   - Copy or <Ctrl>+<C> — to copy the selected fragment onto the clipboard
   - Paste or <Ctrl>+<V> — to insert the contents of the clipboard into the formula

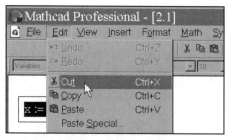

FIGURE 2.18. Editing formulae using the **Edit** command from the main menu

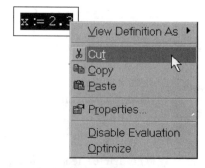

FIGURE 2.19. Editing formulae using the right-click menu

To move (or copy) a formula fragment between different parts of the document, cut (or copy) it to the clipboard, then go to the desired location and paste the contents of the clipboard there.

Modifying Numbers and Names of Variables or Functions

To edit the existing math expression by changing the specific number or name of a variable or function, proceed as follows:

1. Click the name of the required variable or function with the mouse. If necessary, move the editing lines with the mouse or by using the arrow keys and the space bar.
2. Type the required numbers and characters from the keyboard. If necessary, delete the existing numbers or characters by moving the insertion line before them and pressing the <BackSpace> key.

Sometimes it might be more convenient to delete the existing formula fragment and to insert a new name or number into the placeholder.

Modifying Operators

To delete an operator, move the insertion line before it and then press the <BackSpace> key. As a result, the operator will either be deleted (and the operands to the left and to the right of the deleted operator will be merged) or (in complex formulae) replaced by the operator placeholder (black rectangular frame). If necessary, the operator placeholder can also be deleted by repeating the <BackSpace> keystroke.

2.2.5. Entering Symbols, Operators, and Functions

Now it's time to summarize the materials covered up to this point. Usually, math expressions contain various specific symbols, which in Mathcad are entered using procedures different from the ones used in most text processors. Mathcad provides the following techniques of inserting such characters into your documents:

- Most symbols, such as Latin characters or digits in variable or function names, are typed from the keyboard.
- The easiest way to insert Greek characters is by using the Greek toolbar (Figure 2.20). As well, you can insert an appropriate Latin character and then

Figure 2.20. The **Greek** toolbar

press the keyboard shortcut <Ctrl>+<G>. The Latin character will then be transformed to the corresponding Greek character (for example, the letter "a" will be replaced by α).

■ Operators can be inserted either by using various math toolbars or by pressing an appropriate keyboard shortcut. For example, the most frequently used operators (see Figure 2.16) are grouped in the **Calculator** toolbar.

■ Function names can either be typed from the keyboard or inserted by using the Insert | **Function** menu commands. Notice that the latter method is more reliable, since it helps to prevent you from making possible typing *errors (see the Section 1.2, "First Acquaintance with Mathcad," Chapter 1).*

■ Parentheses can be inserted by pressing the appropriate keys on the keyboard. However, a more convenient way of enclosing the existing formula fragment in parentheses is by placing the fragment between the editing lines and pressing the <'> key (apostrophe).

Particular features of the operators and functions usage will be covered in detail in *Chapter 3*.

2.2.6. Managing the Display of Specific Operators

Some operators, such as multiplication or assignment operators, can be represented in different ways in Mathcad documents. Mainly, this was done in order to simplify the process of preparing reports (since, for example, the := symbol looks natural in the Mathcad program, but is often unacceptable in report documentation).

The multiplication operator may be represented by using the following characters (Figure 2.21):

■ Dot
■ Narrow Dot
■ Large Dot

■ ×
■ Thin Space
■ No Space

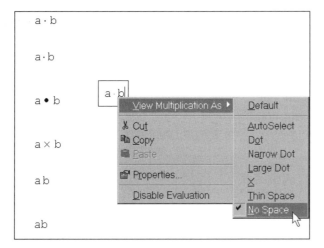

FIGURE 2.21. Various representations of multiplication operator and appropriate options from the right-click menu

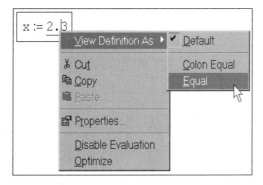

FIGURE 2.22. Modifying representation of the assignment operator

The assignment operator can be represented either by := (combination of colon and equal sign — **Colon Equal**), or simply by the equal character (**Equal**), as shown in Figure 2.22.

To change representation of the aforementioned operators in your documents, do the following:

1. Right-click the operator to display the context menu.
2. Select the first menu item to display the submenu.
3. Select the preferred representation for the operator as shown in Figures 2.21 and 2.22.

Take note of the fact that when editing the formula, the operator representation temporarily changes to that which is set to default, even if you have selected a different representation.

Modify the representation of operators manually only when it is really necessary. Remember that an unusual display of the operator might cause confusion and even lead to errors, especially when studying programs.

You can always reset the operator representation to the default. To do so, select the **Default** command from the right-click menu. Moreover, you can select the **AutoSelect** command, which will instruct the Mathcad editor to display the operator depending on the context.

Selection of the default representation for the two above-mentioned and some other operators is performed on the **Display** tab of the **Math Options** window (see Figure 2.23). To open this window, select the **Math I Options** commands from the main menu. One can specify default operator representations by selecting the required options from appropriate drop-down lists. For example, to set the default view of the multiplication operator go to the **Multiplication** list, and to select representation of the assignment operator set the desired option from the **Definition** list.

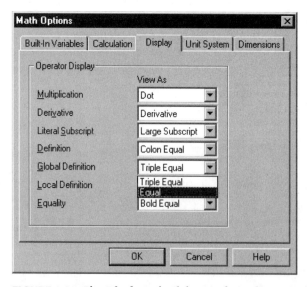

FIGURE 2.23. The **Display** tab of the **Math Options** window

2.3. TYPING AND EDITING TEXT

Mathcad is a math editor mainly intended for typing in math formulae and performing calculations according to these formulae. At the same time, Mathcad provides advanced capabilities for formatting text. The function of text regions within Mathcad documents might vary from task to task and depend on a user. It is recommend that you use a different approach for text fragments used for the following purposes:

■ As comments
■ As an element of document layout intended for creating descriptive reports both in printed and electronic representation

In this section, we will concentrate on the main techniques of working with text. The extended text processing capabilities provided by the Mathcad text processor will be covered in detail in *Chapter 16*.

2.3.1. Text Input

A *text area* (or *text region*) can be placed in any unoccupied location within a Mathcad document. However, when the user places the cursor in an unoccupied position and simply starts entering symbols, Mathcad, by default, interprets them as the initial symbols of a formula. To instruct the program to create a text region rather than a math region, press the <"> key before you begin text input. As a result, a new text region will appear at the cursor's location, which has a characteristic appearance (Figure 2.24). The cursor will look like a vertical red line, known as the *text input line,* which is somewhat similar to the editing lines in formulae.

You can also create a new text region by selecting the Insert | Text Region commands from the main menu.

Now you can start typing your text into the text region. Each new character you type will be inserted into the position designated by the text input line.

FIGURE 2.24. The newly created text region

2.3.2. Editing Text

To edit the existing text within a document, proceed as follows:

1. Click the text area with the mouse. The selected text region will appear in a box (Figure 2.25).
2. If necessary, move the text input line along the text area to the characters that you are going to modify. This can be done by clicking the required location within a text area with the mouse or by moving the text input line with arrow keys and <Home>/<End> keys on the keyboard.
3. Edit the text.

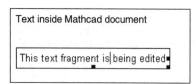

FIGURE 2.25. Text regions within Mathcad documents (the second one is being edited)

To edit the text, you can use the same tools for modifying formulae:

- Select the text fragment by dragging with the mouse or by pressing and holding the <Shift> key and pressing arrow keys.
- Cut, copy, and paste text fragments, either by using keyboard shortcuts such as <Ctrl>+<X>, <Ctrl>+<C>, <Ctrl>+<V>, respectively, or by using appropriate commands from the **Edit** menu, from the right-click menu, or by clicking buttons on the **Standard** toolbar.

*Mathcad also provides advanced text formatting capabilities, such as changing font type and size, specifying text alignment options, etc. Most of these formatting capabilities are implemented by the tools available on the **Formatting** toolbar. They will be covered in detail in Chapter 16.*

2.3.3. Importing Text

Mathcad can import text fragments from other applications, such as Notepad or Microsoft Word. The easiest way to perform this operation is by using the clipboard:

1. When working with another application that you need to import text from, copy the fragment that you want to import.

2. Then go to the Mathcad window and select the required location for the imported text by clicking it with the mouse.

3. Select one of the following techniques:
   - In a Mathcad document, create a new text region by pressing the <"> key. Insert the contents of the clipboard into the new text region by pressing the <Ctrl>+<V> keys. The fragment will be imported as a normal text area (Figure 2.26), which can be edited using standard Mathcad tools.
   - Insert the clipboard contents into the desired location (this is done by pressing <Ctrl>+<V>) without creating a new text region. The text will be inserted as an OLE object; i.e., the application used to create this text will begin to edit that object (Figure 2.27).

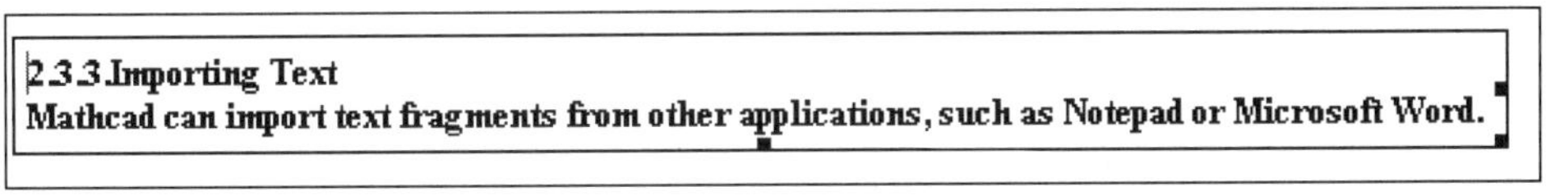

2.3.3.Importing Text
Mathcad can import text fragments from other applications, such as Notepad or Microsoft Word.

FIGURE 2.26. Text fragment imported into the text region

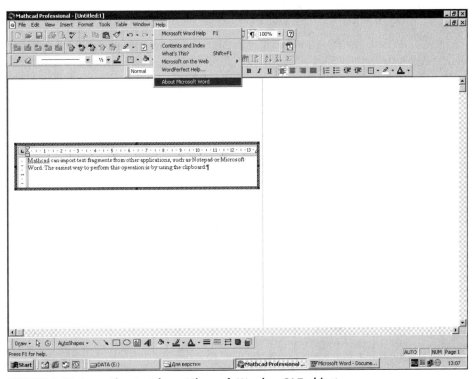

FIGURE 2.27. Importing text from Microsoft Word as OLE object

Import text as an OLE object only in cases where special formatting for that text can't be performed using standard Mathcad functionality. To start editing such objects, double-click them with the mouse. After that, the appropriate tools provided by the program used to create the imported text will appear in the Mathcad window (notice, for example, the Microsoft Word **Help** menu displayed in Figure 2.27). Using these tools, you'll be able to edit the object and then return to the normal Mathcad editing mode by clicking somewhere outside the imported OLE object.

2.3.4. Math Symbols within the Text

To produce high-quality documents, most often you'll need text regions containing math expressions. To create such regions, proceed as follows:

1. Click the required location within a text area.
2. Select the **Insert | Math Region** commands from the main menu or press <Ctrl>+<Shift>+<A> to create a blank placeholder within your text (Figure 2.28).
3. Insert a math expression to replace the placeholder, proceeding just the same way as you do when entering normal formulae (see Section 2.2).

When inserting math formulae into your text, remember that these formulae have the same effect on your calculations as if they were placed into the math region directly in the document. Figure 2.29 shows (from top to bottom) two math regions followed by the text region (which is being edited), where the x variable is assigned a new value; then another math region and another text region, where the x value is displayed. Notice that the x variable has changed its value after being redefined within the first text region.

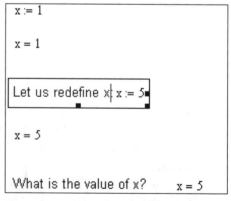

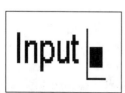

Figure 2.28. Inserting math symbols into the text region

Figure 2.29. Math regions inserted into the text region influence your calculations

If you don't want math regions within your text regions to influence your calculations, you can disable them. To achieve this, switch to the formula-editing mode, and select the Format | Properties commands from the main menu. When the Properties window appears, go to the Calculations tab, set the Disable Evaluations checkbox, and click OK.

2.3.5. Hyperlinks

Sometimes you might need to format the text area in such a way as to make it a hyperlink that moves the cursor to some other location within an active document, to another Mathcad document, or even to the specific URL on the Internet. To insert a hyperlink, use the Insert | Hyperlink commands from the main menu (this topic will be covered in more detail in *Chapter 16*).

2.4. EDITING DOCUMENTS

In the previous sections we discussed the procedures for editing stand-alone text or math regions. Now you will learn how to apply standard editing techniques to other parts of documents, either blank portions or those containing one or more regions. Let's consider in brief the most typical document-editing techniques that are standard for Windows applications.

Selecting a Document Fragment

To select one or more fragments in succession, point the mouse cursor to the first region, click the left mouse button and drag the mouse pointer to the last region that must be selected. All selected regions will be highlighted with a dashed line (Figure 2.30).

You can also select several adjacent regions by clicking the first one, pressing and holding the <Shift> key and then clicking the last region.

To select several non-adjacent regions (Figure 2.31), click the first region to be selected, press and hold the <Ctrl> key and sequentially click all the other regions.

To select all the contents of a document, use the Edit | Select All commands from the main menu or press the <Ctrl>+<A> keys on the keyboard.

To deselect the document fragment(s), click anywhere within a document.

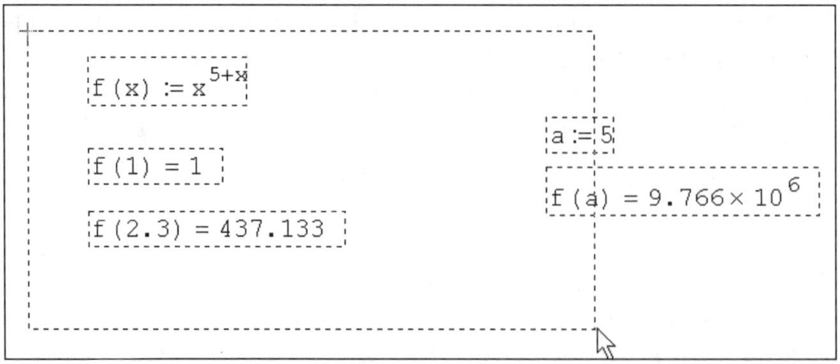

$$f(x) := x^{5+x}$$

$$a := 5$$

$$f(1) = 1$$

$$f(a) = 9.766 \times 10^6$$

$$f(2.3) = 437.133$$

FIGURE 2.30. Selecting a document fragment comprising more than one adjacent region

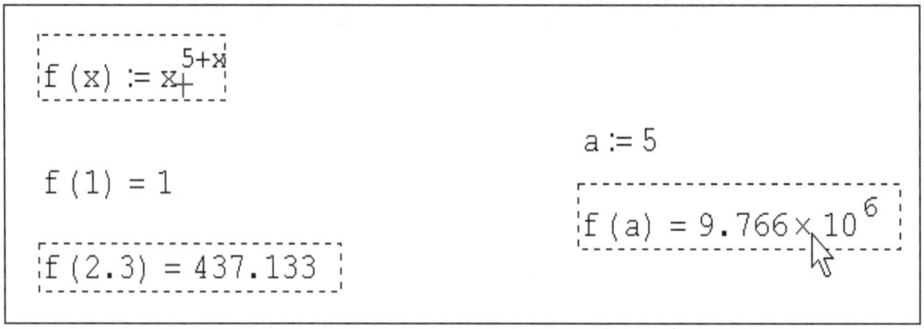

$$f(x) := x^{5+x}$$

$$a := 5$$

$$f(1) = 1$$

$$f(a) = 9.766 \times 10^6$$

$$f(2.3) = 437.133$$

FIGURE 2.31. Selecting several non-adjacent regions

Deleting a Document Fragment

The selected regions can be deleted by pressing <Del> or <Ctrl>+<D>.

The currently selected region can be deleted by pressing <Ctrl>+<D> or by selecting the Edit | Delete commands from the main menu.

Blank rows can be deleted from a document by clicking their upper part with the mouse and pressing the <Del> key.

To enter a blank row below the crosshair cursor click <Enter>.

Cutting, Copying, Inserting, and Moving Document Fragments

To cut, copy, or paste the selected document fragments, use one of the following standard capabilities:

- The Edit command from the main menu
- The right-click menu
- Standard toolbar buttons
- The following keyboard shortcuts: <Ctrl>+<X>, <Ctrl>+<C>, and <Ctrl>+<V>

The easiest way to move and copy the selected document regions is provided by the Drag-and-Drop technology:

- To move the selected region, point to it with the mouse cursor (it will change its appearance to the shape of a hand), then click the left mouse button and drag it with the mouse to the desired location (Figure 2.32). When you release the mouse button, selected regions will be moved to the new location.
- To copy the selected regions, press and hold the <Ctrl> key and drag them with the mouse to the desired location.

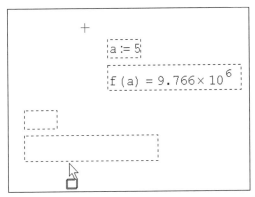

FIGURE 2.32. Dragging document fragment to the desired location

Aligning Regions

For a better display of the documents, Mathcad lets you align both math and text regions vertically at the left edge and horizontally along the upper edge. To align several regions, select one or more regions and then click one of the two available alignment buttons on the toolbar (Figure 2.33) or select the Format I Align Regions commands from the main menu and select the desired option (Across or Down) from the submenu (Figure 2.34).

FIGURE 2.33. The region alignment toolbar buttons

The result of aligning regions is shown in Figure 2.35. To position the regions in a geometrically correct order, you will probably need to repeat the alignment operation several times.

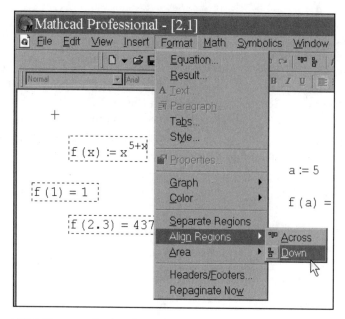

FIGURE 2.34. Aligning the selected regions using menu options

$$f(x) := x^{5+x} \qquad a := 5$$

$$f(1) = 1 \qquad a = 5$$

$$f(2.3) = 437.133 \qquad f(a) = 9.766 \times 10^6$$

FIGURE 2.35. The result of horizontal and vertical alignment of the selected regions

*When you attempt to align regions, they will probably start overlapping. In such a case, Mathcad will display the dialog containing the following question: "Selected regions may overlap. Align selected regions?" If you cancel the operation by pressing the **Cancel** button, the alignment operation won't be performed.*

Refreshing the Display

The Mathcad editor is rather complex, and, therefore, from time to time, the document display might become cluttered with "garbage" — extra characters that actually aren't present in the document. If you suspect that you have encountered such a situation, perform the View | Refresh commands from the main menu or press the keyboard shortcut <Ctrl>+<R>. As a result, all the garbage will disappear.

Searching and Replacing

When working with Mathcad, you can easily search for characters, fragments, or words within a document (Figure 2.36):

1. Select the Edit | Find commands from the main menu or press <Ctrl>+<F>. The Find dialog will appear.
2. Type the text that you want to find into the Find what field.
3. Specify the search options by setting or clearing the following checkboxes as required:
   - Match whole word only
   - Match case
   - Find in Text Regions
   - Find in Math Regions
4. When necessary, specify the search direction by setting the Up or Down radio buttons.
5. Click the Find Next button to start searching the next location within a document where the specified symbol occurs.
6. To close this dialog, click Cancel. You'll move to the location within a document found as a result of the searching procedure.

You can automatically replace symbols within a document by proceeding in a the similar way (Figure 2.37):

1. From the main menu, select the Edit | Replace commands or press the keyboard shortcut <Ctrl>+<H>. The Replace dialog will appear.
2. Fill the Find what field with the text to be replaced.

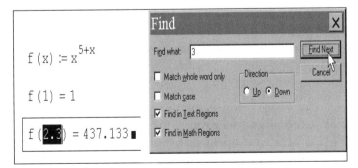

FIGURE 2.36. The results of the searching procedure

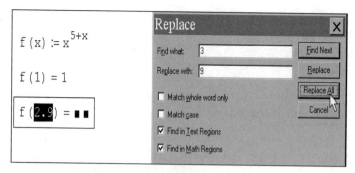

FIGURE 2.37. The result of replacing symbols in a document

3. Enter the text for replacement in the Replace with field.
4. If necessary, specify the search options described above.
5. Press one of the buttons:
    *   Find Next — To find the next occurrence of the specified symbol
    *   Replace — To replace the next symbol found by the search procedure
    *   Replace All — To replace all the characters within a document that satisfy the search criteria
    *   Cancel — To close the Replace dialog

Checking the Spelling

To check English spelling, select text regions that need to be checked and select the **Edit | Check Spelling** commands from the main menu or click the **Standard** toolbar button labeled by the checkmark. If you want to perform spell checking for the whole document, don't select any text regions, but rather, place the input line at the point at which you want to start the check.

Mathcad checks the spelling only within text regions.

If during the check Mathcad finds a word that is missing from its dictionary, it will highlight the word and display the Check Spelling window (Figure 2.38).

The Check Spelling dialog contains the following components:

*   Not Found — The word specified in this field is not found in the dictionary. You'll have to check the spelling yourself and then enter the corrected version.
*   Change To — Displays a list of words from the dictionary that most closely resemble the word displayed in the **Not Found** field. Select the correct suggestion from this list.

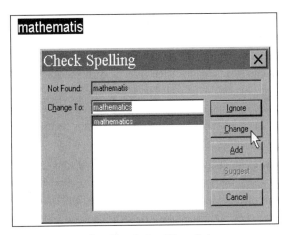

FIGURE 2.38. The **Check Spelling** dialog

- Change — Click this button to replace the word in the document by the corrected one.
- Ignore — Don't replace the word and ignore its incorrect spelling.
- Add — Don't replace the word in a document and add it to the Mathcad dictionary. Later on, this spelling will be interpreted as correct.
- Cancel — Don't perform any replacements, and close the Check Spelling dialog.

2.5. PRINTING DOCUMENTS

To print a copy of the currently active document, press the keyboard combination <Ctrl>+<P> or press the **Print** button on the **Standard** toolbar, labeled with the printer icon.

If you need more advanced capabilities of managing the printing process, use the following commands from the menu File menu:

- Page Setup — Specifies the page setup options for the currently active document, such as page size, margins, source, etc.
- Print Preview — Allows you to preview the currently active document before sending it to the printer.
- Print — Allows you to print the currently active document. Here you can select a printer (if there are several printers), and change printer settings (such as print quality, resolution, number of copies to be printed, and print range).

When you select any of these commands from the File menu, Mathcad opens a dialog with the same name, where you need to specify the options according to your requirements, and then when you are finished, it starts printing the document. Mathcad implements all these capabilities in a way that is standard for all Windows applications.

Notice that when you press the Print toolbar button, the active document will be printed immediately, using the currently selected print options and printer settings.

2.6. SENDING DOCUMENTS VIA E-MAIL

You can send a document via e-mail using any mail client application, such as Microsoft Outlook, by attaching a Mathcad document to the message in the normal way. You can also send the currently active document without exiting Mathcad. To do so, select the File | Send commands from the main menu. The New Message window will appear (Figure 2.39) with the active Mathcad file automatically attached to the message. Now the user only needs to specify the recipient's e-mail address, the message subject, and the text (notice that the latter two items are optional), and to send the message.

To use this option, you need to have an Internet connection. As well, the required mail client application must be installed.

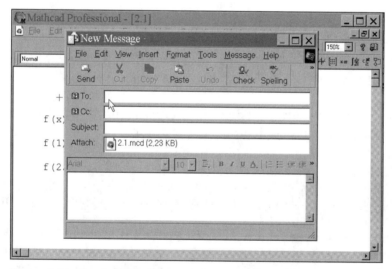

Figure 2.39. Creating a mail message for sending a document via e-mail

3 | Calculations

This chapter covers the basics of Mathcad calculations. It provides all the necessary information on how to use variables, functions, assignment operators, numeric input, and symbolic output *(see Section 3.1)*, and it also considers other operators *(see Section 3.2)*. Finally, we will discuss the basic tools intended for managing calculations in Mathcad *(see Section 3.3)* and briefly consider the error message display *(see Section 3.4)*.

3.1. VARIABLES AND FUNCTIONS

The main tools any mathematician uses are operations involving variables and functions. In Mathcad variables, operators and functions are implemented in an intuitive manner, which means that you input expressions in the Mathcad editor in just the same way as you would on a sheet of paper. The order of calculations in a Mathcad document is also obvious: math expressions and operations are interpreted by the processor from left to right and from top to bottom.

Let us review, in brief, all the main actions that the user can perform for defining variables and functions and performing their output.

3.1.1. Defining Variables

To define a variable, it is sufficient to specify its name and assign some value to it (the assignment operator that will be discussed in the next section is used for this purpose).

3.1.2. Assigning Values to Variables

To assign a new value to a variable (for example, to make the x variable take the 10 value), do the following:

1. Go to the desired location within a document and enter the variable name, for example, x.
2. Insert the assignment operator. To do so, you can use the <:> key on the keyboard or click the Definition button on the Calculator toolbar or the Evaluation toolbar, as shown in Figure 3.1.
3. Enter the variable value (10) to replace the placeholder.

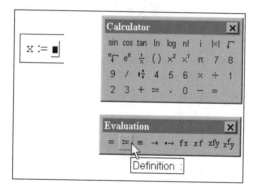

FIGURE 3.1. The result of using the assignment operator

Listing 3.1 shows the result of these actions described.

For the sake of convenience, the assignment operator button is included in two toolbars: Calculator and Evaluation.

NOTE

LISTING 3.1. Assigning Numeric Value to a Variable

```
x := 10
```

The new value assigned to a variable might represent a number, a math expression containing other variables (Listing 3.2) and functions *(see subsequent chapters)*, or a string expression (Listing 3.3). In the latter case, we are dealing with the s variable of the string data type.

LISTING 3.2. Assigning a Numeric Expression to a Variable

$$x := 10$$

$$y := (x - 3)^2 + 1$$

LISTING 3.3. Assigning a String Value to a Variable

$$s := \text{"Hello, "}$$

If this is the first time that a specifically named variable occurs in the current document, you can use the equals sign ("=") instead of the assignment operator, since Mathcad will automatically replace it with the assignment operator (:=).

Sometimes this is not possible (for example, when you are assigning a value to the variable with the name reserved by Mathcad for specific purposes). For example, to assign a value to the variable named N, you must always use the := assignment operator, since, by default, this name is reserved by Mathcad to designate the dimensions of force (newtons, N).

To redefine the value of a variable defined in the document, use the := assignment operator rather than the equal sign, or, alternately, use the toolbar.

This difference of the assignment operator from the commonly accepted math style (:= rather than =) is actually a compromise due to the fact that Mathcad is a programming system. In contrast to other operators, the assignment operator evaluates the expression on the right of the ":=" and assigns the result to the variable on the left of the ":=". Don't let the unusual look of this operator deceive you. From Mathcad's point of view, this operator designates that the variable value *is not displayed on the screen* (this action is designated by the =), but, rather, the value on the right part is *assigned* (:=) to this variable.

On the other hand, when preparing reports, you might need to change the display of the assignment operator from the default (":=") to the equal sign. To change the default display of a single assignment operator, right-click

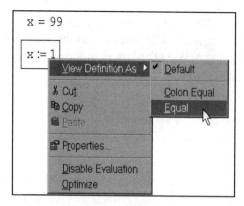

FIGURE 3.2. Changing the display of the assignment operator via the right-click menu

the specific operator and select the **View Definition As** command from the right-click menu (Figure 3.2). To change the assignment operator display for the whole document, select the **Math | Options | Display** commands from the main menu *(see Section. 2.2.6, "Managing the Display of Specific Operators," in Chapter 2).*

Besides the assignment operator that we have just considered (it is used most frequently), Mathcad also provides the capability of *global assignment or global definition.*

Global assignment will be covered in detail in *Section 3.2.5.*

3.1.3. Functions

In Mathcad, functions are written in the form that is commonly accepted by mathematicians:

- ◾ f(x, . . .) — The function:
    - ● f — Function name
    - ● x, . . . — List of arguments

The easiest way to enter function names is by typing the name from the keyboard. In Mathcad, the functions are formally classified by the following two types:

- ◾ Built-in functions
- ◾ User-defined (custom) functions

Both function types in calculations are used in practically the same way, with the single exception that any built-in function can be used immediately at any location within a document (information on the insertion of built-in functions was provided in *Section 1.2, "First Acquaintance with Mathcad"*). In contrast to built-in functions, a user-defined function must first be defined in a document before you can use it in your calculations.

3.1.4. Creating User-Defined Functions

To create a user-defined function, for example, $f(x,y) = x^2 \cdot \cos(x+y)$, do the following:

1. Go to the desired location within a document and enter the function name (f).
2. Type in the left (opening) parenthesis "(", then the names of the function arguments separated by a comma (x, y) and then the right (closing) parenthesis ")". When you type the left parenthesis and commas, Mathcad will display the appropriate placeholders.
3. Enter the assignment operator from the toolbar or by pressing the <:> key.
4. The placeholder will appear. Enter the expression to determine your user-defined function (for example, $x^2 \cdot \cos(x+y)$), using the keyboard or available toolbars.

The result of this operation is illustrated in Listing 3.4.

LISTING 3.4. User-Defined Function

$$f(x,y) := x^2 \cdot \cos(x+y)$$

All variables present in the right part of the function definition must either be included in the list of function arguments (in parentheses, next to the function name) or defined earlier. If they are not, Mathcad will display an error message, and display the name of the undefined variable in red (Figure 3.3).

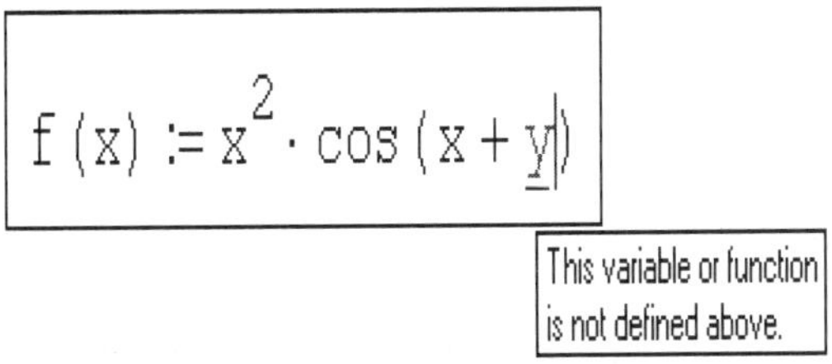

$$f(x) := x^2 \cdot \cos(x + \underline{y})$$

This variable or function is not defined above.

FIGURE 3.3. Error message informing the user of the presence of an undefined variable

3.1.5. Displaying Variable and Function Values

To calculate some math expression that might include variables, operators, and functions (both built-in and user-defined), proceed as follows:

1. Enter the expression to be calculated, for example: x^y.
2. Press the <=> key.

The calculated value of the expression that you have entered will appear to the right next to the equal sign (Listing 3.5, the last line). You can't edit the value to the right of the equal sign, since it represents the result produced by Mathcad's numeric processor, which is totally hidden from the user. Sometimes (when the expression contains functions that implement numeric methods, often in rather complex combinations) calculation algorithms are very complicated and time-consuming. Expressions within a document that are in the process of calculation are framed by a green border. Furthermore, you can't perform any action in Mathcad when the calculation is in progress.

LISTING 3.5. Calculating an Expression

$$x := 10$$

$$y := (x - 3)^2 + 1$$

$$x^y = 1 \times 10^{50}$$

$$x = 10$$

Notice that before you start calculating a math expression, you must define the value of each variable included in it (notice the first two lines in Listing 3.5). The expression can contain any number of variables, operators, and functions. The last line in Listing 3.5 demonstrates the output of the current value of specific variables, while the output of function values is illustrated in Listings 3.6 and 3.7.

LISTING 3.6. Function Value Output

$$f(x, y) := x^2 \cdot \cos(x + y)$$

$$f(2, 5.99) = -0.542$$

$$f(1.3, 7) = -0.729$$

LISTING 3.7. Function Value Output *(Listing 3.6, Continued)*

x := 1.3

y := 7

f (x, y) = −0.729

NOTE

When creating custom functions using miscellaneous variables, it is important to make sure that the names of the variables used in the function definition are present in the argument list or defined earlier in the document text. For example, the output results of the f(x,y) function (Listing 3.6) would remain the same, even if x and y variables are assigned specific values before or after the function definition. This happens because the argument values are specified directly in the line performing the calculation of the function value. However, if we define the f(x) function as shown in Listing 3.8, it will depend on the value of the y variable at the moment when the f(x) value is calculated (i.e., y = 5), since y is not included in the list of f(x) arguments. Actually, the function can be written as follows: $f(x) = x^2 \cdot cos(x + 5)$. Even if the user redefines the y somewhere later in the program, Mathcad will "remember" the f(x) function as the $x^2 \cdot cos(x + 5)$ expression Listing 3.9).

LISTING 3.8. Example Illustrating User-Defined Functions

y := 5

$f (x) := x^2 \cdot cos (x + y)$

f (1) = 0.96

LISTING 3.9. Example Illustrating User-Defined Functions *(Listing 3.8, Continued)*

y := 0

x := 1

f (x) = 0.96

$x^2 \cdot cos (x + y) = 0.54$

$x^2 \cdot cos (x + 5) = 0.96$

Notice that the number of arguments must be the same when defining the function and performing the output of function values. Compare, for example, Listings 3.6 and 3.8, where two different functions are created — f(x,y) and f(x), respectively, despite the fact that the expressions in the right parts are the same.

By entering the equal sign for calculating math expressions in Mathcad, you are applying the evaluation operator. To insert this operator, you can use the toolbar button labeled by the equal sign (such buttons are available on both the **Calculator** and **Evaluation** toolbars) (see Figure 3.1). The numerical evaluation operator ensures that all calculations are performed with numbers, and that all built-in algorithms are implemented by appropriate numerical methods.

3.1.6. Symbolic Output

Besides numerical calculations, Mathcad makes it possible to perform symbolic (or analytical) calculations. For symbolic calculations, Mathcad provides a range of special tools that will be covered in detail later in this book *(see Chapter 5)*. The simplest of these tools is the symbolic evaluation operator. It is designated by the → symbol and in most cases, is used in a way similar to that of the numerical evaluation operator. However, the internal difference between these two operators is tremendous. Numerical output in the normal sense of this term is simply a "programmed" calculation, according to the formulae and numerical methods, with technical details hidden from the user. In contrast, symbolic output is the result produced by Mathcad's built-in Artificial Intelligence system (AI), known as the symbolic processor. The details of the symbolic processor's work are also hidden from the user (and, most often, even difficult to imagine). Actually, the symbolic processor parses and analyses the text of mathematical expressions themselves. Of course, the range of formulae that can be calculated symbolically is narrower, since, generally speaking, a relatively small part of math problems has analytical solutions.

To calculate a math expression analytically (for example, B·sin(arcsin(C · x)), where B, C, x — are some variables), proceed as follows:

1. Enter the expression that you need to calculate: B · sin(asin(C · x)).
2. Insert the symbolic evaluation operator by pressing the <Ctrl>+<.> keys or by clicking an appropriate button (Figure 3.4), which is available on the **Symbolic** or **Evaluation** toolbar.

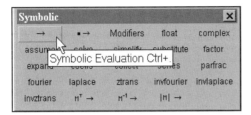

FIGURE 3.4. Toolbar button for insertion of the symbolic evaluation operator

After this is completed, either the analytically determined value of the expression will appear to the right of the symbolic evaluation operator (Listing 3.10), or the following error message will be displayed: "No answer found." If Mathcad's symbolic processor can't analytically simplify the expression, it will display the expression to the right of the $\rightarrow$ sign without changing it.

LISTING 3.10. Symbolic Output of the Expression

$$B \cdot \sin(\text{asin}(C \cdot x)) \rightarrow B \cdot C \cdot x$$

LISTING 3.11. Symbolic Output of the Expression That Can't Be Simplified

$$x^2 \cdot \cos(x + y) \rightarrow x^2 \cdot \cos(x + y)$$

Pay special attention to Listings 3.10 and 3.11 and take note of the fact that for symbolic output you don't need to define variables included in the left part of the expression in advance! If variables were previously assigned some values, the symbolic processor will simply substitute them in the simplified formula and produce a result that takes these values into account. Consider, for example, the following two listings — 3.12 and 3.13).

Function values can also be calculated using the symbolic processor. This is done in exactly the same way as when you are using a numeric processor. For example, compare the results provided in Listings 3.12 and 3.13 (of course, the symbolic and numeric results are equal: $9 \cdot \cos(8) = -1.31$). Similarly, it is also possible to perform the symbolic output of variable values. For example, you can assign some variable a function value or complex expression (Listing 3.13, the second line) and then output the value of that variable in symbolic representation.

LISTING 3.12. Numeric and Symbolic Output of Function Value

$$f(x) := x^2 \cdot \cos(x + 5)$$

$$f(3) = -1.31$$

$$f(3) \rightarrow 9 \cdot \cos(8)$$

LISTING 3.13. Numeric and Symbolic Output of Variable

$$f(x) := x^2 \cdot \cos(x + 5)$$

$$a := f(3)$$

$$a \rightarrow 9 \cdot \cos(8)$$

$$a = -1.31$$

As is obvious from the preceding examples, the advantage of symbolic calculations is the fact that they produce analytical results, which are more valuable from the mathematician's point of view. Therefore, based on characteristics of a specific task, you must decide if you should try to obtain an analytical solution in addition to a numeric one.

3.1.7. Valid Names for Variables and Functions

To conclude this topic, let us consider the characters that can and can't be used in variable and function names and provide a list of limitations that must be observed when assigning names. Valid characters include the following:

- Upper- and lowercase letters — Mathcad distinguishes the case of the character; thus, x and x define different variables. In addition, Mathcad also determines the font; therefore, the names x and x will be considered different names.
- Numbers from 0 to 9.
- Infinity characters (keyboard shortcut <Ctrl>+<Shift>+<Z>).
- The prime symbol (keyboard shortcut <Ctrl>+<F7>).
- Greek letters — they are inserted using the Greek toolbar.
- Underscore character.
- Percent character.
- Subscripts.

Be careful when using subscripts in the names of variables and functions; don't confuse them with array subscripts. To enter a name containing subscript characters, such as K_{max}, for example, enter the "K" letter, then the period ("."), after which the insertion lines will extend down somewhat lower, and then enter the subscript max.

Now let us consider the limitations for the names of variables and functions:

- The name can't start with the digit (0 to 9), underscore (_), prime symbol (') or a percent (%).
- The infinity symbol (∞) can only be the first character of the name.
- All characters in the name must have the same font, point size, and be of the same style.
- The names can't coincide with the names of the built-in functions, constants, or units, such as sin or TOL. You can redefine these reserved built-in names, but keep in mind that the similarly named built-in function will no longer be available after the definition.
- Mathcad doesn't distinguish between variable names and function names. For example, if you first define the function f(x), and you later define the variable f, you'll find out that you can't access the f(x) function anywhere in the document below the definition of the f variable.

Names Containing Operators and Special Characters

Sometimes it might be desirable to use function and variable names containing symbols of Mathcad operators or other characters that can't be inserted into the name directly. Mathcad provides the following capabilities to do so.

First, the name composed from any characters and enclosed in square brackets will be interpreted correctly (Figure 3.5, the first line). For example, to enter a name such as [a + b], do the following:

1. Press <Ctrl>+<Shift>+<J> keys — a pair of square brackets with the placeholder enclosed within them will appear.
2. Enter any sequence of characters to replace the placeholder, for example: a + b.

If you don't like the presence of the square brackets in the name, you can use a somewhat more complex method to insert special characters. For example, to enter the a+b name, proceed as follows:

1. Enter the first character (a), which must be a valid character for Mathcad names.

$$[a+b] := 7$$

$$\$:= 5$$

$$a+b := 0$$

FIGURE 3.5. Using special characters in variable names

2. Press <Ctrl>+<Shift>+<K> to switch to the special "text" editing mode.
3. Enter any sequence of characters (+).
4. Press <Ctrl>+<Shift>+<K> once again in order to return to the normal editing mode. Now you can continue entering valid characters into the name (b).

The result of these actions is shown in the last line in Figure 3.5. If you need to enter the name starting with a special character (like the one in the second line of Figure 3.5), it is necessary to perform all the steps from 1–4, starting the name with any valid character, and to delete this character upon completion of the input.

3.2. OPERATORS

Each operator in Mathcad designates some math operation by representing it as a character. In complete accordance with the terminology that is conventional in mathematics, some actions (such as addition, division, matrix transposition, etc.) are implemented in Mathcad as built-in operators, while other actions (for example, sin, erf, etc.) are carried out as built-in functions. Each operator is applied to one or two numbers (variable or function), which are known as *operands*. If one or both operands are missing at the moment the operator is inserted, the missing operands will be displayed as placeholders. To insert any operator into the required position in your document, use one of the following methods:

■ Press an appropriate key (or key combination) on the keyboard.
■ Click an appropriate button on one of the available mathematical toolbars.

I'd like to remind you that most mathematical toolbars contain toolbar buttons grouped according to the math operators. To display these toolbars, click the appropriate button on the **Math** toolbar.

In this section, and later on, we will only consider the second method of inserting an operator. Those of you who prefer to use the keyboard can refer to the list of keyboard shortcuts in Mathcad's online Help system.

Earlier in this chapter we considered specific features that are characteristic for using the following three operators: the assignment operator *(see Section 3.1.2)*, numeric *(see Section 3.1.5)* and symbolic output *(see Section 3.1.6)*. Now let us consider the principles of using other Mathcad operators and the capabilities of creating user-defined operators.

3.2.1. Arithmetic Operators

Operators designating main arithmetic operations are entered from the Calculator toolbar shown in Figure 3.6:

- Addition and subtraction: + and − (Listing 3.14)
- Multiplication and division: ·, / and ÷ (Listing 3.15)
- Factorial: ! (Listing 3.16)
- Modulus of a number: |x| (Listing 3.16)
- Square root: $\sqrt{}$ (Listing 3.17)
- n-th root: $\sqrt[n]{}$ (Listing 3.17)
- Raising x to the power of y: x^y (Listing 3.17)
- Priority change: () (Listing 3.18)
- Numeric output: = (all listings)

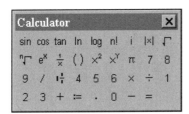

FIGURE 3.6. The **Calculator** toolbar

LISTING 3.14. Addition, Subtraction, and Negation Operators

$1 + 3 - 7 = -3$

$-(-2) = 2$

LISTING 3.15. Division and Multiplication Operators

$$\frac{5}{2} = 2.5$$

$$5 \div 2 = 2.5$$

$$2\frac{3}{4} = 2.75$$

$$1 \cdot 2 \cdot 3 \cdot 4 \cdot 5 = 120$$

LISTING 3.16. Factorial and Magnitude (Modulus) Operators

$$5! = 120$$

$$\left|-10\right| = 10$$

LISTING 3.17. Root and Exponentiation Operators

$$\sqrt{4} = 2$$

$$\sqrt[3]{125} = 5$$

$$e^{\ln(3)} = 3$$

$$3^2 = 9$$

$$10^{0.2} = 1.585$$

LISTING 3.18. Priority Change Operator

$$(1 + 2) \cdot 3 = 9$$

$$1 + 2 \cdot 3 = 7$$

As you can see from the illustration, this toolbar allows you to enter not only the operators listed here but also their most frequently used combinations, such as raising the exponent to a power, mixed multiplication and division, along with an imaginary unit and π. Notice that the division operator can be written on either one or two lines (notice the two respective buttons on the Calculator toolbar).

Also notice that the Mathcad editor allows you to select the display of the multiplication operator *(see Section 2.2.6, "Managing the Display of Specific Operators," in Chapter 2)*. To change the display mode of the multiplication operator, do the following:

1. Right-click the expression containing the multiplication operator.
2. Select the first command (View **Multiplication As**) from the right-click menu.
3. Select the item that corresponds to the required representation style for multiplication from the submenu: as a normal dot (**Dot**); a dot with a narrow distance between multipliers and the operator (**Narrow Dot**); a large dot (**Large Dot**); a cross (**Ч**); a narrow space separating the multipliers (**Thin Space**); or without any operator symbol or space between multipliers (**No Space**). To view the expression, as it would look when using the last two representations, you need to remove the selection from it. To return to the default representation, select the **Default** command from the right-click menu.

Some operators, such as the complex conjugate operator, are not on the toolbars (Listing 3.19). The only way to insert this operator is by pressing the <"> key on the keyboard (you must perform this operation with the required expression highlighted within the math region since striking the <"> key in a blank position within a document will create a new text area).

LISTING 3.19. Complex Conjugate Operator

$$\overline{(3 + i)} = 3 - i$$

3.2.2. Calculus Operators

Calculus operators are inserted into the document using the Calculus toolbar. When you click on any of these toolbar buttons, the symbol designating an appropriate math operation will appear in the document, having several placeholders. The number and position of the placeholders is determined by the operator type and precisely corresponds to the conventional mathematical notation. For example, when you insert the summation operator (Figure 3.7), you need to specify the following four values: the variable according to which summation must be performed, lower and upper limits, and the expression under

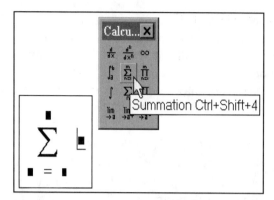

FIGURE 3.7. Insertion of the summation operator

the summation character (the example of the summation operator is shown in Listing 3.22). To calculate an indefinite integral, you need to fill two placeholders: the integrand and integration variable.

After you insert any calculus operator, you have the option of evaluating its value either numerically, by pressing the <=> key, or symbolically, by using the symbolic output operator.

Let us consider the calculus operators in brief and provide the simplest examples of their usage:

- Differentiation and integration:
  - Derivative (Listing 3.20)
  - N-th derivative (Listing 3.20)
  - Definite integral (Listing 3.21)
  - Indefinite integral (Listing 3.21)
- Summation and calculation of the product:
  - Summation (Listing 3.22)
  - Iterated product (Listing 3.22)
  - Summation with range variables (Listing 3.23)
  - Iterated product with range variables (Listing 3.23)
- Limits (Listing 3.24):
  - Two-sided limit
  - Left-handed limit
  - Right-handed limit

LISTING 3.20. Operators for Evaluating Derivatives

$$\frac{d}{dx}\sin(x) \to \cos(x)$$

$$\frac{d^2}{dx^2}\sin(x) \to -\sin(x)$$

LISTING 3.21. Integration Operators

$$\int_a^\infty \frac{1}{x^3} \, dx \to \frac{1}{2 \cdot a^2}$$

$$\int \ln(x) \, dx \to x \cdot \ln(x) - x$$

LISTING 3.22. Summation and Iterated Product Operators

$$\sum_{i\,=\,1}^{10} i = 55 \qquad \sum_{i\,=\,1}^{10} i \to 55$$

$$\prod_{i\,=\,1}^{30} i = 2.653 \times 10^{32}$$

LISTING 3.23. Summation and Iterated Product with Range Variables

$$i := 1 .. \ \ 5$$

$$\sum_i i^2 \cdot i! = 3.447 \times 10^3$$

$$\prod_i e^i = 3.269 \times 10^6$$

Range variables and specific features of their usage will be covered in detail in the next chapter (see Section 4.3.2, "Range Variables," in Chapter 4).

LISTING 3.24. Operators for Symbolic Evaluation of Limits

$$\lim_{x \to \infty} \frac{1 + 3 \cdot x}{x} \to 3$$

$$\lim_{x \to 0^+} \frac{1}{x} \to \infty$$

$$\lim_{x \to 0^-} \frac{1}{x} \to -\infty$$

In contrast to other operators, limit operators can only be evaluated symbolically (see Chapter 5).

Summation and iterated product operators are actually more convenient forms of denoting summation (+) and product (×) operators with a large number of operands. In contrast to these operators, differentiation and integration operators are significantly different from summation and product evaluation operators. The most significant difference is found in the fact that they are implemented based on specific numeric methods that are run by Mathcad's numeric processor in a form hidden from the user. When performing a numeric evaluation of derivatives and integrals, it is essential that you understand the basics of respective algorithms in order to avoid errors or unexpected events when obtaining the results (Chapter 7 concentrates on the numeric methods of differentiation and integration).

It is important to notice that Mathcad is capable of evaluating integrals with one or both infinite limits, and of finding analytic solutions when working out the values of infinite limits, sums, and products. For convenience, the infinity button is available on the same toolbar (**Calculus**). Figure 3.8 illustrates an example of inserting the infinity symbol when evaluating the sum of an infinite series.

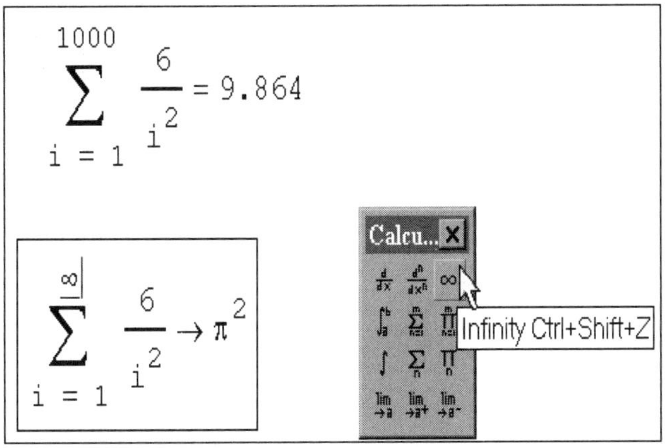

$$\sum_{i=1}^{1000} \frac{6}{i^2} = 9.864$$

$$\sum_{i=1}^{\infty} \frac{6}{i^2} \to \pi^2$$

FIGURE 3.8. Evaluating the sum of an infinite series

3.2.3. Logical Operators

The results of logical (Boolean) operators are 0 (if the logical expression is false) or 1 (if the logical expression is true). To evaluate a logical expression, such as 1 = 1, proceed as follows (Figure 3.9):

1. Insert the required operator (=) from the **Boolean** toolbar.
2. Fill in two placeholders with the operands (two values of 1).
3. Press the <=> key on the keyboard.

You'll get a solution that might seem absurd at first: 1 = 1 = 1. Actually, however, everything is correct. There is a logical expression (1 = 1) to the left of the evaluation operator. Notice that a logical equal sign looks somewhat different from an ordinary one. The logical expression in the left part is true. Therefore, the value of this logical expression is equal to 1, and this value appears to the right from the equal sign.

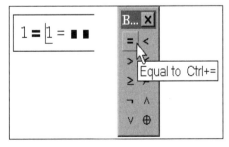

FIGURE 3.9. Insertion of the logical operator

Logical operators are listed below:

- Greater Than — $x > y$
- Less Than — $x < y$
- Greater Than or Equal — $x \geq y$
- Less Than or Equal — $x \leq y$
- Equal — $x = y$
- Not Equal to — $x \neq y$
- And — $x \wedge y$
- Or — $x \vee y$
- Exclusive or — $x \oplus y$
- Not — $\neg x$

The operands in logical expressions can take the form of any number. However, if the operator is applicable to 0 or 1 only, then any nonzero value is, by default, set to 1. Thus, the result produced by Boolean operators can be equal either to 0 or to 1. For example, $\neg(-0.33) = 0$.

The examples illustrating the principles of logical operators are provided in Listings 3.25 and 3.26.

LISTING 3.25. Comparison Operators

$2 \equiv 3 = 0$ $\qquad$ $5 > 1 = 1$ $\qquad$ $3 > 3 = 0$

$7 \equiv 7 = 1$ $\qquad$ $3 < \infty = 1$ $\qquad$ $3 \geq 3 = 1$

$0 \neq 0 = 0$

LISTING 3.26. Boolean Operators

$1 \vee 0 = 1$ $\qquad$ $1 \wedge 0 = 0$ $\qquad$ $1 \oplus 0 = 1$ $\qquad$ $\neg 1 = 0$

$0 \vee 0 = 0$ $\qquad$ $0 \wedge 0 = 0$ $\qquad$ $0 \oplus 0 = 0$ $\qquad$ $\neg 0 = 1$

$1 \vee 1 = 1$ $\qquad$ $1 \wedge 1 = 1$ $\qquad$ $1 \oplus 1 = 0$

Logical operators are very important when writing algebraic equations and inequalities in a form acceptable for Mathcad.

3.2.4. Vector and Matrix Operators

Vector and matrix operators are intended for performing various operations with vectors and matrices. Since most of them implement numeric algorithms, they will be covered in detail in Part III *(see Chapter 9)*.

3.2.5. Evaluation Operators

Nearly all evaluation operators were discussed above *(see Section. 3.1)*. They are grouped on the **Evaluation** toolbar.

■ Evaluate Numerically — = *(see Section 3.1.5)*
■ Evaluate Symbolically — → *(see Section 3.1.6)*
■ Definition — := *(see Section 3.1.2)*
■ Global Definition — ≡

Now let us discuss the difference between normal definition and *global definition* operators (the process of inserting a global definition operator into your document is illustrated in Figure 3.10). To evaluate an expression containing some variable or function, it is necessary to assign some value to this variable or function somewhere earlier in the document. If this requirement is not satisfied, you'll get an error message (Figure 3.11). However, if you insert a global definition operator into any part of the document (for example, somewhere at the bottom), the variable defined by this operator will be defined in any part of your document (Listing 3.27).

LISTING 3.27. Definition and Global Definition Operators

$$x = 5$$

$$x := 10$$

$$x = 10$$

$$x \equiv 5$$

$$x = 5$$

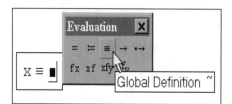

FIGURE 3.10. Global Definition button on the **Evaluation** toolbar

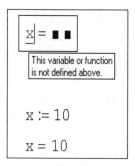

FIGURE 3.11. The normal definition operator only influences the subsequent part of the document

As is obvious from Listing 3.27, normal or *local* definition influences the x variable only from the instance $x := 10$ to the global definition $x \equiv 5$. Generally speaking, Mathcad analyzes documents for variable definition in a two-pass algorithm: first, it recognizes all global definitions and evaluates all expressions from top to bottom and from left to right according to these global definitions. On the second pass, it analyzes all local definitions in the same order and re-evaluates all expressions according to this correction. Let us provide an important example illustrating the interaction between local and global definitions (Listing 3.28).

LISTING 3.28. Interaction between Global and Local Definitions

$x \equiv 5$

$x = 5$

$x := 10$

$x = 10$

$y \equiv x^2$

$y = 25$

Pay special attention to the fact that the x variable is defined locally in the third line of this listing ($x := 10$), and that the value of y is still calculated according to the global definition $x \equiv 5$, since the variable y itself is globally defined using the x variable.

Be very careful when dealing with global variable definitions. To avoid confusion, don't redefine them locally. Use global definition only for defining constants. If possible, avoid situations where the evaluation operator precedes the global definition, since this will improve readability of your documents.

You can also use global definitions for functions. This is done in a way similar to that used for variable global definitions (Listing 3.29).

LISTING 3.29. Global Definition of the User-Defined Function

$$f(2) = 128$$

$$f(x) \equiv x^7$$

The global definition operator can be displayed both as an identical equal (triple equal) sign and as a normal equal sign. To specify this option, right-click the global definition operator, then select the View Definition As command from the context menu, and select the Equal command from the submenu.

3.2.6. Creating User-Defined Operators

An advanced user might not be satisfied by the set of built-in Mathcad operators. To insert predefined user operators into your documents, use the **Evaluation** toolbar.

Naming Operators

User-defined operator can have absolutely any name (refer to the "Names Containing Operators and Special Characters" section earlier in this chapter). However, based on the mathematical sense of the operators, it would be logical to name them using symbols. The most convenient way to perform this operation is by using the collection of symbols in the Resource Center. Open the **Resource Center** window, select the **QuickSheets and Reference Tables** link and then open the section labeled **Extra Math Symbols**. You'll find quite a large collection of symbols, each of which can be simply dragged with the mouse to the required position within your document.

To assign a specific action to the user-defined operator, proceed the same way as you would with user-defined functions.

Creating a Custom Binary Operator

To create a custom binary operator (for example, one implementing the following operation: $x \cdot y^2$), proceed as follows:

1. Enter the name for the operator, for example, `bin`.
2. Enter the left parenthesis sign ("("), then the list of two operands separated by commas — "x", ",", "y" — then the right parenthesis (")").
3. Enter the assignment operator ":".
4. Type in an expression that will be assigned to the operator (in our example, this will be $x \cdot y^2$).

Creating a Custom Unary Operator

Unary operators must be created in a similar way. However, in this case, instead of two comma-separated operands, you need to type in a single operand. For example, to create an operator named % while implementing a conversion of fractions to percentages (Listing 3.30), proceed as follows:

1. Type in the name of the custom operator. To do so, press the following keys: <a>, <Ctrl>+<Shift>+<K>,<%>; then once again press <Ctrl>+<Shift>+<K>; then delete the letter "a" in the name of the custom operator.
2. After the % sign, enter the left parenthesis "(", followed by "x", then the right parenthesis ")".
3. Enter the assignment operator by pressing the <:> key.
4. Enter the following expression: $x \cdot 100$.

LISTING 3.30. Creating a Custom Unary Operator

```
% (x) := x·100
```

Using the Binary Operator

There are two ways to insert a custom binary operator into the document, differing only in terms of their display in the document. To insert an operator using graph (tree) notation, proceed as follows:

1. Click the **Tree Operator** button on the **Evaluation** toolbar (Figure 3.12, left part).
2. Fill in the placeholders with the operator name (the middle placeholder) and values of the operands (the remaining two "branch" placeholders).
3. Enter the assignment operator by pressing the <=> key.

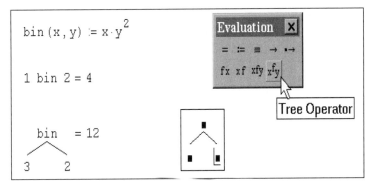

FIGURE 3.12. The usage of the custom binary operator

The result of this operator's action is shown in Figure 3.12, in the bottom-left corner.

In addition to the above-described notations, also known as "treefix", the operator can be used in so-called "infix" notation — i.e., as the following sequence: "first operand — operator name — second operand" (Figure 3.12, the second line in the left part). To enter the operator in "infix" notation, click the **Infix Operator** toolbar button labeled as **xfy**.

Using the Unary Operator

Insertion of custom unary operators is similar to that of custom binary operators. The only difference is that instead of two operands, you need to enter only one (Figure 3.13). To insert a custom unary operator, click the **fx** (Prefix operator) or the **xf** (Postfix operator) button on the **Evaluation** toolbar. The right part of Figure 3.13 (at the point of insertion) and the result of the action (at the left) illustrate the first notation. The second notation is illustrated on the lower-left line in the same picture.

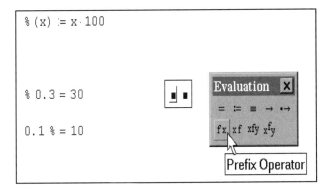

FIGURE 3.13. The usage of the custom unary operator

3.3. MANAGING CALCULATIONS

A Mathcad document is a full-featured computer program. The Mathcad system itself represents a fully functional programming environment, however, it's oriented toward a mathematician rather than a professional programmer. In contrast, most other programming environments (such as implementations of popular programming languages including C, FORTRAN, Basic, etc.) separate the processes of editing code and running programs. In Mathcad, the program code and the result produced by the program are joined within a document. Still, the functions of editing formulae and calculations according to these formulae are separate, and the user is capable of managing all the important aspects of their calculations.

3.3.1. Calculation Modes

All the examples that we consider in this book implicitly suggest that you are working in automatic calculation mode. This mode is enabled by default when the user creates a blank document. Therefore, when the user enters expressions containing evaluation operators, the calculations are performed immediately. Generally speaking, there are two modes of calculations:

- *Automatic mode* — All calculations are performed automatically, as the user enters formulae.
- *Manual mode* — The user manually starts calculations with each formula or the whole document.

You can set the calculation node using the Math I Automatic Calculation commands, as shown in Figure 3.14. If this command is checkmarked, automatic mode is enabled. Otherwise the document is edited in the manual mode. To change the calculation mode, simply select this menu item (for example, if you click the mouse button in the situation shown in Figure 3.14, you'll switch to the manual mode).

The calculation mode is set independently for each document. Thus, you can have several documents opened simultaneously, and calculate them using different modes.

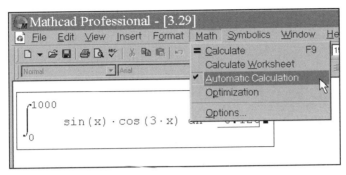

FIGURE 3.14. Selecting the calculation mode

Evidently, there are advantages and drawbacks to each mode. Automatic calculations simplify the process of working with the document, since the results of calculations appear in real time, and the user can analyze them immediately. On the other hand, if you are working with rather complex calculations, they might be time-consuming (this is rather noticeable if your computer can't be classified as a fast and powerful one). Consequently, quite often you'll need to wait for calculations to be completed before you can proceed with your work. Especially if you edit some expression in the starting lines of a large document, particularly the one influencing further calculations, your whole document will need to be recalculated. If this is the case, it is much more convenient to perform editing in the manual mode and enable calculations as necessary.

3.3.2. Interrupting Calculations

Mathcad calculates documents in a way common to most programming environments: from top to bottom and from left to right. As long as an expression is being calculated (by numeric or symbolic processor), it is highlighted by the green frame (Figure 3.15), and any user actions in the document being edited are disabled. If you are performing complex calculations, and your computer is not especially fast, you'll be able to notice how the green frame moves from expression to expression.

If you need to interrupt what has become a time-consuming calculation, press the <Esc> key. The dialog shown in Figure 3.16 will appear, prompting you to confirm your intention to interrupt the calculations by clicking OK. After you confirm the interruption, all expressions that were not calculated will be highlighted in red. To resume the interrupted calculation, press <F9> or select the **Math | Calculate** commands from the main menu.

$$\frac{d^2}{dt^2}y(t) + 0.1 \cdot \frac{d}{dt}y(t) + 1 \cdot y(t) = 0$$

$$y'(0) = 0$$

$$y(0) = 0.1$$

$$y := \text{Odesolve}(t, 50)$$

FIGURE 3.15. The calculation is in progress

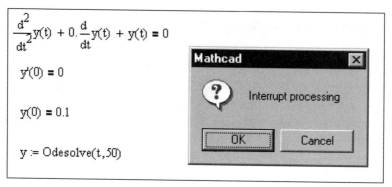

FIGURE 3.16. Interrupting the calculation process

3.3.3. Manual Calculations

If the automatic calculation mode is disabled (the Math I Automatic Calculation command in the menu has no checkmark), the user must start the calculations manually.

- ■ To evaluate all formulae in the whole document, select the Math I Calculate Worksheet command.
- ■ To evaluate all the formulae within the visible part of your document, select the Math I Calculate commands from the main menu, or press <F9>, or click the Calculate button on the standard toolbar.
- ■ If you need to interrupt the calculations, press the <Esc> key.

To manage the size of the visible part of your document, you might wish to zoom in.

TIP

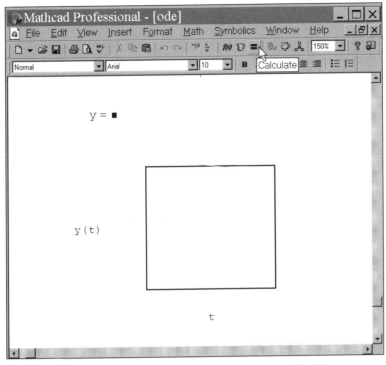

FIGURE 3.17. To start calculations, click the **Calculate** toolbar button

When editing the text in manual mode, Mathcad doesn't perform any calculations and doesn't plot any graphs. Placeholders formally designate appropriate positions within expressions (Figure 3.17).

3.3.4. Disabling Calculations for Specific Formulae

Mathcad allows you to disable the calculation of any specific formula. When you do this, that formula won't be able to influence any further calculations. To disable calculations for a specific formula within your document, do the following:

1. Right-click the required formula.
2. Select the **Disable Evaluation** command from the right-click menu, as shown in Figure 3.18.

To disable calculations for a single formula, you can also use an equivalent method. To use this method, open the **Properties** dialog by selecting the command

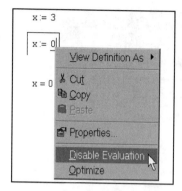

FIGURE 3.18. Disabling calculations for a selected formula using the right-click menu

with the same name from the right-click menu (see Figure 3.18) or by choosing the **Format** command from the main menu. When the **Properties** dialog appears, go to the Calculations tab and set the **Disable Evaluation** checkbox.

Listing 3.31 illustrates the result of disabling evaluations for a formula. In this listing, the second definition is disabled, which is designated by the presence of the black rectangular placeholder directly after the formula. Respectively, the value of the x variable displayed in the last line doesn't "sense" the disabled definition and retains its value (3).

LISTING 3.31. Disabling Evaluation for the Second Assignment Operator

$$x := 3$$

$$x := 0 \; \blacksquare$$

$$x = 3$$

3.3.5. Optimizing Calculations

Newer releases of Mathcad are distinguished by their enhanced and improved capabilities of speeding up numeric evaluations due to the use of symbolic math. Before performing a numeric evaluation, Mathcad automatically tries to simplify the expression using the symbolic processor. This process is known as *optimization*. Since the symbolic processor itself becomes more and more powerful with the release of each new version, symbolic conversion often makes the process of performing calculations significantly faster. The optimization mode can be enabled both for the whole document and for specific formulae.

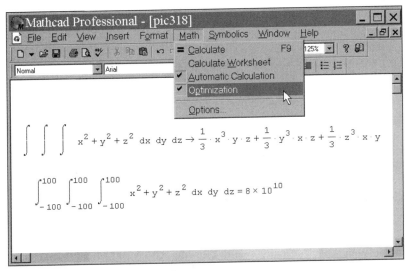

FIGURE 3.19. Optimization mode

To enable or disable the optimization mode for the active document, select the Math | Optimization commands from the main menu, as shown in Figure 3.19. The document contents displayed in this illustration are intended to help you understand the mathematical idea behind the optimization mode. For example, to boost the process of evaluating the lower (definite) integral, it is beneficial to use its analytical solution found by the symbolic processor (the upper indefinite integral).

To change the optimization mode for a separate formula without changing the selected mode for other formulae of your document, right-click the required formula and select the **Optimize** command from the right-click menu. The same task can also be performed using the **Properties** dialog discussed in the previous section. To open this dialog, select the **Format | Properties | Calculations** commands from the main menu or select a similar command from the right-click menu **Properties | Calculations**.

You can check to see if the optimization mode is enabled for a specific formula by right-clicking that formula and viewing the context menu. If the checkmark is present to the left of the Optimize command, the optimization mode is enabled; otherwise it is disabled. Thus, for the example shown in Figure 3.18, the optimization mode is disabled. If you view the formula carefully with the optimization mode enabled, you'll be able to see if the symbolic processor has managed to simplify the expression. The expression is followed by an asterisk symbol (Listing 3.32). If this symbol is red, the symbolic processor has successfully simplified the expression; otherwise it has failed.

LISTING 3.32. Optimization of the Integral Evaluation

$$J := \int_{-10}^{10} \int_{-10}^{10} \int_{-1}^{1} \sin(x + y + z) \ dx \ dy \ dz \ _{*}$$

3.3.6. The *Math Options* Dialog

In addition to other methods of specifying the calculation modes, there is another convenient method of specifying these options on the Calculations tab of the Math Options dialog. To open this window, select the Math | Options commands from the main menu. There are three checkboxes that specify the calculation mode (Figure 3.20).

- Recalculate automatically — Enables the automatic calculation mode.
- Use strict singularity checking for matrices — New option introduced with Mathcad 2001i. It is important for some operations with matrices. The option provides an additional test for matrix singularity before using numeric algorithms, and it allows you to avoid the inappropriate use of a numeric method and the display of an error message if the matrix is singular.
- Optimize expressions before calculating — Enables optimization mode.

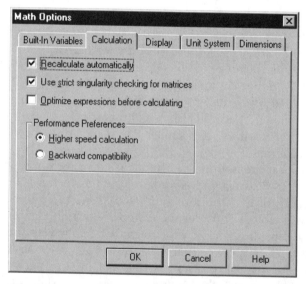

Figure 3.20. Managing calculation mode in the **Math Options** dialog

Besides checkboxes, Mathcad 2001i introduces a couple of switches that enable the user to implement the new mode of *higher speed calculation*. To enable this mode, set the **Higher speed calculation** radio button (Figure 3.20). To disable this mode, set the **Backward compatibility** option. In this case, calculations will be performed without additional optimization, in the same way it was done in the previous version of the product (Mathcad 2000). It might be necessary to use this mode if you encounter error messages when processing documents created in earlier Mathcad versions and working correctly with those versions.

3.4. ERROR MESSAGES

When the Mathcad processor can't evaluate an expression for some reason, it displays an error message instead of displaying the result (Figure 3.21). If the cursor is somewhere outside the erroneous formula, the name of the variable that caused the error is highlighted in red. When you click such a formula, the error message appears (Figure 3.21, at the bottom).

If an expression causes an error, it is ignored, while subsequent expressions of the document are still calculated. Certainly, if the formulae causing the error influence the values of subsequent formulae, those formulae will also be interpreted as erroneous. Thus, when you encounter error messages in your document, find and correct the first one. Quite often, this will help you to eliminate all subsequent errors.

Even advanced Mathcad users encounter error messages in their documents. These error messages might be caused both by orthographic errors and by more serious internal reasons that require a sound understanding of the numeric algorithms. The art of the mathematician comprises the skill of analyzing an erroneous situation and finding an adequate way of correcting it.

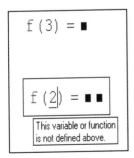

FIGURE 3.21. Error message

4 Data Types

In this chapter we will consider data types used in Mathcad documents and the principles of their input and output in the simplest numeric form. Normal data types (real and complex numbers, constants, and string data) are listed in the first section of this chapter *(see Section 4.1)*, while the principles of their input into your documents are covered at the same time. The final section of this chapter considers formatting capabilities for representing the results of numeric calculations *(see Section 4.4)*. Besides traditional functionality for processing numeric data, Mathcad provides a powerful toolkit for working with arrays. Arrays are implemented in the form of vectors and matrices, which allows one to approach the calculation style in the most common mathematical form *(see Section 4.3)*.

A distinguishing feature of the Mathcad environment is its capability to process dimensional variables supplied with physical dimensional units *(see Section 4.2)*. These tools are intended to significantly simplify engineering and scientific calculations.

Other forms of data input and output, such as graphs, animation, and file input/output are covered in the last part of this book *(see Chapter 15)*.

4.1. DATA TYPES

The simplest and the most common form of data input and output is implemented in Mathcad by assignment and output (either numeric or symbolic) directly into the document. Variables and functions used for data input and output can have different types of values (numeric, string, etc.). Listed below are the main data types processed by Mathcad processors:

- ■ *numbers* (including real and complex numbers and built-in constants) — Mathcad stores all numbers in the floating-point, double precision format (without classifying them to integer, Boolean, etc.).
- ■ *strings* — Any quoted text.
- ■ *arrays* (including range variables, vectors, and matrices) — Ordered sequences of numbers or strings.

Let us consider these data types and their direct input into the document by means of assigning values to variables in more detail.

4.1.1. Real Numbers

Mathcad interprets any expression that starts with a digit as a number. Therefore, to input a number, simply start typing it. Although Mathcad stores all numbers in the same format, you can use any notation that you consider to be the most suitable for your document, including the following:

- ■ As an *integer.*
- ■ As a *decimal number,* using decimal notation with any number of digits after the decimal point.
- ■ Using the *exponential notation* — The so-called *scientific notation.* To enter a number in scientific notation, enter the number, followed by a multiplication sign, and then enter 10 raised to the required power.
- ■ As a number to another base.

The first three notations are illustrated by the contents of the respective rows in Listing 4.1.

When entering numbers greater than or equal to 1000, do not use either a comma or a period to separate digits into groups of three. Simply type the digits one after another, as shown in the first line of Listing 4.1. It is invalid to enter 1000 as 1,000 or 1.000.

LISTING 4.1. Entering Real Numbers

$a := 10000$

$b := 2.57285$ $c := 312.1$

$d := 4.17 \cdot 10^{-23}$ $e := 345.1 \cdot 10^{3}$

If you continue Listing 4.1 by sequential output of all variables, you might be surprised by the fact that some numbers look different than expected (for example, $d=0$). The reason for this will be explained in Section 4.2.

To enter the number in forms other than a decimal, for example, as a *binary*, an *octal*, or a *hexadecimal*, proceed as follows:

1. Enter the digits that make up the number in the appropriate notation, using only valid characters (only 0 and 1 are valid for binary numbers; only digits from 0 to 7 are valid for octal numbers; and only digits from 0 to 9 and letters from a to f are valid for hexadecimal numbers). For example, the number 34 in binary representation will be represented as 100010.
2. Enter b (for binary numbers), o (for octal numbers), or h (for hexadecimal numbers) immediately after the last digit of the number.

Listing 4.2 illustrates the usage of non-decimal numbers. Notice that nonetheless, output is still performed in decimal notation.

LISTING 4.2. Entering Non-Decimal Numbers

$a := 100010b$ $a = 34$

$b := 37o$ $b = 31$

$c := 0af0h$ $c = 2.8 \times 10^{3}$

Logical functions use bit numbers (true or false). Mathcad designates them by real numbers 0 (false) and 1 (true).

4.1.2. Complex Numbers

Most operations in the Mathcad environment are, by default, performed over *complex numbers*. A complex number is a sum of a *real part* (real number) and an *imaginary part* (imaginary number, which is obtained by multiplying any real number by an *imaginary unit* i, according to the definition, $i = \sqrt{-1}$ or $i^2 = -1$). To enter an imaginary number, such as 3i, proceed as follows:

1. Enter the real multiplier (3).
2. Directly following the multiplier, type in "i" or "j" characters, both of which can be used to represent an imaginary unit.

To enter an imaginary unit, press sequentially <1>, <i>. If you simply enter the "i" character, Mathcad will interpret it as a variable named i. Furthermore, an imaginary unit will be represented as 1i only when the respective formula is selected. In all other cases, an imaginary unit is represented simply by i (Figure 4.1).

$$a := i + 10$$

$$x := 1i$$

$$x = i$$

FIGURE 4.1. Entering an imaginary unit

Complex numbers can be entered as an ordinary sum of real and imaginary parts or as any expression containing an imaginary number. Listing 4.3 provides several examples illustrating input and output of complex numbers.

LISTING 4.3. Complex Numbers

```
x := 2i + 4

y := 19.785j + 0.1

z := 23 · e^{0.1i}
```

```
x = 4 + 2i

y = 0.1 + 19.785i

z = 22.885 + 2.296i
```

Mathcad provides several simple built-in functions and operators for working with complex numbers *(see Section 10.2, "Functions for Complex Numbers," in Chapter 10)*. These functions and operators are illustrated in Listing 4.4.

LISTING 4.4. Functions for Working with Complex Numbers

```
y := 19.785j + 0.1

Im (y) = 19.785        Re (y) = 0.1

z := 23 · e^0.1i

|z| = 23               arg (z) = 0.1
```

An imaginary unit can be represented in the calculation results as j rather than as i. To change the representation of an imaginary unit, select the required option from the Imaginary Value list in the Result Format dialog, which you can open by selecting the Format I Result I Display Options commands from the main menu.

4.1.3. Built-In Constants

Some names in Mathcad are reserved for system variables, which are also known as *built-in constants*. Built-in constants are classified by two types: *math constants*, storing the values of some common math symbols, and *system constants*, which determine the functioning of most numeric algorithms implemented in Mathcad.

Math Constants

- ∞ — Infinity symbol (entered by the <Ctrl>+<Shift>+<z> keystroke)
- e — Natural logarithm base (entered by the <e> keystroke)
- π — The value of "pi" (entered by the <Ctrl>+<Shift>+<p> keystroke)
- i, j — Imaginary unit (entered by the <1>, <i> or <1>, <j> keystrokes)
- % — Percentage symbol, <%>, equivalent to 0.01

In numeric and symbolic calculations, math constants are interpreted differently. The numeric processor simply interprets them as specific numbers (Listing 4.5), while the symbolic processor recognizes each of them based on the math context. Furthermore, the symbolic processor is capable of returning math constants as the results of calculations.

LISTING 4.5. Values of Math Constants

$$\infty = 1 \times 10^{307}$$

$$e = 2.718$$

$$\pi = 3.142$$

$$i = i$$

$$j = i$$

$$\% = 0.01 \qquad 100 \cdot 25 \cdot \% = 25$$

If desirable, you can change the value of each of the above listed constants or use them as variables in your calculations (see, for example, Listing 4.1, where the e constant is redefined). Notice, however, that if you redefine a constant, its previous value becomes unavailable.

System Variables

- TOL — Controls the precision of numeric methods *(see Part III)*.
- CTOL — Controls the precision of evaluating expressions used in some numeric methods *(see Part III)*.
- ORIGIN — Specifies the starting index in arrays *(see Section 4.3.1)*.
- PRNPRECISION — Specifies the number of significant digits when performing output into a file *(see Chapter 15)*.
- PRNCOLWIDTH — Specifies the width of columns during output into a file *(see Chapter 15)*.
- CWD — This is a string representation of the path to the current working directory that Mathcad is using.

Listing 4.6 lists predefined values of the system variables. These values can be changed in any part of your document by simply assigning an appropriate new

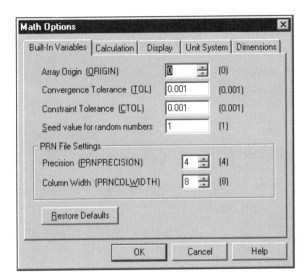

FIGURE 4.2. The **Built-In Variables** tab of the **Math Options** dialog

value to a specific variable. You can perform this task for the whole document in the Math Options dialog (Figure 4.2), which is opened by selecting the Math | Options | Built-In Variables commands from the main menu. To reset all system variables to their default values, press the Restore Defaults button.

LISTING 4.6. Default Values of the System Variables

$$\text{TOL} = 1 \times 10^{-3}$$

$$\text{CTOL} = 1 \times 10^{-3}$$

$$\text{ORIGIN} = 0$$

$$\text{PRNPRECISION} = 4$$

$$\text{PRNCOLWIDTH} = 8$$

$$\text{CWD} = \text{"C:\Dima\MCAD\MathCad 2001\4 Data\"}$$

4.1.4. String Expressions

Variables and functions can take not only numeric, but also string values. String values are strings comprising any quoted sequence of characters (Listing 4.7).

Mathcad provides several built-in functions intended for working with strings *(see Section 10.7, "String Functions," in Chapter 10).*

LISTING 4.7. String Input and Output

```
s := "Hello,"

s = "Hello,"

concat (s, " world!") = "Hello, world!"
```

Custom string functions can be defined in just the same way.

4.2. DIMENSIONAL VARIABLES

In Mathcad, numeric variables and functions can have dimensions. This is done in order to simplify engineering and physical calculations. Mathcad provides a wide range of dimension units, whose use enables you to create dimensional variables.

4.2.1. Creating a Dimensional Variable

To create a dimensional variable (for example, one defining the current of 10 A), proceed as follows:

1. Enter the expression assigning the value of 10 to the I variable, for example: 10: I:=10.
2. Directly after the value (10), enter the multiplication symbol: "*".
3. Then in the placeholder area, select the Insert | Unit commands from the main menu, click the Insert Unit toolbar button (the one with the measuring glass icon), or press the <Ctrl>+<U> keyboard shortcut (Figure 4.3).
4. The Insert Unit dialog will appear (Figure 4.4). Select the required measurement unit (Ampere (A), in our example) from the Unit list.
5. Click OK.

If you have difficulties selecting the appropriate measurement unit but know the dimension of the variable (in our example the dimension is electric current), try to select it from the Dimension list of the Insert Unit dialog. After you accomplish this, the Unit list will display valid measurement units for this dimension, which will simplify your selection (Figure 4.5).

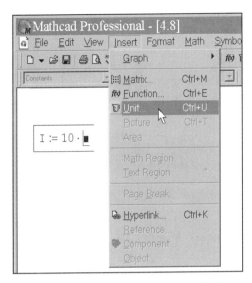

FIGURE 4.3. Assigning a dimensional unit to a dimensional value

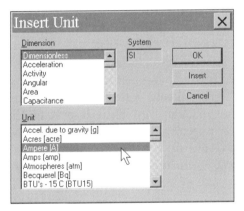

FIGURE 4.4. The **Insert Unit** dialog

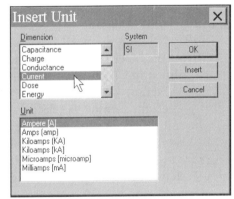

FIGURE 4.5. Selecting the dimension

You can view the inserted measurement units without closing the Insert Unit dialog. To do so, click the Insert button instead of OK. In this case, you'll notice that the measurement unit has appeared at the required position of the document, while retaining the capability to change it without leaving the Insert Unit window.

NOTE

Most units can be represented using conventional symbols, for example, ampere can be represented as A *or* amp, *ohm as* ohm *or* Ω, *etc.*

4.2.2. Working with Dimensional Variables

When working with dimensional variables, be prepared for Mathcad to constantly check the accuracy of the calculations. For example, you are not allowed to sum the values of different dimensions, otherwise you'll get the following error message: "The units in this expression do not match" (Figure 4.6). Still, you are allowed to add, for example, amperes and kiloamperes (Figure 4.8).

Any calculations valid from the physical point of view are allowed when performing operations over dimensional variables. Listing 4.8 represents an example of calculating the resistance via the ratio of voltage to current.

LISTING 4.8. Calculations Using Dimensional Variables

$$I := 10 \cdot A$$

$$U := 12 \cdot V$$

$$R := \frac{U}{I}$$

$$R = 1.2 \, kg \, m^2 \, s^{-3} \, A^{-2}$$

Pay special attention to the fact that the result in Listing 4.8 is not returned in ohms. Still, it can be easily converted both to ohms and to other valid units. To do so, simply double-click the placeholder directly following the calculated value when the formula is selected (Figure 4.7). As a result, the Insert Unit dialog will appear, where you can change the measurement unit for the calculated result. As a result of this action, the calculated result will be converted and re-calculated according to the newly selected measurement unit (using the uppermost formula shown in Figure 4.7, for example).

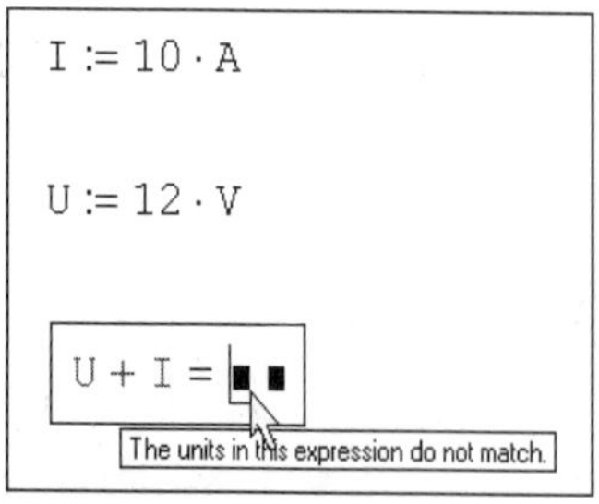

FIGURE 4.6. Adding variables of different dimensions is not allowed

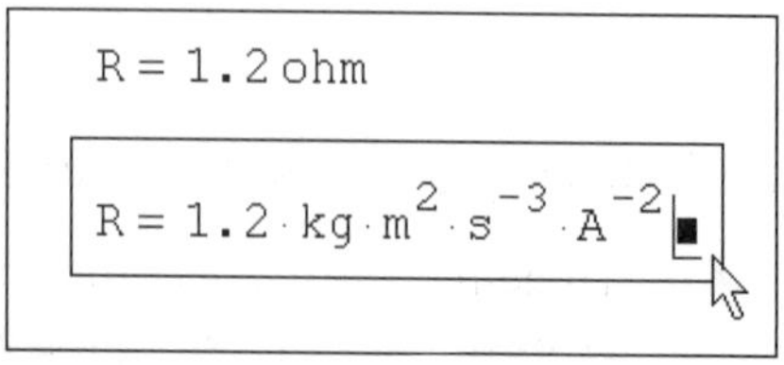

FIGURE 4.7. Notice the change of the measurement units in the result

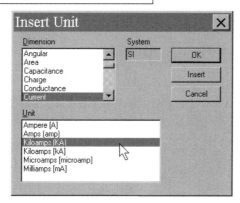

FIGURE 4.8. Adding variables of the same dimension but in different units

You can enable automatic conversion to simpler measurement units. To do so, open the Result Format window by selecting the Format | Result | Unit Display commands and check the Simplify units when possible checkbox.

The process of changing measurement units during their selection in the Insert Unit dialog is shown in Figure 4.8. As a result of your selection, the current value in amperes will be changed to the value in kiloamperes: 1.01 кА.

4.2.3. Selecting a System of Units

As can be easily noticed, we have used the SI units in all examples presented in this section. Both the list of available units and the display-only **System** list in the **Insert Unit** dialog (with the selected SI item) serve as evidence of this fact. To change the system of units for the whole document, select the **Math | Options** commands from the main menu, and then go to the **Unit System** tab of the **Math Options** dialog. Here you can select one of the radio buttons corresponding to the desired unit system.

4.2.4. Defining a Custom Dimension

To define a new (custom) unit for a specific dimension of even a new dimension, express it via available dimensions and assign it to a variable with its respective name. An example of the creation of a new measurement unit named nanoampere is shown in Listing 4.9.

LISTING 4.9. Defining a New Unit

$$nA := 10^{-9} \cdot A$$

$$3 \cdot A = 3 \times 10^9 \, nA$$

Custom units are not available in the **Insert Unit** dialog. Therefore, you'll need to type them directly from the keyboard (as this was done for the nA unit in the second line of Listing 4.9).

4.3. ARRAYS

Arrays are ordered sequences of numbers, known as array elements. You can access any element of the array by its *index*, i.e., ordinal number in the sequence of numbers (in Listing 4.10 a is an array and a_1 is an element of the array). Array usage is especially efficient in math calculations.

LISTING 4.10. One-Dimensional Array (Vector)

$$a := \begin{pmatrix} 14 \\ 1.4 \\ 4.7 \end{pmatrix}$$

$$a_0 = 14$$

$$a_1 = 1.4$$

$$a_2 = 4.7$$

Mathcad provides the following types of arrays:

- *Vectors* (one-index arrays, Listing 4.10), *matrices* (two-index arrays, Listing 4.11), and *tensors* (multiple-index arrays)
- *Range variables* — Vectors, whose elements depend on the index in a specific manner

LISTING 4.11. Two-Dimensional Array (Matrix)

$$a := \begin{pmatrix} 0.1 & 2.8 \\ 3.7 & 0 \end{pmatrix}$$

$a_{0,0} = 0.1$

$a_{1,0} = 3.7$

$a_{1,1} = 0$

4.3.1. Accessing Array Elements

Access to the whole array is implemented by naming the vector variable. For example, the sequence of "a", "=" symbols in Listings 4.10 and 4.11 will perform the output of the respective vector or matrix. Mathcad provides both operators and built-in functions that act on vectors and matrices as a whole (they will be considered in detail in Chapter 9), such as transposition, matrix multiplication, etc.

It is also possible to operate within particular elements of an array, the same way as you would with numbers. To do so, you only need to correctly specify an appropriate index or combination of the array indexes. For example, to access the 0-th element of the a vector (Listing 4.10), proceed as follows:

1. Enter the array variable name (a).
2. Click the Subscript button labeled by the x_n icon on the Matrix toolbar or type [.
3. Enter the desired index (0) into the array subscript placeholder that will appear directly to the right of the array name.

Now, if you type the numeric output symbol (=), the value of the 0-th array element will appear to the right, as shown by the second line of Listing 4.10.

To access a particular element of the multiple-index array (for example, the $a_{1,0}$ element of the a matrix, see Listing 4.11), do the following:

1. Enter the array variable name (a).
2. Switch to the input of the subscript by typing [.
3. Enter the first subscript (1) to the placeholder, then type "," and fill the second placeholder by the value of the second index (0).

As a result, you'll get access to the array element, as shown in the next to last line of the Listing 4.11.

In the listings provided above, array element numbering starts with zero. In other words, the first array element has the index value of 0. The starting index of an array is specified by the ORIGIN system variable, which, by default, is set to 0. If you need to number vector or matrix elements starting with 1, assign this value to the ORIGIN variable (Listing 4.12). Notice that in this case an attempt to access 0-th element of a vector will result in an error, since its value becomes undefined.

Besides access to specific array elements, you have the capability of operating within its subarrays (for example, column vectors that make up a matrix). Just click the $x^{<>}$ button on the **Matrix** toolbar *(see Chapter 9)*.

LISTING 4.12. Changing the Starting Index of an Array

ORIGIN := 1

$$a := \begin{pmatrix} 14 \\ 1.4 \\ 4.7 \end{pmatrix}$$

$a_0 = \blacksquare$

$a_1 = 14$

$a_2 = 1.4$

$a_3 = 4.7$

4.3.2. Range Variables

Range variables in Mathcad are types of vectors. Mainly, they are intended for creating *loops* or *iterative calculations*. The simplest example of a range variable is an array of sequential numbers belonging to a specified range.

For example, to create a range variable s with the elements $0,1,2,3,4,5$, proceed as follows:

1. Point the cursor to the required position within a document.
2. Enter the variable name (s) and insert an assignment operator ":".
3. Click the **Range Variable** button on the **Matrix** toolbar (Figure 4.9) or type the semicolon character from the keyboard.
4. Fill the placeholders (Figure 4.9) with the left and right range limits (0 and 5).

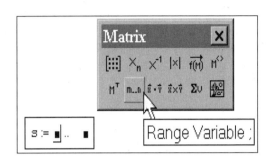

FIGURE 4.9. Creating a range variable

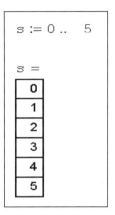

FIGURE 4.10. Output of the range variable

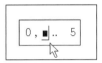

FIGURE 4.11. Creating a range variable with a step other than 1

The resulting range variable is shown in Figure 4.10.

To create a range variable that goes up or down with a step other than 1, for example, 0,2,4,6,8, proceed as follows:

1. Create a range variable in the range from 0 to 8 (see Figure 4.9).
2. Place the insertion lines at the starting value of the range (0).
3. Type a comma.
4. Fill the placeholder that will appear (Figure 4.11) with the value of the step (2).

The resulting range variable will range from 0 to 8 inclusively, in increments of 2.
Range variables are most frequently used for the following tasks:

- Parallel calculations (Listings 4.13 and 4.14)
- Assigning values to the elements of other arrays (Listings 4.14 and 4.15)

Pay special attention to a typical example of the usage of range variable provided in Listings 4.13 and 4.14. Most math operations implemented in Mathcad are performed with range variables in just the same way as they are with ordinary numbers. In this case, the same operation is performed in parallel with all elements of a range variable.

LISTING 4.13. Range Variable in Parallel Calculations

$$i := 0, 2 \ldots 8$$

$$s(i) := i^2 + 1$$

i =	s(i) =	sin(s(i))
0	1	0.841
2	5	-0.959
4	17	-0.961
6	37	-0.644
8	65	0.827

Parallel calculations with arbitrary vectors, which aren't necessarily range variables, are performed in the same way. For example, in Listing 4.14 you can define vector i, similar to the vector defined in Listing 4.10, and perform parallel calculations with its elements.

NOTE

LISTING 4.14. Range Variable in Parallel Calculations

$$i := 0 \ldots 5$$

$$s_i := i^2 + 1$$

$$s = \begin{pmatrix} 1 \\ 2 \\ 5 \\ 10 \\ 17 \\ 26 \end{pmatrix} \qquad \sin(s) = \begin{pmatrix} 0.841 \\ 0.909 \\ -0.959 \\ -0.544 \\ -0.961 \\ 0.763 \end{pmatrix}$$

LISTING 4.15. Range Variable Usage for Matrix Definition

$$i := 0 .. \quad 3$$

$$j := 0 .. \quad 5$$

$$c_{i,j} := i + j$$

$$C = \begin{pmatrix} 0 & 1 & 2 & 3 & 4 & 5 \\ 1 & 2 & 3 & 4 & 5 & 6 \\ 2 & 3 & 4 & 5 & 6 & 7 \\ 3 & 4 & 5 & 6 & 7 & 8 \end{pmatrix}$$

NOTE

When defining an array using range variables (Listings 4.14 and 4.15), make sure that these range variables take the sequence of values corresponding to all required array indexes. For example, if the range variable changes in steps of 2, half of the vector elements will be undefined.

Remember that a range variable is simply a type of vector with a simplified form of element definition. Quite frequently, one must do the same calculations cyclically, many times. For example, you might need to evaluate the f(x) functions within specific x range in order to plot its graph. Manually specifying all argument values (similar to the vector in Listing 4.10) is a tedious task. On the other hand, if using a range variable, this task can be solved with a single line of code.

4.3.3. Creating Arrays

There are several methods used to create arrays:

- Entering all elements manually using the **Insert Matrix** command
- Defining specific elements of the array
- Creating a data table and filling it with numbers
- Using built-in functions for creating arrays *(see Chapter 9)*
- Interacting with another application, such as Excel or MATLAB
- Reading data from an external data file
- Importing data from an external data file

Let us discuss the main methods of creating arrays, considering the fact that the latter two capabilities will be covered in detail in the last part of this book *(see Chapter 15).* Use the method that is optimal for a specific document in terms of simplicity and readability, or, alternately, one that is the most convenient from your own point of view.

Creating a Matrix with the *Insert Matrix* Command

The simplest and the most illustrative method of creating a vector or matrix is as follows:

1. Click the **Matrix** or **Vector** button on the **Matrix** toolbar (Figure 4.12). Alternately, you press the <Ctrl>+<M> keyboard shortcut or select the Insert | Matrix commands from the main menu.
2. When the Insert Matrix dialog appears, specify an integer number or rows and columns for the matrix to be created. For example, to create a 3×1 vector, enter the values as shown in Figure 4.12.
3. Click OK or Insert to insert the matrix. As a result, a new array of empty placeholders with the specified number of rows and columns will appear in your document (Figure 4.13).
4. Fill the placeholders with the values of the matrix elements. Use the mouse cursor or arrow keys to go from placeholder to placeholder.
5. To add new rows or columns to the existing matrix, proceed in the same way:
6. Move the insertion lines to the matrix element below and to the right of which you are going to insert new columns and/or rows.
7. Insert a new matrix, as was described above. A zero value is acceptable when specifying the number of rows or columns (Figure 4.14).
8. Fill the placeholders with matrix elements.

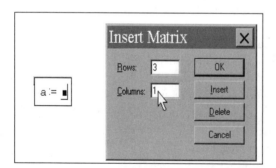

FIGURE 4.12. Inserting a matrix

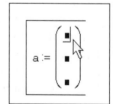

FIGURE 4.13. Filling the matrix with elements

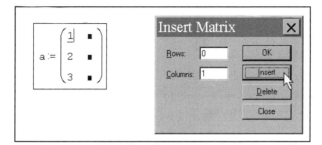

FIGURE 4.14. Inserting a column into the existing matrix

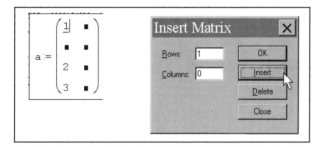

FIGURE 4.15. Inserting a row into the existing matrix

Figures 4.14 and 4.15 show the results of the sequential insertion of a column and row into the matrix after specifying the appropriate number of rows and columns in the **Insert Matrix** dialog and confirming user input by clicking the **Insert** button.

Matrix placeholders can be filled with numbers (real or complex) or any math expressions containing variables, operators, built-in functions and custom functions (Listing 4.16, the second line).

LISTING 4.16. Using Variables and Functions When Defining a Matrix

$$x := 3$$

$$A := \begin{pmatrix} \sin(x) \\ x \end{pmatrix}$$

$$A = \begin{pmatrix} 0.141 \\ 3 \end{pmatrix}$$

Creating an Array by Defining its Specific Elements

The simplest way to create an array is to define any number of its elements. This can be done using the following methods:

- Assigning values directly to the array elements
- Using range variables (see Listing 4.15)

Any of these methods allow you to assign the required value(s) both to all array elements (see Listing 4.15) and to a specific part of the array elements or even to a single element. In the latter case, Mathcad will create an array, whose dimension is specified by the indexes of the inserted element (Listing 4.17), while undefined elements are, by default, assigned zero values.

LISTING 4.17. Creating a Matrix by Defining a Single Element

$$s_{3,2} := 99$$

$$s = \begin{pmatrix} 0 & 0 & 0 \\ 0 & 0 & 0 \\ 0 & 0 & 0 \\ 0 & 0 & 99 \end{pmatrix}$$

In any position of the document, you can redefine any of the array elements (Listing 4.18, the first line) or change its dimensions. To change the dimension of the whole array, simply assign any value to the new elements whose indexes fall outside the range of the previous dimension (Listing 4.18, the second line).

Just like with the normal assignment operator, you can insert any functions into the matrix placeholders.

NOTE

LISTING 4.18. Changing the Matrix *(Listing 4.17, Continued)*

$$s_{1,2} := 1$$

$$s = \begin{pmatrix} 0 & 0 & 0 \\ 0 & 0 & 1 \\ 0 & 0 & 0 \\ 0 & 0 & 99 \end{pmatrix}$$

$$s_{4,4} := -7$$

$$s = \begin{pmatrix} 0 & 0 & 0 & 0 & 0 \\ 0 & 0 & 1 & 0 & 0 \\ 0 & 0 & 0 & 0 & 0 \\ 0 & 0 & 99 & 0 & 0 \\ 0 & 0 & 0 & 0 & -7 \end{pmatrix}$$

Creating a Tensor

Defining specific elements is very convenient for creating tensors (multiple-index arrays). Mathcad provides the means for direct editing of only vectors and matrices. However, you can create a tensor by defining a *nested array*. To achieve this, you need to assign each matrix element the value in the form of another vector or matrix (Listing 4.19). The user must only make sure that the tensor indexes are correct and avoid confusion when indexing nested matrices (see the last line of the listing).

LISTING 4.19. Creating a Tensor and Accessing Its Elements

$$s_{0,0} := \begin{pmatrix} 1 \\ 2 \end{pmatrix} \qquad s_{1,0} := \begin{pmatrix} 3 \\ 4 \end{pmatrix}$$

$$s_{0,1} := \begin{pmatrix} 5 \\ 6 \end{pmatrix} \qquad s_{1,1} := \begin{pmatrix} 7 \\ 8 \end{pmatrix}$$

$$s = \begin{pmatrix} \{2,1\} & \{2,1\} \\ \{2,1\} & \{2,1\} \end{pmatrix}$$

$$\left(s_{1,0} \right)_0 = 1 \qquad \left(s_{1,0} \right)_1 = 4$$

The process of creating a tensor can be automated by using range variables.

TIP

Notice that, by default, Mathcad doesn't display the three-dimensional structure of a tensor (the next to last line of the Listing 4.19). Rather, it displays information on the dimensions of each element of the s matrix. To expand nested arrays, select the Format I Result I Display Options commands from the main menu and set the Expand Nested Arrays checkbox on the Display Options tab.

Creating the Input Table

A Mathcad document can contain various objects created in other applications. For example, if you have copied a fragment of a table created in another application to the clipboard (Microsoft Excel, for example), this fragment can be used when creating a matrix using the input table. To do so, proceed as follows:

1. Click the mouse anywhere in the unoccupied area within a document.
2. From the main menu, select Insert I Component I Table.
3. The Component Wizard window will appear. Select the Input Table component from the list and click the Finish button (Figure 4.16).
4. Insert the matrix name into the placeholder, and then insert the data into the cells of the input table that will appear in the document (Figure 4.17). Notice that you can fill in the cells of this table with the fragments pasted from the clipboard, and you can insert any real or complex numbers, but you can't use formulae.

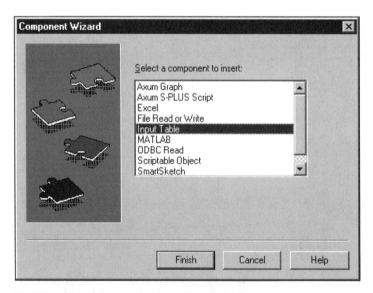

FIGURE 4.16. The **Component Wizard** window

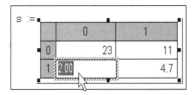

FIGURE 4.17. Entering numbers into the input table

The matrix defined using the input table looks the same as the matrix defined normally (Listing 4.20).

LISTING 4.20. The Matrix Defined Using the Input Table (Figure 4.17)

$$s = \begin{pmatrix} 23 & 11 \\ 2 & 4.7 \end{pmatrix}$$

Creating Arrays by Importing Data from Other Applications

Using the Component Wizard, which can be started by selecting the Insert | Component | Table commands from the main menu, it is easy to establish a connection with other components by importing data. Detailed information on this topic can be found in the Mathcad on-line Help system in the following sections:

- Excel Component
- MATLAB Component

4.3.4. Displaying Vectors and Matrices

You have probably already noticed that matrices, vectors, and range variables have been displayed differently in different examples. This is due to the presence of automatic settings for matrix display, accepted in Mathcad by default. Actually, there are two styles of displaying the array: in matrix form and in table form (Figure 4.18).

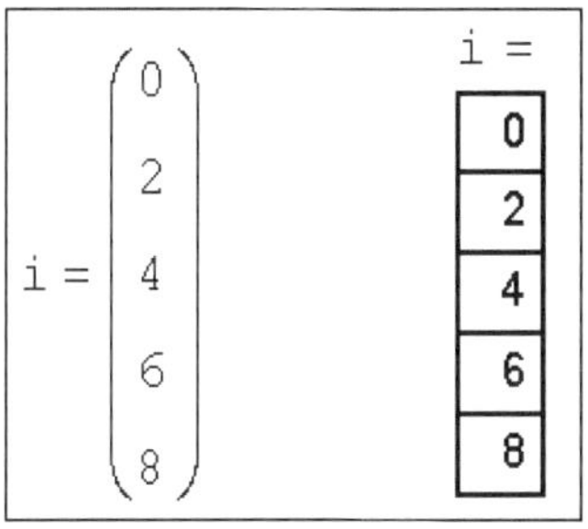

$$i = \begin{pmatrix} 0 \\ 2 \\ 4 \\ 6 \\ 8 \end{pmatrix}$$

$$i =$$

0
2
4
6
8

FIGURE 4.18. Displaying arrays in matrix form (left) and in table form (right)

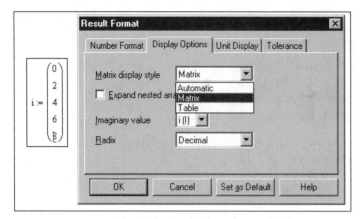

FIGURE 4.19. Changing the display style of the array

To change the style of display for a specific array, select the **Format | Result** command from the main menu. The **Result Format** dialog will appear (Figure 4.19). Go to the **Display Options** tab and select the required style from the **Matrix display style** list. The following options are available:

- **Automatic** — Lets Mathcad select the display style
- **Matrix**
- **Table**

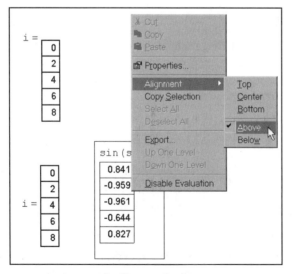

FIGURE 4.20. Matrix alignment options

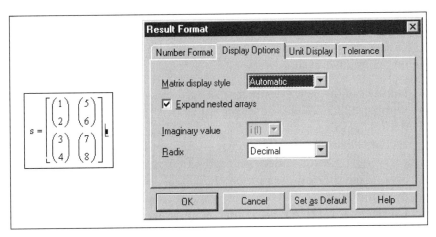

FIGURE 4.21. Expanding nested arrays

The table display provides various options for aligning the matrix relative to the expression on the left side of the output operator (Figure 4.20). To change the currently selected alignment option, right-click the table, select the **Alignment** command from the context menu, and then select the desirable alignment option from the submenu.

In addition to the matrix display style, the Result Format dialog allows you to set the tensor (nested array) display style. To display tensors in a style similar to the one shown in Figure 4.21, set the **Expand nested arrays** checkbox. To display tensors in collapsed style (see Listing 4.19), clear this checkbox if it is set.

The most illustrative form of displaying vectors is done through creating their graph *(see Chapter 15)*.

4.4. NUMERIC DATA OUTPUT FORMAT

Although it is impossible to influence the result displayed to the right of the output operator that outputs the values of variables, functions, and expressions, you can change the display format of numeric output. As was already mentioned *(see Section 4.1.1)*, both data input and output can be performed using the following two notations:

- *Decimal,* for example: 13478.74559321
- *Exponential,* for example: 1.348×10^4

Selection of the numeric output format is performed in the **Result Format** dialog, which can be opened by selecting the **Format | Result** commands from the main menu.

4.4.1. Formatting the Result

To manage the numeric output format (in decimal or exponential notation), you can set the following parameters:

- Number of displayed decimal places after the decimal point. For example, the number 122.5587 will be displayed as 122.56, if you choose to display two decimal places.
- Display or suppress trailing zeros — this option allows you to display or suppress trailing zeros in decimal representation of a number. If you opt to suppress trailing zeros, the number 1.500 will always be displayed as 1.5 (even if the number of decimal places after the decimal point is set to 3).
- Exponential threshold — if this value is exceeded, the number will be displayed in exponential notation. For example, if exponential threshold is set to 3, the number 122.56 will be displayed in decimal notation, if this value is set to 2, it will be displayed as "1.23×10^2".

When you are using exponential notation, the number of decimal places of the first multiplier is controlled by the first parameters from the list provided above.

- In addition, the number presented in exponential notation can be represented using equivalent formats, "1.23×10^2", or in engineering format, "1.23E+002".

Mathcad provides several types of formats, each of which allows one to change the various parameters of number representation. To select the format, go to the **Number Format** tab of the **Result Format** dialog (Figure 4.22).

General Format

This format is selected for numeric output by default. Here you can specify both the number of decimal places (in the **Number of decimal places** field) and the exponential threshold (in the **Exponential threshold** field). When the exponential threshold value is exceeded, the number will be displayed in exponential notation (as shown in Figure 4.22). Listing 4.21 provides several examples of representing the same number using the general format. The left column contains numbers with the exponential threshold set to 3 and number of decimal places (from top to bottom) set to 3, 4, 4, and 5 respectively. For the number displayed in the last line, the **Show trailing zeros** checkbox is set. The right column contains numbers with an exponential threshold from 1 to 4 (from top to bottom).

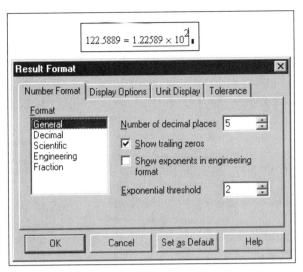

FIGURE 4.22. Selecting number format for numeric output

LISTING 4.21. General Format of the Result

$$152.5889 = 152.589$$ $$152.5889 = 1.526 \times 10^{2}$$

$$152.5889 = 152.5889$$ $$152.5889 = 1.5 \times 10^{2}$$

$$152.5889 = 152.5889$$ $$152.5889 = 1.5 \times 10^{2}$$

$$152.5889 = 152.58890$$ $$152.5889 = 152.58890$$

Decimal Format

Numbers are displayed in decimal representation only, while exponential format is never used.

Scientific Format

Numbers are displayed in exponential notation only, and both the number of decimal places of the left multiplier and the display or suppression of trailing zeros are defined by the user.

Engineering Format

Numbers are displayed in exponential notation only, and exponents are always multiples of 3; as with scientific format, the user can change the number of decimal places.

Fraction Format

This format is different from the previous ones, since it represents numbers as fractions (Figure 4.23). You can manage the accuracy of the number representation using the Level of accuracy field and control the number display by specifying if the number can be displayed in a mixed format (as shown in Figure 4.23). To specify the option of displaying a number in a mixed format, check the Use mixed numbers checkbox.

Listing 4.22 provides examples of the display of the same number in a different format. The first line illustrates the decimal format; the second, the scientific format with three decimal places; and the third, the engineering format (also with three decimal places). The last two lines show the fraction format: the next to last line shows the number displayed with the accuracy level of 5, and for the number displayed in the last line, this parameter is set to 10. Furthermore, the Use mixed numbers checkbox is set for the last line.

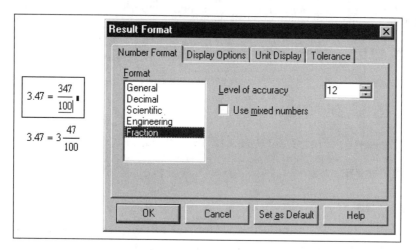

FIGURE 4.23. Fraction format

LISTING 4.22. Formats for Representing the Calculation Results

$$12340.56789 = 12340.568$$

$$12340.56789 = 1.234 \times 10^4$$

$$12340.56789 = 12.341 \times 10^3$$

$$12340.56789 = \frac{999586}{81}$$

$$12340.56789 = 12340 \frac{56789}{100000}$$

4.4.2. Rounding Small Numbers to Zero

Mathcad automatically rounds small numbers to zero (Listing 4.23). One can set the threshold value (in powers of 10) separately for real and imaginary parts of a number. Numbers whose absolute value is smaller than this threshold value are displayed as zeros.

Remember that this relates only to the display of the numeric results. All numbers are stored in RAM correctly.

LISTING 4.23. Representation of Small Numbers

$$2.15 \cdot 10^{-23} = 0$$

$$3.4 + i \cdot 10^{-11} = 3.4$$

$$-0.0000000000000001 = 0$$

To change the threshold values, do the following:

1. Click anywhere in an unoccupied position within a document.
2. Open the **Result Format** dialog by selecting the **Format | Result** commands.
3. Go to the **Tolerance** tab.
4. Set the threshold value for real numbers in the **Zero threshold** field and for imaginary numbers in the **Complex threshold** field.
5. Click **OK**.

You can change the threshold value for an imaginary zero even when editing the formula (Figure 4.24), however, you can't change the real threshold value when working in this mode.

To view the numeric output at the maximum precision level, press <Ctrl>+<Shift>+<N>. In this case, the status bar (at the bottom left of the Mathcad window) will display the result with maximum precision for a short time (Figure 4.25).

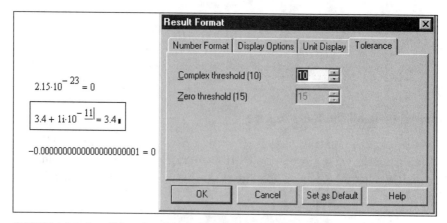

FIGURE 4.24. Specifying the threshold value for an imaginary zero

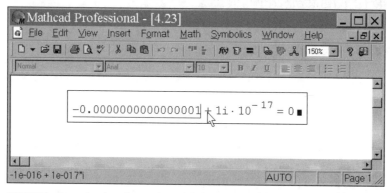

FIGURE 4.25. Viewing the result with maximum precision

4.4.3. Displaying Output in Other Notations

Akin to the usage of other notation during input *(see Section 4.1.1)*, you can also perform numeric output in decimal, binary, octal, or hexadecimal notations (Listing 4.24, from top to bottom).

LISTING 4.24. Numeric Output Using Different Notations

```
47 = 47

47 = 101111b

47 = 57o

47 = 2fh
```

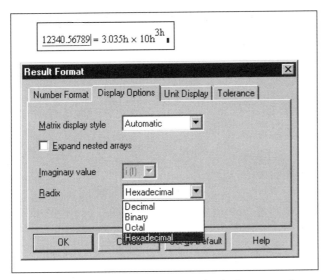

FIGURE 4.26. Selecting notation for displaying numeric output

To specify the notation for numeric output, select the Format | Result | Display Options commands from the main menu, and then select the required option from the Radix list (Figure 4.26). When displaying numbers using other notations, you'll also have formatting capabilities, which are provided on the Number Format tab of the Result Format dialog. Listing 4.25 provides several examples of how to format binary numbers.

LISTING 4.25. Formatting Capabilities for Binary Numbers

```
47 = 101111b
```

$$47 = 1.1b \times 10b^{101b}$$

$$47 = 1.100b \times 10b^{101b}$$

In this chapter, we have considered the most basic principles of numeric input and output. More advanced options of data input and output, such as graphs, animation, and file input/output will be covered later in this book (see Chapter 15).

PART

II

Calculations

5 Symbolic Calculations

I n this chapter we are going to consider the capabilities of the Mathcad symbolic processor. It enables you to solve most math problems analytically, without having to use numeric methods and, consequently, without introducing any calculation errors. In the initial section of this chapter we will briefly consider the ways of performing symbolic calculations in the Mathcad editor *(see Section 5.1)*, and then we will proceed with a discussion of how to use symbolic calculations for solving specific tasks. Mathcad lets you perform a wide range of analytical conversions, such as algebraic and matrix operations *(see Section 5.2)*, main operations of mathematical analysis *(see Section 5.3)*, and calculations of integral function conversions *(see Section 5.4)*.

It should be noticed that the techniques of most symbolic calculations are also covered in the third and the fourth parts of this book, during the discussion on solutions of specific calculation tasks. Finally, we will conclude the discussion by describing several practical examples of how to perform efficient math symbolic calculations *(see Section 5.5)*.

5.1. METHODS OF SYMBOLIC CALCULATIONS

Symbolic calculations in Mathcad can be performed using the following two methods:

- Using menu commands
- Using the symbolic output operator →, symbolic processor keywords, and normal formulae (in the Mathcad on-line Help system this method is described as *live symbolic evaluations*)

The first method is more convenient, and it is employed when it is necessary to quickly obtain an analytical result that is to be used only once, without saving the process of calculations. The second method is more illustrative, since it provides you with the capability to record symbolic calculations in the traditional math form and to save symbolic calculations in Mathcad documents. Furthermore, analytical conversions performed using the menu relate only to a single expression, that which is currently selected. As a result, any formulae located in a Mathcad document above the currently selected expression (for example, assignment operators) have no effect on this conversion. The symbolic output operator, on the other hand, takes account of all previous contents of the document and produces the result with respect to them.

In symbolic calculations one is able to use most Mathcad built-in functions.

For symbolic calculations using menu commands, use the **Symbolics** menu, which groups math operations that Mathcad can perform analytically (Figure 5.1). To implement the second method, you can use all Mathcad tools suitable for numeric calculations (for example, the **Calculator**, **Evaluation** toolbars, etc.) and a special math toolbar that can be displayed on the screen by clicking the **Symbolic Keyword Toolbar** button on the **Math** toolbar. The **Symbolic** toolbar contains buttons corresponding to specific commands used in symbolic conversions, such as factorization of an expression, Laplace transform, and other operations, which can't be performed numerically in Mathcad and for which, incidentally, there are no built-in functions.

Let us consider both types of symbolic calculations with a simple example — namely, with the expansion of the following expression: `sin(2·x)`.

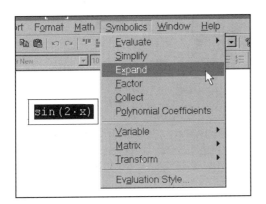

FIGURE 5.1. The **Symbolics** menu

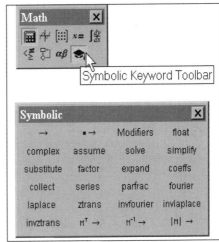

FIGURE 5.2. The **Symbolic** toolbar

Here is the first method (using the menu).

1. Enter the expression that you need to expand: `sin(2·x)`.
2. Select it (Figure 5.1).
3. From the main menu, select Symbolics | Expand.

After you accomplish this, the result of the expansion (factorization in this case) will appear directly below as an extra line (Figure 5.3).

Symbolic operations using menu commands are only possible for a specific object (for example, an expression, some fragment of an expression, or some specific variable). To obtain the correct result for the desired analytical conversion, select the object to which the operation must be applied. In the case we are considering here, the conversion was applied to the whole `sin(2·x)` expression. If you select some fragment of the formula, as shown in Figure 5.4, the respective conversion will be performed on the selected fragment only (consider the lower line in that illustration).

```
sin(2·x)


2·sin(x)·cos(x)
```

FIGURE 5.3. The result of applying the **Symbolics | Expand** menu commands

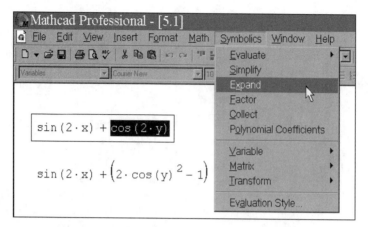

FIGURE 5.4. Symbolic expansion of the selected fragment

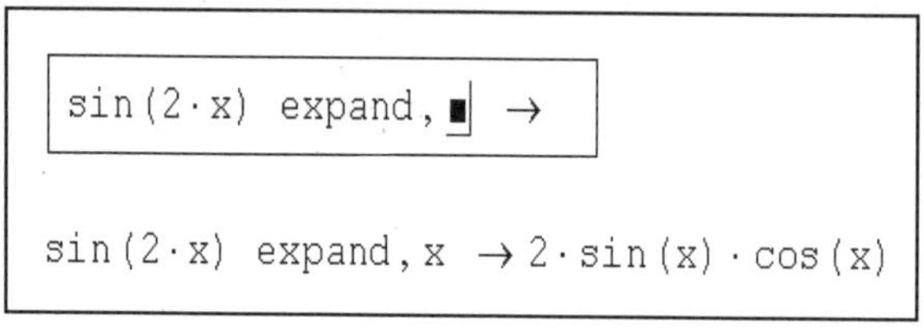

$$\sin(2 \cdot x) \text{ expand,} \blacksquare \rightarrow$$

$$\sin(2 \cdot x) \text{ expand,} x \rightarrow 2 \cdot \sin(x) \cdot \cos(x)$$

FIGURE 5.5. Symbolic expansion of the expression

Here is the second method of symbolic conversions (using the → operator).

1. Enter the expression to be transformed: $\sin(2 \cdot x)$.
2. Click the **Expand** button on the **Symbolic** toolbar.
3. Fill the placeholder next to the right of the expand keyword (Figure 5.5, top line) with the name of the variable (x) or press the <Del> key to delete the placeholder.
4. Insert the symbolic output operator →.
5. Press <Enter> or simply click the mouse somewhere outside the expression.

As you probably remember, the symbolic output operator can be entered using several methods: by clicking the → button available on the **Evaluation** toolbar or on the **Symbolic** toolbar or, alternately, by pressing the <Ctrl>+<.> keys. The result of symbolic expansion will appear next to the right (see Figure 5.5, on the bottom line).

If symbolic calculations are performed using the second method, the symbolic processor will take into account all the formulae inserted into the document before the current one (Figure 5.6, bottom). If the same conversions are performed using the menu, the symbolic processor won't "see" anything except for the selected formula and, consequently, will interpret all its variables analytically, even if they were previously assigned some values (Figure 5.6, top). As a result, predefined custom functions are unavailable when performing symbolic conversions via menu commands.

If you can select a method of performing symbolic calculations, we recommend that you use the second method — using the → operator — since it allows you to cut down on user actions in a Mathcad document. The presence of the special menu for symbolic calculations is retained for backward compatibility with previous Mathcad versions where analytical conversions were not so perfectly integrated and were available, mainly, via the menu.

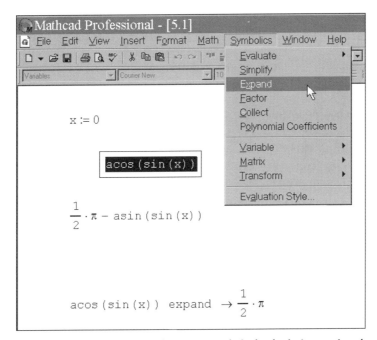

FIGURE 5.6. The difference between analytical calculations using the menu and the → operator (the last line)

Not all expressions can be converted analytically. If this is the case (either because the task has no analytical solution or because it proves to be too complex for the Mathcad symbolic processor), the displayed result represents the expression itself (Listing 5.1).

LISTING 5.1. Symbolic Conversions

$$\cos(2 \cdot x) \; \text{expand}, x \; \rightarrow 2 \cdot \cos(x)^2 - 1$$

$$\cos(x) \; \text{expand}, x \; \rightarrow \cos(x)$$

Later on in this chapter, when considering symbolic calculations using the menu, we will illustrate the results with the screenshots. As for symbolic calculations using the → operator, they will be illustrated by listings.

5.2. SYMBOLIC ALGEBRA

Mathcad's symbolic processor can perform main algebraic conversions, such as simplifying, factoring, collecting terms, and calculating symbolic sums and products.

5.2.1. Simplifying Expressions

Simplifying expressions is the most common operation. The Mathcad symbolic processor tries to transform the expression in such a way as to represent it in a simpler form. When it is performing this operation, the processor uses various arithmetic formulae, collecting terms, trigonometric identities, calculating inverse functions, etc. To simplify an expression using the menu (Figure 5.7), proceed as follows:

1. Enter the expression to be simplified.
2. Select the whole expression or its fragment that needs to be simplified.
3. From the menu, select Symbolics I Simplify.

To simplify the expression using the symbolic output operator, use the `simplify` keyword (Listing 5.2). Don't forget that if some variables were assigned values earlier, these values will be substituted into the final expression when the symbolic output is performed (Listing 5.3).

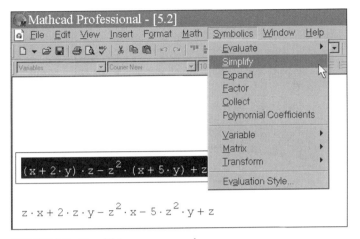

FIGURE 5.7. Simplifying an expression

LISTING 5.2. Simplifying an Expression

$$(x + 2 \cdot y) \cdot z - z^2 \cdot (x + 5 \cdot y) + z \ \text{simplify} \ \rightarrow z \cdot x + 2 \cdot z \cdot y - z^2 \cdot x - 5 \cdot z^2 \cdot y + z$$

LISTING 5.3. Simplifying an Expression with Variable Substitution

$$x := 10 \quad y := 1$$

$$(x + 2 \cdot y) \cdot z - z^2 \cdot (x + 5 \cdot y) + z \ \text{simplify} \ \rightarrow 13 \cdot z - 15 \cdot z^2$$

When numeric expressions are being simplified, the procedure is performed differently, depending on whether there is a decimal point in the numbers. If there is, the numeric expression will be calculated directly (Listing 5.4).

LISTING 5.4. Simplifying Numeric Expressions

$$\sqrt{3} \ \text{simplify} \ \rightarrow \sqrt{3}$$

$$\sqrt{3.01} \ \text{simplify} \ \rightarrow 1.7349351572897472412$$

5.2.2. Expanding Expressions

The symbolic expansion operation (or expansion) is the inverse of the simplification operation. During this operation, all sums and products are expanded, and trigonometric dependencies are factored using trigonometric identities. To expand an expression, use the Symbolics I Expand menu commands or insert the symbolic output operator with the `expand` keyword. The usage of the expansion operation was covered in detail in Section 5.1 (see Figures 5.3–5.6 and Listing 5.1).

5.2.3. Factoring

To factor an expression, select the Symbolics I Factor commands from the main menu (Figure 5.8) or use the symbolic output operator with the `factor` keyword (Listing 5.5). This operation enables you to transform a polynomial into a product of simpler polynomials and an integer number into a product of primes. Before using these menu commands, don't forget to select the whole expression or the part that you need to transform into a product.

LISTING 5.5. Factoring Examples

$$x^4 - 16 \text{ factor } \rightarrow (x - 2) \cdot (x + 2) \cdot \left(x^2 + 4\right)$$

$$28 \text{ factor } \rightarrow 2^2 \cdot 7$$

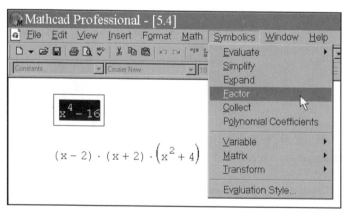

FIGURE 5.8. Factoring an expression

5.2.4. Collecting Terms

To simplify an expression by representing it as a polynomial in a particular variable (Figure 5.9), proceed as follows:

1. Enter an expression to be simplified by collecting terms.

2. Within that expression, select the variable in relation to which it is necessary to collect the terms (in the example shown in Figure 5.9 this is the y variable).
3. From the main menu, select the **Symbolics | Collect** commands.

As a result, another line will appear directly below the expression, representing the resulting polynomial (the lower line in Figure 5.9).

To collect the terms using the symbolic output operator (Listing 5.6), proceed as follows:

1. Enter the expression to be simplified.
2. Click the Collect button on the Symbolic toolbar.
3. The collect keyword will appear, followed by the placeholder. Enter the variable name on which to collect the terms (in the first line of the Listing 5.6 this is the x variable, in the second it is y).
4. Insert the symbolic output operator →.
5. Press <Enter>.

The collect *keyword may be followed by a comma-separated list of variables. In this case, Mathcad will sequentially collect the terms on all listed variables, as shown in the last line of Listing 5.6.*

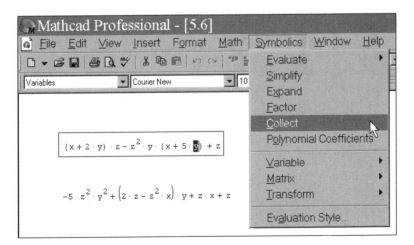

FIGURE 5.9. Collecting terms

LISTING 5.6. Collecting Terms on Different Variables

$$(x + 2 \cdot y) \cdot z - z^2 \cdot y \cdot (x + 5 \cdot y) + z \; \text{collect}, x \; \rightarrow \left(z - z^2 \cdot y\right) \cdot x + 2 \cdot z \cdot y - 5 \cdot z^2 \cdot y^2 + z$$

$$(x + 2 \cdot y) \cdot z - z^2 \cdot y \cdot (x + 5 \cdot y) + z \; \text{collect}, y \; \rightarrow -5 \cdot z^2 \cdot y^2 + \left(2 \cdot z - z^2 \cdot x\right) \cdot y + z \cdot x + z$$

$$(x + 2 \cdot y) \cdot z - z^2 \cdot y \cdot (x + 5 \cdot y) + z \; \text{collect}, x, y, z \; \rightarrow \left(z - z^2 \cdot y\right) \cdot x + 2 \cdot z \cdot y - 5 \cdot z^2 \cdot y^2 + z$$

5.2.5. Polynomial Coefficients

If an expression is a polynomial in a specific variable x written as a product of other simpler polynomials rather than in canonic form $(a_0 + a_1 x + a_2 x^2 + \ldots)$, the Mathcad symbolic processor simply calculates the coefficients $a_0, a_1, a_2 \ldots$. Coefficients themselves can be functions on other variables (and sometimes rather complex ones).

To calculate polynomial coefficients using menu commands (Figure 5.10), proceed as follows:

1. Enter the expression.
2. Within the expression, select a variable or another expression, for which you need to calculate polynomial coefficients (in the example shown in Figure 5.10, the z variable has been selected).
3. From the main menu, select the **Symbolic | Polynomial Coefficients** commands.

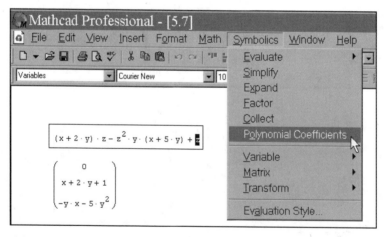

FIGURE 5.10. Calculating polynomial coefficients

As a result, the vector composed of polynomial coefficients will appear directly below the expression. The first element of the vector is the free term a_0, the second is a_1, and so on.

A specific task that requires calculating polynomial coefficients will be discussed in the section dedicated to the numeric calculation of the roots of a polynomial (see Section 8.2, "The Roots of a Polynomial," in Chapter 8).

To calculate polynomial coefficients using the symbolic output operator, do the following:

1. Enter the expression.
2. Click the **Coeffs** button on the **Symbolic** toolbar.
3. Enter the polynomial argument into the placeholder directly following the `coeffs` keyword.
4. Insert the symbolic output operator →.
5. Press <Enter>.

Examples of calculating polynomial coefficients are provided in Listings 5.7 and 5.8. Listing 5.7 illustrates the calculation of coefficients for different arguments. The latter listing demonstrates the capability of defining coefficients not only of specific variables, but also for more complex expressions included in the formula under consideration.

LISTING 5.7. Calculating Polynomial Coefficients

$$(x + 2 \cdot y) \cdot z - z^2 \cdot y \cdot (x + 5 \cdot y) + z \text{ coeffs}, z \rightarrow \begin{pmatrix} 0 \\ x + 2 \cdot y + 1 \\ -y \cdot x - 5 \cdot y^2 \end{pmatrix}$$

$$(x + 2 \cdot y) \cdot z - z^2 \cdot y \cdot (x + 5 \cdot y) + z \text{ coeffs}, x \rightarrow \begin{pmatrix} 2 \cdot z \cdot y - 5 \cdot z^2 \cdot y^2 + z \\ z - z^2 \cdot y \end{pmatrix}$$

LISTING 5.8. Calculating Polynomial Coefficients for a Simple Variable and for an Expression

$$(x - 4) \cdot (x - 7) \cdot x + 99 \quad \text{coeffs}, x \rightarrow \begin{pmatrix} 99 \\ 28 \\ -11 \\ 1 \end{pmatrix}$$

$$(x - 4)^3 + (x - 4) \cdot (x - 7) \cdot x + 99 \quad \text{coeffs}, x - 4 \rightarrow \begin{pmatrix} 99 \\ x^2 - 7 \cdot x \\ 0 \\ 1 \end{pmatrix}$$

5.2.6. Series and Products

To perform a symbolic calculation of a finite or infinite sum or product, proceed as follows:

1. Enter the expression using the **Calculus** toolbar for inserting appropriate summation or iterated product symbols *(see the Section 3.2.2, "Calculus Operators," in Chapter 3)*. When necessary, insert the infinity symbol as a row limit by pressing <Ctrl>+<Shift>+<Z>.
2. Depending on the desired style of symbolic calculations, select the **Symbolics | Simplify** commands from the main menu or enter the symbolic output operator →.

The examples of numeric and symbolic calculations of series and iterated products are shown in Listings 5.9 and 5.10.

LISTING 5.9. Symbolic and Numeric Calculations of Series

$$\sum_{i=0}^{10} 2^i = 2.047 \times 10^3 \qquad\qquad \sum_{i=0}^{10} 2^i \rightarrow 2047$$

$$\sum_{i=0}^{\infty} a^i \rightarrow \frac{-1}{(a-1)}$$

$$\sum_{n=0}^{\infty} \frac{x^n}{2^n \cdot n!} \rightarrow \exp\left(\frac{1}{2} \cdot x\right) \qquad \sum_{n=0}^{\infty} \frac{1^n}{2^n \cdot n!} \rightarrow \exp\left(\frac{1}{2}\right) = 1.649$$

$$\sum_{n=0}^{100} \frac{1^n}{2^n \cdot n!} = 1.649$$

LISTING 5.10. Symbolic Calculation of the Iterated Product

$$\prod_{n=1}^{\infty} \frac{1}{n^3 + 1} \rightarrow 0 \qquad \prod_{n=1}^{\infty} \sqrt{n} \rightarrow \infty$$

5.2.7. Converting to Partial Fractions

To convert a complex fraction into a partial fraction, you must either select the Symbolics | Variable | Convert to Partial Fraction commands from the main menu (Figure 5.11), or use the symbolic output operator with the `parfrac` keyword (Listing 5.11). If you are using the first method (via the menu), don't forget to select the variable before running menu commands. If you are using the second method (via the symbolic output operator), then you'll need to specify the variable name after the `parfrac` keyword. Generally, the sequence of actions when converting to partial fractions is the same as in standard practice *(see Section 5.2.4).*

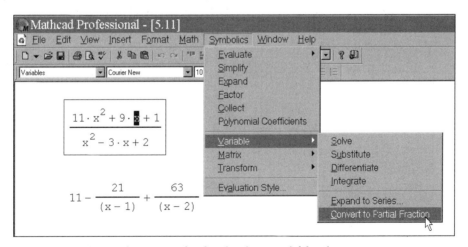

FIGURE 5.11. Converting a complex fraction into partial fractions

LISTING 5.11. Converting to Partial Fractions

$$\frac{11 \cdot x^2 + 9 \cdot x + 1}{x^2 - 3 \cdot x + 2} \text{ convert , parfrac , x } \rightarrow 11 - \frac{21}{(x-1)} + \frac{63}{(x-2)}$$

5.2.8. Substitution

The operation of substituting a variable value into an expression is a rather convenient feature of symbolic calculations. To use the menu for variable substitution, proceed as follows (Figure 5.12):

1. Select the variable value that you need to substitute into a specific expression. The variable value can be any expression containing any variables (in Figure 5.12 the first line of the document is taken as the value for substitution).
2. Copy the variable value onto the clipboard. This can be done either by pressing the <Ctrl>+<C> keyboard shortcut or by clicking the **Copy** button on the **Standard** toolbar.
3. Within the expression into which you need to substitute the value from the clipboard, select the variable that must be replaced (in the example shown in Figure 5.12, the x variable in the second line is selected).
4. Select the **Symbolics | Variable | Substitute** commands from the main menu.

The result of these actions is shown on the bottom line in the document shown in Figure 5.12.

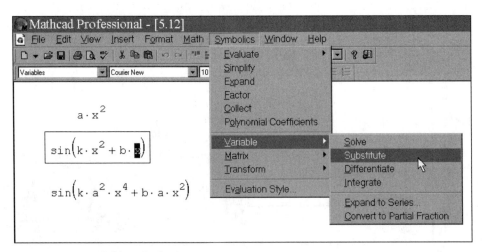

FIGURE 5.12. Variable substitution

To perform the same actions using the symbolic output operator, use the `substitute` keyword, which can be inserted by the **Substitute** button on the Symbolic toolbar. Directly after the `substitute` keyword, fill in the placeholder with the logical expression showing which variable must be substituted by which formula (Listing 5.12).

LISTING 5.12. Variable Substitution

$$\sin\left(k \cdot x^2 + b \cdot x\right) \text{ substitute }, k = a \cdot x^2 \ \rightarrow \sin\left(a \cdot x^4 + b \cdot x\right)$$

5.2.9. Matrix Algebra

The Mathcad symbolic processor allows you to perform analytically varied matrix calculations. Since most operations and built-in functions are performed with matrices in the same way as with ordinary numbers, the simplification menu command (**Simplify**) described above can also be applied to matrices.

Also, there is a range of specific matrix operations that can be implemented by using either the **Symbolics | Matrix** menu commands or by using several buttons available on the **Symbolic** toolbar (see Figure 5.2). These matrix operations are listed below:

- Transpose
- Invert
- Determinant

Matrix operations are performed using the same sequence as symbolic operations over scalar variables. Before applying these operations, don't forget to select the matrix with which you are going to perform your operation.

5.3. MATHEMATICAL ANALYSIS

The most illustrative demonstration of the functional capabilities of the Mathcad symbolic processor can be produced by analytical calculations of limits, derivatives, integrals, series expansions, and by solving algebraic equations. All these operations can be performed via the **Symbolics** menu (by selecting an appropriate command from the **Variable** submenu). Naturally, before performing an operation, you need to select a variable within an expression on which you are going to perform an operation. To select a variable, move the insertion lines to it (however, to make the procedures more illustrative, it is recommended that you highlight the variable by dragging the mouse above the required part of the expression).

All operations listed above can also be performed using the symbolic output operator. This method will be covered in detail in the respective chapters of Part III (except for the series expansion, which will be discussed in Section 5.3.3). Later on in this section, we will describe the procedures involved in performing mathematical analysis operations using the menu.

Symbolic calculation of the function limit is described in Section 3.2.2, "Calculus Operators," in Chapter 3.

NOTE

5.3.1. Differentiation

To analytically differentiate an expression by a specific variable, select the required variable within the expression, and then select the Symbolics I Variable I Differentiate commands from the main menu (Figure 5.13).

As a result, the derivative value will appear in the next line. To find the second derivative, re-apply the aforementioned sequence of procedures to the differentiation result that you have obtained. Higher-order derivatives are calculated in the same way.

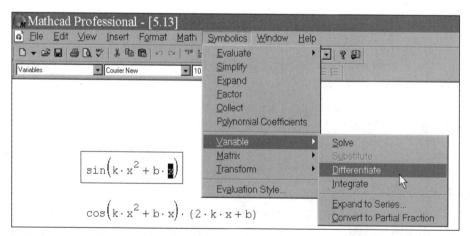

FIGURE 5.13. Differentiation using the menu

5.3.2. Integration

To calculate an indefinite integral of an expression with a specific variable, select the required variable within that expression and then run the Symbolics I Variable I Integrate commands from the main menu (Figure 5.14). Analytical representation of the indefinite integral will appear directly below. The result may contain both Mathcad built-in functions *(see Chapter 10 and Appendix A)* and other special functions that can't be directly calculated in Mathcad. Despite the fact that these functions can't be calculated, the symbolic processor can produce them as a result of some symbolic operations.

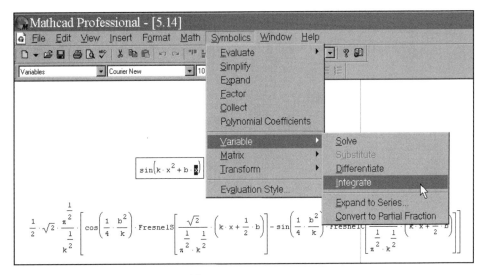

Figure 5.14. Integration by a variable

More detailed information about symbolic solutions of algebraic equations, differentiation and integration using the symbolic output operator, including calculation of higher-order derivatives and definite multiple integrals, will be provided in Part III of this book (see Chapter 7).

5.3.3. Expansion to Series

Using the Mathcad symbolic processor, it is possible to expand an expression to *Taylor series* by any variable (x) around the x = 0 point (i.e., to represent the expression in the vicinity of the point x by the following sum:

$a_0 + a_1x + a_2x^2 + a_3x^3 + \ldots$). Here a_i — some coefficients independent from x, but, possibly, representing functions of some other variables contained within the initial expression. If the expression has a singularity in the $x = 0$ point, its respective expansion is known as *Laurent series*.

To expand an expression into a series, proceed as follows:

1. Enter the required expression.
2. Select the variable that you want to expand around.
3. Run the **Symbolics I Variable I Expand to Series** commands from the main menu (Figure 5.15).
4. The **Expand to Series** dialog will appear (Figure 5.16). Here you'll need to enter the desired order of approximation into the **Order of Approximation** field. After having accomplished this, click OK.

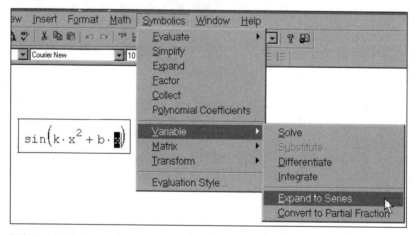

FIGURE 5.15. Preparing an expression for expanding it to a series with the x variable

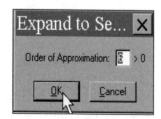

FIGURE 5.16. Expanding an expression to the Taylor series

$$\sin\left(k \cdot x^2 + b \cdot x\right)$$

$$b \cdot x + k \cdot x^2 + \frac{-1}{6} \cdot b^3 \cdot x^3 + \frac{-1}{2} \cdot k \cdot b^2 \cdot x^4 + \left(\frac{1}{120} \cdot b^5 - \frac{1}{2} \cdot k^2 \cdot b\right) \cdot x^5 + O\left(x^6\right)$$

FIGURE 5.17. The result of expansion to the Taylor series

The result of the operation will be displayed directly below the expression (Figure 5.17).

Don't forget that the expansion is performed only around the x=0 point. To create an expansion for another point (x = a), for example, substitute the x variable with the x - a value (see Section 5.2.8).

To use an alternative method of expanding an expression to a series, use the symbolic output operator with the series keyword, which can be inserted by the appropriate button on the **Symbolic** toolbar. Specify the variable name and approximation order (Listings 5.13 and 5.14) directly after the series keyword, separated by a comma. Figure 5.18 provides a comparison of the function and its expansions to a series with different orders of approximations (for k = b = 1). From this illustration, it is obvious that the expansion to a series approximates a function rather well around x = 0 and becomes rougher and rougher as the distance from this point grows.

LISTING 5.13. Expanding an Expression to a Series with a Different Approximation Order

$$\sin\left(k \cdot x^2 + b \cdot x\right) \text{ series}, x, 2 \;\rightarrow b \cdot x$$

$$\sin\left(k \cdot x^2 + b \cdot x\right) \text{ series}, x, 3 \;\rightarrow k \cdot x^2 + b \cdot x$$

$$\sin\left(k \cdot x^2 + b \cdot x\right) \text{ series}, x, 4 \;\rightarrow b \cdot x + k \cdot x^2 - \frac{1}{6} \cdot b^3 \cdot x^3$$

$$\sin\left(k \cdot x^2 + b \cdot x\right) \text{ series}, x, 5 \;\rightarrow b \cdot x + k \cdot x^2 - \frac{1}{6} \cdot b^3 \cdot x^3 - \frac{1}{2} \cdot k \cdot b^2 \cdot x^4$$

LISTING 5.14. Expanding an Expression to a Series around Different Variables

$$\sin\!\left(k \cdot x^2 + b \cdot x\right) \text{ series, } k, 3 \;\rightarrow\; \sin(b \cdot x) + \cos(b \cdot x) \cdot x^2 \cdot k - \frac{1}{2} \cdot \sin(b \cdot x) \cdot x^4 \cdot k^2$$

$$\sin\!\left(k \cdot x^2 + b \cdot x\right) \text{ series, } b, 3 \;\rightarrow\; \sin\!\left(k \cdot x^2\right) + \cos\!\left(k \cdot x^2\right) \cdot x \cdot b - \frac{1}{2} \cdot \sin\!\left(k \cdot x^2\right) \cdot x^2 \cdot b^2$$

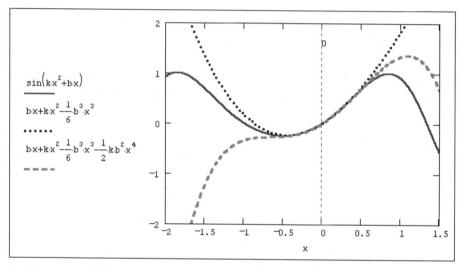

FIGURE 5.18. Initial function and the results of its expansions to the Taylor series

5.3.4. Solving Equations

The symbolic processor allows you to find the values of a variable analytically, at which point the expression turns to zero. To achieve this, do the following:

1. Enter the expression.
2. Select a variable you want to solve for.
3. From the **Symbolics** menu, select the **Variable I Solve** commands (Figure 5.19).

Detailed information on solving algebraic equations symbolically will be provided in Part III (see Chapter 8). In particular, we will cover the capabilities of solving systems of algebraic equations and writing equations in the common form of logical equality.

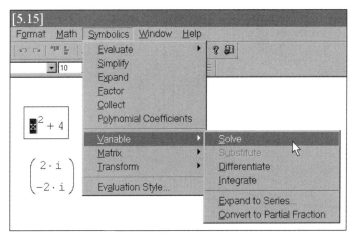

FIGURE 5.19. Solving an equation symbolically

5.4. INTEGRAL TRANSFORMS

By definition, integral transforms map the `f(x)` function to another function of another argument `F(ω)`. The `f(x) → F(ω)` mapping is specified by an integral dependence. The Mathcad symbolic processor is capable of performing the following three types of integral transforms: Fourier transform, Laplace transform and Z-transform. Besides direct transforms, it also has the ability to perform an inverse transform for any of these transforms (i.e., `F(ω) → f(x)`).

All symbolic integral transforms are performed in a way similar to the operations that we have already considered. To evaluate a transform for a specific expression, select the variable for which the transform must be performed and then select an appropriate menu item. Transforms using the symbolic output operator are used with appropriate keywords, next to which you need to specify the name of the required variable.

Let us consider some examples of symbolic evaluation using all these integral transforms.

5.4.1. Fourier Transform

The Fourier transform represents the `f(x)` function as an integral of harmonic functions, known as *Fourier integral*:

$$F(\omega) \;=\; \int_{-\infty}^{\infty} f(x) \cdot \exp(-i\omega x)\, dx .$$

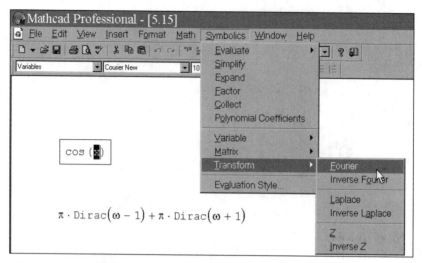

FIGURE 5.20. Calculating Fourier transform using the menu

Analytical calculation of the Fourier transform using the menu is shown in Figure 5.20. Listing 5.15 provides two examples of how to calculate the direct Fourier transform using the `fourier` keyword and symbolic output operator →. Listing 5.16 illustrates inverse Fourier transform for one of the functions from the previous listing.

Mathcad also allows you how to calculate the Fourier transform by using the numeric processor by means of the popular Fast Fourier Transform (FFT) algorithm (see Section 14.4.1, "The Fourier Transform," in Chapter 14).

LISTING 5.15. Direct Fourier Transform

$$\cos(x) \text{ fourier}, x \;\to\; \pi \cdot \text{Dirac}(\omega - 1) + \pi \cdot \text{Dirac}(\omega + 1)$$

$$\left(x^2 + 4\right) \text{ fourier}, x \;\to\; -2 \cdot \pi \cdot \text{Dirac}(2, \omega) + 8 \cdot \pi \cdot \text{Dirac}(\omega)$$

LISTING 5.16. Inverse Fourier Transform

$$-2 \cdot \pi \cdot \text{Dirac}(2, \omega) + 8 \cdot \pi \cdot \text{Dirac}(\omega) \text{ invfourier}, \omega \;\to\; t^2 + 4$$

5.4.2. Laplace Transform

Laplace transform is an integral of the f(x) function in the following form:

$$F(s) = \int_0^\infty f(x) \cdot \exp(-sx)\, dx.$$

Laplace transform can be calculated in a way absolutely similar to Fourier transform (see the previous section). Examples of Laplace transform are presented in Listing 5.17.

LISTING 5.17. Direct and Inverse Laplace Transforms

$$x^2 + 4 \ \text{laplace}, x \ \rightarrow \frac{2}{s^3} + \frac{4}{s}$$

$$\frac{2}{s^3} + \frac{4}{s} \ \text{invlaplace}, s \ \rightarrow t^2 + 4$$

5.4.3. Z-Transform

Z-transform of the f(x) function is determined via an infinite sum represented as follows:

$$F(z) = \sum_{n=0}^\infty f(n) \cdot z^{-n}.$$

An example of Z-transform is shown in Listing 5.18.

LISTING 5.18. Direct and Inverse Z-Transforms

$$x^2 + 4 \ \text{ztrans}, x \ \rightarrow z \cdot \frac{\left(-7 \cdot z + 5 + 4 \cdot z^2\right)}{(z-1)^3}$$

$$z \cdot \frac{\left(-7 \cdot z + 5 + 4 \cdot z^2\right)}{(z-1)^3} \ \text{invztrans}, z \ \rightarrow 4 + n^2$$

5.5. EXTENDED CAPABILITIES OF THE SYMBOLIC PROCESSOR

In the previous sections of this chapter we considered the main techniques involved in performing symbolic calculations in Mathcad. As a general rule, these techniques were illustrated by simple examples of how to perform specific symbolic operations. Still, when performing various calculations (including numeric calculations) in Mathcad, you can use the capabilities of the symbolic processor more efficiently. In this section, we are going to discuss several enhanced capabilities of the symbolic processor.

5.5.1. Using Custom Functions

When performing symbolic calculations using the symbolic output operator, the Mathcad symbolic processor correctly interprets both built-in and custom functions and variables defined previously in a Mathcad document. Thus, it provides a powerful toolkit for including symbolic calculations into user programs. Listings 5.19 and 5.20 provide examples illustrating how to use custom functions. Compare the last lines of these listings. Notice that they produce different results despite the identity of the expressions in the left parts. This is due to the fact that in Listing 5.20, the x variable was previously assigned the value of 4. Since variable values influence symbolic calculations, the result will be produced on account of the substitution of the number 4 instead of x.

LISTING 5.19. Custom Function in Symbolic Calculations

$$f(k,x) := \cos(k \cdot x) + 4 \cdot x^{2-k}$$

$$f(k,x) \quad \text{substitute}, k = \sqrt{x} \quad \rightarrow \cos\left(x^{\left(\frac{3}{2}\right)}\right) + 4 \cdot x^{\left(2-x^{\frac{1}{2}}\right)}$$

$$f(k,x) \quad \text{series}, k, 2 \quad \rightarrow 1 + 4 \cdot x^2 - 4 \cdot x^2 \cdot \ln(x) \cdot k$$

LISTING 5.20. Variable Values Influence the Result of Symbolic Calculations

$$f(k,x) := \cos(k \cdot x) + 4 \cdot x^{2-k}$$

$$x := 4$$

$$f(k,x) \quad \text{series}, k, 2 \quad \rightarrow 65 - 64 \cdot \ln(4) \cdot k$$

In contrast, when you perform symbolic operations using the **Symbolics** menu, the symbolic processor doesn't "see" anything except the expression within the insertion lines. For this reason, neither custom functions nor predefined variables will be taken into account during calculation.

If you need to perform analytic operations "on the fly" and get the result in a generic form without taking into account any predefined values of the variables included into the expression, use the Symbolics menu.

5.5.2. Calculating Numeric Value of the Expression

The symbolic processor is capable of calculating numeric values of expression (both real and complex). Sometimes it is more convenient than using the numeric processor (i.e., the normal equals sign). To calculate the numeric value of a specific expression (Figure 5.21), use the **Symbolics | Evaluate | Symbolically** or **Symbolics | Evaluate | Floating Point** commands. In the latter case, you'll be prompted to specify the output precision in the **Floating Point Evaluation** dialog. As a result, whenever possible, Mathcad will replace symbolic calculations with floating-point values.

The **Symbolics** menu provides just another item, namely, **Symbolics | Evaluate | Complex**, that allows you to represent the result in a complex form $(a + b \cdot I)$.

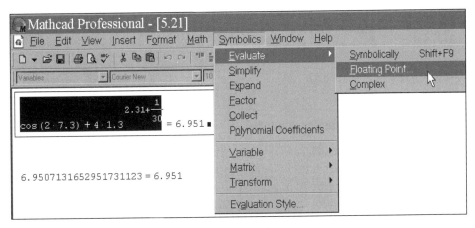

FIGURE 5.21. Calculating a floating-point expression

Similar actions can be performed using the float and complex keywords, which you can insert from the **Symbolic** toolbar. The float keyword is applied with the output precision value (Listing 5.21). Using the complex keyword, you can transform expressions both symbolically and taking into account the numeric values, if they were previously assigned to variables (some examples are presented in Listing 5.22).

LISTING 5.21. Calculating a Floating-Point Expression

$x := 3$ $\qquad\qquad$ $k := 2.4$

$\cos(k \cdot x) + 4 \cdot x^{2-k}$ float, 3 $\rightarrow 3.19$

$\cos(k \cdot x) + 4 \cdot x^{2-k}$ float, 10 $\rightarrow 3.185927374$

$\cos(k \cdot x) + 4 \cdot x^{2-k}$ float, 20 $\rightarrow 3.1859273744412716730$

LISTING 5.22. Complex Transformations of the Expressions

e^{z+2i} complex $\rightarrow \exp(z) \cdot \cos(2) + i \cdot \exp(z) \cdot \sin(2)$

$4.2 \cdot 2i^{1.8-3i}$ complex $\rightarrow 1193.4523970930846183 + 1107.3477730509390980 \cdot i$

$x := i$

$4 \cdot x^3$ complex $\rightarrow -4 \cdot i$

$4 \cdot x^{3.1}$ complex $\rightarrow .62573786016092347604 - 3.9507533623805509048 \cdot i$

5.5.3. Sequences of Symbolic Commands

Symbolic calculations can be performed using chains of keywords. To make such calculations, insert the keywords corresponding to symbolic operations sequentially from the **Symbolic** toolbar. The principle of organizing chains of keywords is very similar to that of the Mathcad built-in programming language *(see Chapter 6)*. Listings 5.23 and 5.24 provide several examples illustrating the usage of the sequences of symbolic operators.

The sequences of symbolic operations allow you to introduce additional conditions into your calculations, such as limitations for real or complex forms of the result. To achieve this, use the assume *keyword. More detailed information on this topic is provided in the Mathcad on-line Help system.*

LISTING 5.23. Fourier Transform, Expansion to Series and Calculation

$$e^{-x} \text{ fourier}, x \to 2 \cdot \pi \cdot \text{Dirac}(\omega - i)$$

$$e^{-x^2} \left| \begin{array}{l} \text{fourier}, x \\ \text{series}, \omega, 5 \end{array} \right. \to \pi^{\frac{1}{2}} - \frac{1}{4} \cdot \pi^{\frac{1}{2}} \cdot \omega^2 + \frac{1}{32} \cdot \pi^{\frac{1}{2}} \cdot \omega^4$$

$$e^{-x^2} \left| \begin{array}{l} \text{fourier}, x \\ \text{series}, \omega, 5 \\ \text{float}, 3 \end{array} \right. \to 1.77 - .443 \cdot \omega^2 + 5.54 \cdot 10^{-2} \cdot \omega^4.$$

LISTING 5.24. Z-Transform and Conversion to Partial Fractions

$$x^2 + 4 \text{ ztrans}, x \to z \cdot \frac{\left(-7 \cdot z + 5 + 4 \cdot z^2\right)}{(z-1)^3}$$

$$x^2 + 4 \left| \begin{array}{l} \text{ztrans}, x \\ \text{convert}, \text{parfrac}, z \end{array} \right. \to 4 + \frac{2}{(z-1)^3} + \frac{3}{(z-1)^2} + \frac{5}{(z-1)}$$

6 Programming

athcad is a system oriented towards end-users who don't necessarily know anything about programming. Mathcad developers initially aimed to provide professionals in the field of mathematics, physics, and engineering with the capability of performing complex calculations on their own, without needing to contact professional programmers. However, despite a brilliant implementation of this idea, it became evident that without programming Mathcad would be significantly less powerful, mainly due to the complaints of those users who are acquainted with coding techniques and who desire to automate their calculations in a programming style with which they are accustomed. The users of older Mathcad versions were limited to combining several specific built-in functions and range variables, instead of adhering to well-known programming principles *(see Section 6.1)*.

Later releases of Mathcad provide their own built-in programming language, which, despite its limitations, operates rather smoothly *(see Section 6.2)*. Although this language is not very powerful, it still enables the programmer to use software code in Mathcad documents efficiently. Also, on account of its simplicity and because it is easy to use, beginners are able to master it quickly. Finally, programmatic modules within Mathcad documents combine both isolation (which allows them to be easily distinguished from other formulae), and clear perception.

Despite a rather limited number of operators, the Mathcad programming language allows you to solve different (and sometimes rather complex) problems and, therefore, is a serious and helpful instrument for calculations *(see Section 6.3)*.

6.1. PROGRAMMING WITHOUT PROGRAMMING

Earlier versions of Mathcad didn't provide any sort of built-in programming language. Therefore, to implement and use habitual operations to check conditions and organize loops, one had to invent awkward combinations of built-in conditional functions such as if (Listing 6.1) or until and combinations of range variables (Listing 6.2).

Due to generally accepted traditions of the use of the programming language, it is strongly recommended that you avoid using the until function in your future work (this function is still included in Mathcad 2001i, but it is classified as an obsolete function).

LISTING 6.1. Conditional Function

```
f (x) := if (x < 0 , "negative" , "positive" )

f (1) = "positive"

f (−1) = "negative"
```

LISTING 6.2. Organizing a Loop Using a Range Variable

```
i := 0 ..  10
```

$$x_i := i^2$$

Actually, range variables work together to provide a powerful toolkit in Mathcad, similar to the loops in programming. In most cases, it is much more convenient to organize loops (including nested ones) using range variables rather than to write programming code for this purpose. Therefore, it is rather useful to master the techniques of using range variables, vectors, and matrices, since they provide the basis of the most important Mathcad calculations, particularly when preparing graphs. (More detailed information on range variable usage and its related capabilities is provided in *Chapter 4.*)

6.2. THE MATHCAD PROGRAMMING LANGUAGE

Mathcad provides a special toolbar for inserting programming code into Mathcad documents — the Programming toolbar. To display this toolbar on the screen, click the **Programming Toolbar** button on the **Math** toolbar as shown in Figure 6.1. Most buttons on this toolbar are labeled by the text representation of the programming operators (statements), which makes them self-explanatory.

Now, let us proceed with our discussion of the main components of the Mathcad programming language and let's consider examples of their practical sage.

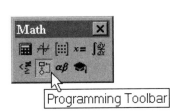

FIGURE 6.1. The **Programming** toolbar

6.2.1. What Is a Program?

The main working tools in Mathcad are math expressions, variables, and functions. Quite often, it is impossible to write a formula using internal logic (such as returning a different value depending on specific conditions) in a single line. Programmatic modules are intended for defining expressions, variables, and functions using several lines, often using specific programming operators.

Compare the definition of the f(x) function from Listing 6.1 to the definition of the same f(x) definition using program module (Listing 6.3).

LISTING 6.3. Conditional Function Defined Programmatically

$$f(x) := \begin{cases} \text{"negative"} & \text{if } x < 0 \\ \text{"positive"} & \text{if } x > 0 \\ \text{"zero"} & \text{otherwise} \end{cases}$$

f(1) = "positive"

f(−1) = "negative"

f(0) = "zero"

Although variable and function definitions using built-in Mathcad functions are principally equivalent to the program modules, in most cases, programming makes documents simpler and improves readability by providing the following advantages:

■ The capability to use loops and conditional statements
■ Simplicity in creating functions and variables requiring several steps (consider, for example, Listing 6.3)
■ The capability to create functions containing the code isolated from other parts of the document, including the advantages of using local variable and exception handling

As can be seen from Listing 6.3, a vertical line denotes the program module in Mathcad. Programming language statements appear directly to the right of this line.

6.2.2. Creating a Program (*Add Line*)

To create a program module, the one represented in the previous section (see Listing 6.3), for example, proceed as follows:

1. Enter the part of the expression that will appear to the left of the assignment operator, then enter the assignment operator itself. In our example this will be the function name: f(x).
2. If necessary, display the **Programming** toolbar on the screen (see Figure 6.1).
3. Click the **Add Line** button on that toolbar.
4. If you know, at least approximately, how many lines of code will make up your program, you can create the required number of lines by sequentially clicking the **Add Line** button the required number of times (Figure 6.2 illustrates the result of clicking the **Add Line** button three times).
5. Fill in the placeholders with the program code that uses statements of the Mathcad programming language. In the example under consideration, each placeholder is filled with the string, such as "positive" (Figure 6.3). Now, click the **If** button on the **Programming** toolbar and fill in the placeholder with the required expression (x > 0, for example), as shown in Figure 6.4.

After the program module is defined completely and all placeholders are filled in, you'll be able to use the function the same way as you normally would, both for numeric and symbolic calculations.

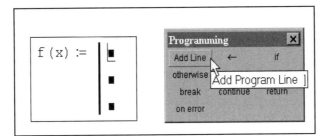

FIGURE 6.2. Starting the procedure of creating a program module

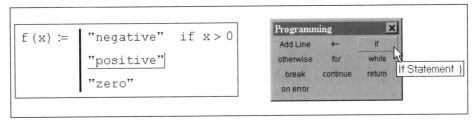

FIGURE 6.3. Inserting a statement

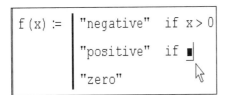

FIGURE 6.4. Inserting a condition into a program

*Don't type the names of program statements from the keyboard. To insert these statements, you should only use **Programming** toolbar buttons or keyboard shortcuts shown by the popup tooltips (Figures 6.2 and 6.3).*

6.2.3. Developing a Program

You can always add a new line of code into the existing program. This can be done at any time using the **Add Line** button of the **Programming** toolbar. To achieve this, first place the insertion lines into the required position within the program module. For example, it you place the insertion lines as shown in Figure 6.5, the new line with the placeholder will be inserted directly before this line. If you move the vertical insertion line from the first (as in Figure 6.5) to the last position of the line, the new line will be inserted directly after this line. If you select a fragment of the row (Figure 6.6) rather than the whole row, this will also influence the position of the new line within the program module (see, for example, Figure 6.7).

TIP

Don't forget that to achieve the desired placement of the insertion lines within a formula, you can use the mouse, the arrow keys, and the spacebar. Sequentially, as you press the spacebar you can select different parts of the formula using the insertion lines.

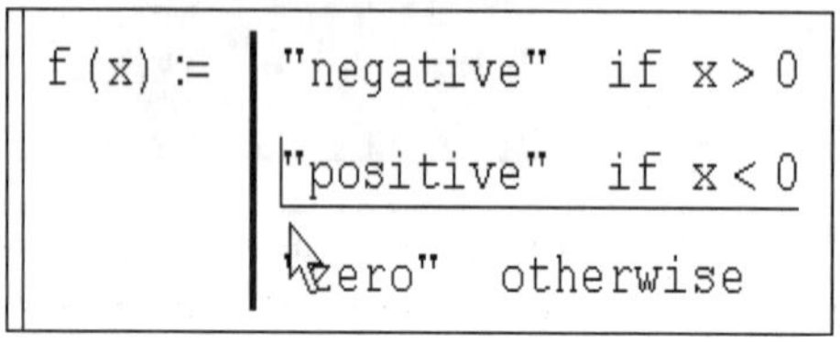

$$f(x) := \begin{array}{ll} \text{"negative"} & \text{if } x > 0 \\ \text{"positive"} & \text{if } x < 0 \\ \text{"zero"} & \text{otherwise} \end{array}$$

FIGURE 6.5. Inserting a new line into the existing program

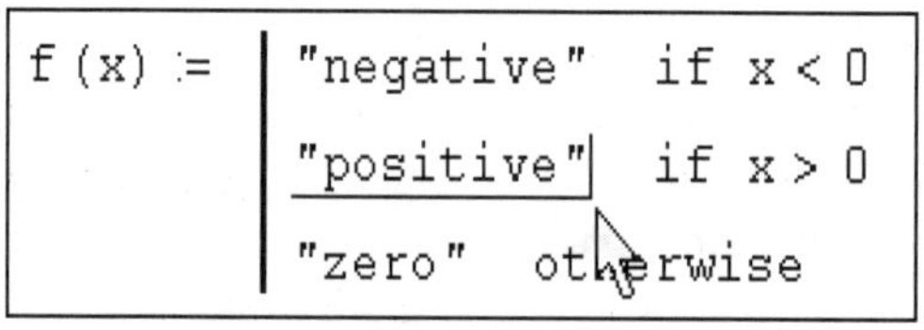

$$f(x) := \begin{array}{ll} \text{"negative"} & \text{if } x < 0 \\ \text{"positive"} & \text{if } x > 0 \\ \text{"zero"} & \text{otherwise} \end{array}$$

FIGURE 6.6. The position of the insertion lines influences the position of the newly inserted line

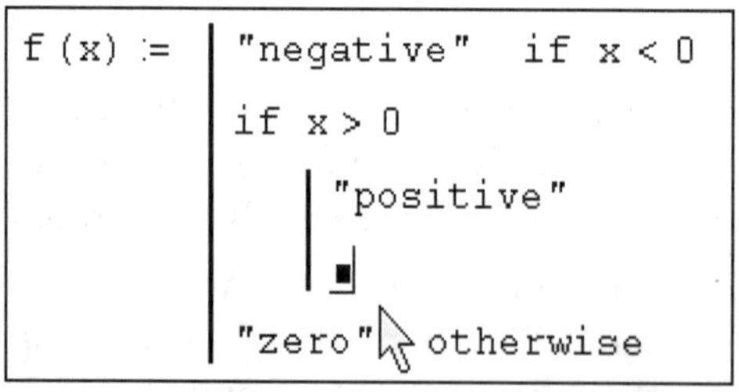

$$f(x) := \begin{array}{l} \text{"negative"} \quad \text{if } x < 0 \\ \text{if } x > 0 \\ \quad \begin{array}{l} \text{"positive"} \\ \blacksquare \end{array} \\ \text{"zero"} \quad \text{otherwise} \end{array}$$

FIGURE 6.7. The result of inserting a new line into an existing program (from the position shown in Figure 6.6)

Why might it be necessary to insert a new line into the position shown in Figure 6.7? The answer is simple – this helps to denote the code fragment that relates to the x > 0 condition that is placed on the fragment's header (notice, that the new vertical line has appeard). An example of further programming is shown in Listing 6.4.

LISTING 6.4. The Improved Program Example

```
f(x) := │ "negative"    if x < 0
        │ if x > 0
        │     │ "positive"
        │     │ "big positive"   if x > 1000
        │ "zero"   otherwise

f(1) = "positive"

f(10^5) = "big positive"
```

When the program executes, and this happens at any attempt to calculate f(x), each line of code is executed sequentially. For example, in the next to last line of Listing 6.4 the program evaluates f(1). Let us consider each line of this listing in more detail.

1. Since x = 1, the x < 0 condition is not satisfied, and the first line is not executed.
2. The condition specified in the second line (x > 0) is satisfied, therefore both lines next to it, which are joined into the common fragment by the short vertical bar, are executed.
3. The f(x) function is assigned the value: f(x) = "positive".
4. The x > 1000 condition is not satisfied, therefore the "big positive" value is not assigned, and f(x) retains the "positive" value.
5. The last line is not executed, since one of the conditions (x > 0) proved to be true, and the otherwise statement is not required.

Thus, the main principle of creating program modules is satisfied through the correct placement of the lines of code. It is not difficult to understand their working principles, since fragments of code of the same level are grouped within a program using vertical bars.

6.2.4. Local Assignment (←)

The Mathcad programming language wouldn't be efficient if it didn't allow you to create local variables within program modules. Local variables are invisible from any other parts of the documents outside the program module to which they belong. In contrast to Mathcad documents, assignment within the program is performed using Mathcad's **Local Definition** operator. To insert this operator, click the **Local Definition** button labeled by the arrow (←) on the **Programming** toolbar.

Both the assignment operator (:=) and the output operator are inapplicable for defining local variables.

NOTE

Local assignment is illustrated by the example in Listing 6.5. The z variable exists only within the program highlighted by the vertical line. It can't be accessed from any other position within the document.

LISTING 6.5. Local Assignment within a Program

$$f(x) := \begin{vmatrix} z \leftarrow 4 \\ z + x \end{vmatrix}$$

$$f(1) = 5$$

6.2.5. Conditional Statements (*if* and *otherwise*)

Action of the if statement comprises two parts. First, it checks the logical expression (conditions) to the right of the statement. If the condition is true, the expression to the left of the if statement is executed. If the condition is false, nothing happens, and the program continues execution with the next line. To insert the conditional statement into the program, proceed as follows:

1. If necessary, enter the left part of the expression and the assignment operator.
2. Create a new line o program code by clicking the **Add Line** button on the **Programming** toolbar.
3. Click the button corresponding to the if statement.
4. Enter the condition to the right of the if statement. Use logical operators by entering them from the **Boolean** toolbar.
5. To the left of the if statement, enter the expression that must be executed when the condition is true.
6. If the program provides additional operations, add a new line into the program code by clicking the **Add Line** button. Enter the code following the preceding steps, using if or otherwise statements.

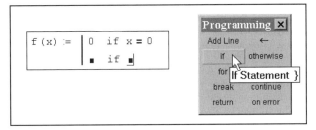

FIGURE 6.8. Inserting a conditional statement

The otherwise statement is used with one or more if statements and specifies the expression that must be executed if no condition proves to be true. The examples illustrating the usage of if and otherwise statements are provided in the next sections (see Listings 6.3 and 6.4).

6.2.6. Loop Statements *(for, while, break, and continue)*

The Mathcad programming language has two loop statements: for and while. The first statement lets the user organize a loop by some variable, making it take the specified range of values. The second statement creates a loop, the body of which keeps executing until a specified logical condition is met. To insert a loop statement into the program module, do the following:

1. Create a new code fragment selected by a vertical line in the program module.
2. Insert the for statement or the while statement by clicking an appropriate button on the **Programming** toolbar.
3. If you have selected the for statement (Figure 6.9), fill in the appropriate placeholders with the name of the loop variable and enter the range of its values (Listings 6.6 and 6.7). If you have inserted the while statement, insert a Boolean expression. When this expression is false, the loop stops execution (Figure 6.8).

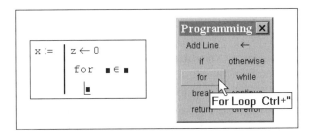

FIGURE 6.9. Inserting the loop statement

4. Fill the lower placeholder with the loop body (i.e., the set of expressions that must keep executing cyclically).
5. When necessary, add additional lines of code to the program.

The range of values of the loop variable in the for *loop can be specified both using range variables (Listing 6.6) or vectors (Listing 6.7).*

LISTING 6.6. The for Loop Specified Using a Range Variable

```
x :=   z ← 0
       for  i ∈ 0 ..  5
         z ← z + i

x = 15
```

LISTING 6.7. The for Loop Specified Using a Vector

```
x :=   z ← 0
       for  i ∈ ( 1   2   3 )
         z ← z + i

x = 6
```

LISTING 6.8. The while Loop

```
x :=   z ← 0
       while   z < 10
         z ← z + 1

x = 10
```

Sometimes it might be necessary to terminate the loop at a specific line within the loop body rather than through a condition specified at the loop header. The break statement is intended specifically to solve this task. Listings 6.9 and 6.10 provide the modified code of Listings 6.6 and 6.8 with the break statement. For example, in Listing 6.9, when the loop variable i reaches the value of 2, the loop halts because of the presence of the break statement in the last line of the program module. Accordingly, the value of the x variable remains equal to $0 + 1 + 2 = 3$.

LISTING 6.9. The break Statement within the for Loop

```
x :=   | z ← 0
       | for  i ∈ 0 .. 5
       |     | z ← z + i
       |     | break   if i = 2
x = 3
```

LISTING 6.10. The break Statement within the while Loop

```
x :=   | z ← 0
       | while  z < 10
       |     | z ← z + 1
       |     | break   if z > 5
x = 6
```

To designate the boundaries where the loop body stops executing, an additional line with the continue *statement can be used at the loop end. This statement can be inserted by clicking the appropriate button on the* **Programming** *toolbar. The Listings 6.11 and 6.12 illustrate examples of modified versions of Listings 6.7 and 6.8, respectively. As you can see, the presence of the* continue *statement has no influence on the result of the program's execution.*

LISTING 6.11. The continue Statement at the End of the while Loop

```
x :=   | z ← 0
       | while  z < 10
       |     | z ← z + 1
       |     | continue
x = 10
```

LISTING 6.12. The continue Statement at the End of the for Loop

$$
x := \begin{array}{|l}
z \leftarrow 0 \\
\text{for } i \in (1 \quad 2 \quad 3) \\
\quad \begin{array}{|l} z \leftarrow z + i \\ \text{continue} \end{array}
\end{array}
$$

$x = 6$

6.2.7. Returning a Value *(return)*

If any variable or function is defined using a program module, then the lines of this module execute in succession when calculating this variable or function within a document. Respectively, the result being calculated changes with program execution. The last assigned value is returned as the final result (examples are presented in Listings 6.3–6.12). To emphasize the fact that a specific value is the return value, you make an assignment on the last line of the program module (Listing 6.13).

LISTING 6.13. Explicitly Specifying the Return Value in the Last Line of the Program Module

$$
f(x) := \begin{array}{|l}
y \leftarrow x^2 \\
z \leftarrow y + 1 \\
z
\end{array}
$$

$f(2) = 5$

On the other hand, you can halt the execution of the program at any point (for example, by using a conditional statement) and return a specific value using the return statement. In this case, when the specific condition has been met (Listing 6.14), the value entered into the placeholder following the return statement will be returned as the final result, and no other code will be executed. To insert the return statement into your program, click the respective button on the **Programming** toolbar.

LISTING 6.14. Using the `return` Statement

$$f(x) := \begin{vmatrix} z \leftarrow x^2 \\ \text{return} \quad \text{"zero"} \quad \text{if } x = 0 \\ \text{return} \quad \text{"i"} \quad \text{if } x = i \\ z \end{vmatrix}$$

$f(-1) = 1$

$f(2) = 4$

$f(0) = \text{"zero"}$

$f(i) = \text{"i"}$

6.2.8. Error Handling *(on error)*

Mathcad programming provides additional error handling capabilities. If the user supposes that the execution of specific code within the program module can cause an error (for example, division by zero), it is possible to handle this error using the `on error` statement. To insert this statement into the program, place the insertion lines into the required position and click the `on error` button on the **Programming** toolbar. As a result, the line of code with two placeholders and the `on error` statement between them will appear (Figure 6.10).

The right placeholder must be filled in with an expression that must be executed in the current line of code. The left placeholder must be filled in with an expression that will be executed instead of the right one, in case it causes an error. Listing 6.15 illustrates the usage of the `on error` statement. The program module presented in this listing calculates the reverse function of n. When $n \neq 0$, the assigned value $z \neq 0$, thus, the last line of this program executes the expression that calculates the $1/z$ value. For example, this happens when calculating the `f(-2)` value. However, if we try to calculate the `f(0)` value, as in the last line of this listing, the execution of the program specified for `f(n)` will result in the division by

$$f(x) := \begin{vmatrix} \blacksquare \quad \text{on error} \quad \blacksquare \\ \blacksquare \end{vmatrix}$$

FIGURE 6.10. Inserting the `on error` statement

zero error on the last line of the program. Thus, instead of the expression next to the right of the on error statement, the left expression will be executed, which, in our case assigns the "user error: cannot divide by zero" sting value to the f(n) function.

LISTING 6.15. Handling the Division by Zero Error

$$f(n) := \begin{vmatrix} z \leftarrow n \\ \\ \text{"user error: can't divide by zero"} \quad \text{on error} \quad \dfrac{1}{z} \end{vmatrix}$$

$$f(-2) \rightarrow \dfrac{-1}{2}$$

f(0) = "user error: can't divide by zero"

It is very convenient to use the error-handling statement with the error(S) built-in function. This function generates an error message in Mathcad's traditional form. Figure 6.11 gives an example of an improved version of Listing 6.15 for this mode of handling the division by zero error. Notice that the text of the error message is passed as an argument to the error function.

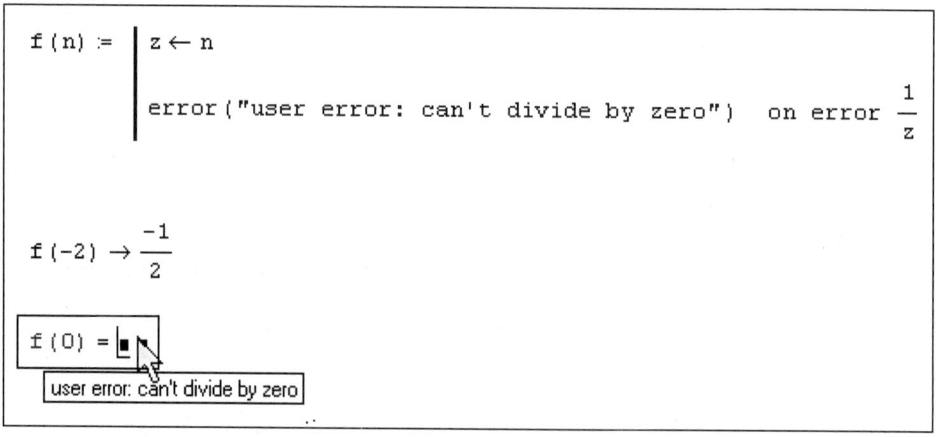

$$f(n) := \begin{vmatrix} z \leftarrow n \\ \\ \text{error("user error: can't divide by zero")} \quad \text{on error} \quad \dfrac{1}{z} \end{vmatrix}$$

$$f(-2) \rightarrow \dfrac{-1}{2}$$

f(0) = ■

user error: can't divide by zero

FIGURE 6.11. Handling the division by zero error

6.3. PROGRAMMING EXAMPLES

Let us consider two simple examples of how to use program modules in Mathcad for numeric (Listing 6.16) and symbolic (Listing 6.17) calculations. These two listings use a large number of the operators and statements considered in this chapter. When you start developing your own custom Mathcad program modules, don't forget that programming statements are inserted into the program text using the **Programming** toolbar buttons. Don't ever" attempt to type their names from the keyboard, since Mathcad won't interpret such input correctly.

Using various programming tools, you can create much more complex programs. Several examples illustrating efficient programming usage will be presented in Section 11.4, "The Phase Portrait of a Dynamic System," in Chapter 11 and Section 12.4.2, "Difference Schemes," in Chapter 12.

LISTING 6.16. Programming Usage in Numerical Calculations

$$
f(n) := \begin{vmatrix} \text{return} & -99 & \text{if } n < 0 \\ z \leftarrow 1 \\ \text{for } i \in 1.. \ n \\ \quad z \leftarrow z \cdot i \\ z \end{vmatrix}
$$

$f(-2) \rightarrow -99$

$f(0) = 0$

$f(3.9) = 6$

$f(3) = 6$

$f(10) = 3.629 \times 10^6$

LISTING 6.17. Programming Usage in Symbolic Calculations

$$f(n) := \begin{vmatrix} -1 & \text{if } n < 0 \\ \\ x & \text{on error } \dfrac{d^n}{dx^n} x^{10} & \text{otherwise} \end{vmatrix}$$

$$f(1) \rightarrow 10 \cdot x^9$$

$$f(10) \rightarrow 3628800$$

$$f(-3) \rightarrow -1$$

$$f(2.1) \rightarrow x$$

7 Integration and Differentiation

In this chapter we are going to consider the main mathematical operations of numeric differentiation and integration.

In terms of calculation, integration *(see Section 7.1)*, and differentiation *(see Section 7.2)* are the simplest operations implemented in Mathcad in the form of operators. Still, if the calculations are performed using a numeric processor, it is necessary to have a sound understanding of the numeric algorithms, specific features and details of which remain invisible for the user. In the same sections *(see Sections 7.1 and 7.2)* we are also going to mention specific features of symbolic differentiation and integration.

7.1. INTEGRATION

Integration in Mathcad is implemented in the form of a numeric operator. It is possible to calculate integrals of scalar functions within specified integration limits (which must also be scalar). Despite the fact that integration limits must be real numbers, the integrand must take complex values, therefore integral values can also be complex. If integration limits have dimension *(see Section 4.2, "Dimensional Variables," in Chapter 4)*, it must be the same for both limits.

7.1.1. Integration Operator

Integration and differentiation operations, like most other mathematical operations, are implemented in Mathcad according to the What You See Is What You
Get (WYSIWYG) principle. To calculate a definite integral, you need to enter its
mathematical formula into the document. This can be done by clicking the button
labeled by the integral sign on the Calculus toolbar, or by pressing the
<Shift>+<7> keyboard shortcut (or the <&> character, which is equivalent). An
integral sign with several placeholders will appear (Figure 7.1), which you must fill
in with the lower and upper integration limits, the integrand, and variable of integration.

*It is also possible to calculate integrals with infinite limits (one or both). To
achieve this, fill the placeholder of the appropriate integration limit with the infinity character using, for example, the Calculus toolbar. To enter the -∞ character, the infinity character should be preceded by the minus sign (like any other
number).*

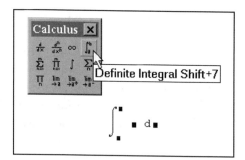

FIGURE 7.1. Integration operator

To obtain the integration result, enter the equal or symbolic equal character.
In the first case, the integration will be performed numerically. In the second case,
if the operation is successful, the symbolic processor will find the precise value of
the integral. These two methods are illustrated in Listing 7.1. Certainly, symbolic
integration is only possible for a limited range of relatively simple integrands.

LISTING 7.1. Numeric and Symbolic Calculation of a Definite Integral

$$\int_0^\pi \sin(x)\ dx = 2 \qquad\qquad \int_0^\pi \sin(x)\ dx \rightarrow 2$$

The integrand might be a function of any number of variables. Thus, to instruct Mathcad which variable it is necessary to calculate the integral with, it is necessary to specify the variable of integration in the appropriate placeholder. Remember that if you need to perform numeric integration with one of the variables, it is necessary to predefine the values of the other variables on which the integrand depends and for which you need to calculate the integral (Listing 7.2).

LISTING 7.2. Integrating the Function of Two Variables with Different Variables

$\alpha := 2$

$$\int_0^\pi \alpha \cdot \sin(x) \ dx = 4$$

$x := 1$

$$\int_0^{10} \alpha \cdot \sin(x) \ d\alpha = 42.074$$

The integration operator can be used just like other operators: for defining functions, in loops, and for calculating range variables. Listing 7.3 provides an example in which the custom function g(x) is assigned a value of a definite integral and several values of this function are calculated.

LISTING 7.3. Usage of the Integration Operator in the Custom Function

$$g(\alpha) := \int_0^\pi \alpha \cdot \sin(x) \ dx$$

$i := 1 .. \ 5$

$g(i) =$

2
4
6
8
10

7.1.2. On the Integration Algorithms

The result of numeric integration is not precise, but rather, it is an approximate value of the integral calculated with an error that depends on the built-in TOL constant. The smaller its value, the more precisely the integral will be evaluated, but at the same time, the more time-consuming the process will be. By default, TOL = 0.001. To increase the speed of the calculation process, it is possible to set a higher value of TOL.

TIP

If the speed of the calculation is of principal importance to you (for example, when it is repeatedly calculating the integral value within a loop), pay special attention when you are selecting the accuracy value. It is recommended that you experiment with the test example using the integrand typical for your calculations. Notice how the decrease of the value of the TOL constant influences the integration error by evaluating the integral with different TOL values and selecting the optimal one based on the precision/calculation speed ratio.

You must also understand that when you insert the integration operator into the Mathcad editor, you are actually creating a real program. For example, the first line of Listing 7.1 is actually a program, most of which MathSoft developers have simply hidden from the user. Most of the time you don't need to pay special attention to this fact, since you can completely rely on Mathcad in this matter. Sometimes, however, you'll need to manipulate the parameters of this program, as was shown in the TOL value selection example. In addition, the user can manually select the algorithm of numerical integration. To achieve this, proceed as follows:

1. Right-click anywhere within the left part of the integral to be calculated.
2. Select one of the available numerical algorithms from the right-click menu (Figure 7.2).

Notice that before you select one of the available algorithms for the first time (Figure 7.2), the **AutoSelect** option in the right-click menu is set by default. This means that Mathcad selects the most appropriate algorithm automatically, based on the results of its analysis of the integration limits and the properties of the integrand. When you select one of the algorithms, this checkmark will be cleared, and the selected algorithm will be marked by a dot.

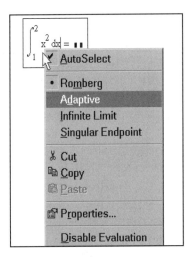

FIGURE 7.2. Selecting an algorithm for numerical integration

Mathcad 2001i has provided the following four numerical integration methods:

- Romberg — Suitable for most functions that have no singularities.
- Adaptive — Intended for functions that rapidly change within the integration interval.
- Infinite Limit — Intended for integrals with infinite integration limits.
- Singular Endpoint — Intended for integrals having a singularity at the endpoints. This is the modified version of the Romberg algorithm, intended for functions undefined at one or both ends of the integration interval.

Whenever possible, always try to rely on Mathcad when selecting the integration algorithm. To enable this option, set the **AutoSelect** option in the right-click menu. Manual selection of another method makes sense when you need to compare the results for specific cases, because you doubt if the obtained result is correct.

If you are dealing with a "good" integrand, which means that it doesn't change too rapidly within the integration limits and doesn't tend toward infinity within the integration interval, numeric evaluation of the integral usually doesn't present any problems or unpleasant surprises. Listed below are the basic concepts of the Romberg algorithm, which is used by default for most such integrands.

- First, the algorithm builds several interpolating polynomials, which replace the integrand $f(x)$ within the integration interval. In the first iteration, the polynomials are calculated by 1, 2, and 4 intervals. For example, the first polynomial, built by one, is simply a straight line drawn via the two endpoints of the integration interval; the second represents a square parabola; and so on.

■ Using each polynomial with the known coefficients as an integrand, it is easy to evaluate the integral analytically. Thus, the algorithm determines the sequence of integrals I_1, I_2, I_4, ... of the interpolating polynomials. For example, according to the trapezoid rule: $I_1 = (b-a) \cdot (f(a) + f(b)) / 2$, and so on.

■ Because of the different number of interpolation points, the calculated integrals I_1, I_2, ... are somewhat different from one another. The larger the number of interpolation points, the better the interpolation polynomial approximates the integral to be evaluated. When the number of interpolation points approaches infinity, the interpolation polynomial gets closer to the integral being calculated. Thus, we extrapolate the I_1, I_2, I_4, ... sequence to the zero width of the elementary interval. The result of this extrapolation, J, is taken for the approximation of the integral being calculated.

■ Using finer partitioning, we proceed with the next iteration by adding a new member of the sequence of interpolating polynomials and by calculating the new (N-th) Romberg approximation, J^N.

■ The larger the number of interpolation points, the closer the Romberg approximation will be to the integral being calculated, and the smaller its difference from the approximation produced by the previous iteration. When the difference between the last two iterations $|J^N - J^{N-1}|$ becomes smaller than the TOL error or less than $TOL \cdot |J^N|$, the iterations halt, and the J^N value is displayed as the integration result.

The polynomial interpolation algorithm will be covered in more detail in Chapter 14.

7.1.3. About Divergent Integrals

If the integral is divergent (tends toward infinity), the Mathcad numeric processor displays an error message and highlights the integration operator in red. Most often, the error message will read: "Found a number with a magnitude greater than 10^307," or "Can't converge to a solution" as, for example, when attempting to calculate the following integral: $\int_0^\infty \frac{1}{\sqrt{x}} \, dx$. Still, the symbolic processor successfully handles this integral by correctly evaluating its infinite value (Listing 7.4).

LISTING 7.4. Symbolic Evaluation of the Divergent Integral

$$\int_0^\infty \frac{1}{\sqrt{x}}\ dx \rightarrow \infty$$

*The symbolic processor provides powerful capabilities of analytical evaluation of integrals, including parametric and indefinite ones, as shown in Listings 7.5 and 7.6. These capabilities of evaluating integrals using the **Symbolics** menu were mentioned in Chapter 5.*

LISTING 7.5. Symbolic Evaluation of the Integral with a Variable Limit

$$\int_0^a \frac{1}{\sqrt{x}}\ dx \rightarrow 2 \cdot a^{\left(\frac{1}{2}\right)}$$

LISTING 7.6. Symbolic Evaluation of the Indefinite Integral

$$\int \frac{1}{\sqrt{x}}\ dx \rightarrow 2 \cdot x^{\left(\frac{1}{2}\right)}$$

If you attempt to obtain a numerical solution of the task shown in Listing 7.4 using a method other than **Infinite Limit**, you'll get an incorrect result (Listing 7.7). Instead of infinity, Mathcad's numerical processor will produce quite a large (but still finite) value – 10^{307} *(see Section 4.1.3, "Built-in Constants," in Chapter 4)*. Notice that Mathcad will select the correct algorithm (**Infinite Limit**), if the AutoSelect option is enabled.

LISTING 7.7. An Incorrectly Selected Numerical Algorithm Produces Incorrect Result for a Divergent Integral

$$\int_0^\infty \frac{1}{\sqrt{x}}\ dx = 6.325 \times 10^{153}$$

7.1.4. Multiple Integrals

To evaluate a multiple integral, proceed as follows:

1. Insert the integration operator, proceeding in the standard fashion.
2. Fill in the appropriate placeholders with the name of the first integration variable and integration limits for this variable.
3. Enter another integration operator into the placeholder intended for the integrand.
4. Proceed in just the same way, and enter the second integration variable, its integration limits and the integrand (if the integral is twofold) or another integration operator, and so on, until you accomplish entering the expression with multiple integrals.

An example of symbolic and numeric calculation of the twofold integrals within infinite integration limits is shown in Listing 7.8. Take note of the fact that the symbolic processor produces the precise value of the integral (π), while the numeric processor approximates this value and displays the result as 3,142.

LISTING 7.8. Symbolic and Numeric Evaluation of the Twofold Integral

$$\int_{-\infty}^{\infty} \int_{-\infty}^{\infty} e^{-\left(x^2+y^2\right)} \, dx \, dy \to \pi$$

$$\int_{-\infty}^{\infty} \int_{-\infty}^{\infty} e^{-\left(x^2+y^2\right)} \, dx \, dy = 3.142$$

NOTE

When entering multiple integrals in the Mathcad editor, be very careful with the integration limits, especially if they are different for different integration variables. Don't confuse the limits for different integration variables. If you are dealing with tasks of this sort, carefully study Listing 7.9, where the symbolic processor calculates a twofold integral. Notice that in the first line of this listing, the integration limits [a, b] relate to the y variable, while in the second line they relate to the x variable.

LISTING 7.9. Symbolic Evaluation of the Twofold Integrals

$$\int_a^b \int_{-1}^1 x + y^3 \ dx \ dy \rightarrow \frac{1}{2} \cdot b^4 - \frac{1}{2} \cdot a^4$$

$$\int_a^b \int_{-1}^1 x + y^3 \ dy \ dx \rightarrow b^2 - a^2$$

7.2. DIFFERENTIATION

Using Mathcad, it is possible to evaluate the derivatives of scalar functions from any number of arguments, from the 0-th to the 5-th order inclusively. Both functions and their arguments might be both real and complex. The only case when differentiation is impossible is differentiation in the vicinity of the singularity points.

The Mathcad numeric processor provides excellent accuracy in numeric differentiation. However, advanced users will certainly also be able to appreciate the rich capabilities of the symbolic processor, which enables them to easily perform routine work when evaluating derivatives of rather awkward functions. This is because, in contrast to other operations, symbolic differentiation can be completed successfully for most analytical functions.

In order to improve both performance and accuracy of numeric differentiation, Mathcad 2001i uses a symbolic processor for numeric differentiation of analytical functions *(see Section 3.3.5, "Optimizing Calculations," in Chapter 3).*

7.2.1. The First Derivative

In order to differentiate the f(x) function at the specific point, do the following:

1. Specify the x point where the derivative must be evaluated, for example, x := 1.
2. Insert the differentiation operator by clicking the **Derivative** button on the **Calculus** toolbar or type the <?> character from the keyboard.
3. Fill in the placeholders (Figure 7.3) with the function depending on the x argument (i.e., f(x), and the argument itself).
4. Insert the numeric <=> or symbolic <→> output operator to display the result.

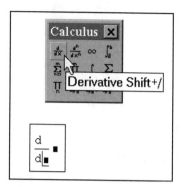

FIGURE 7.3. The differentiation operator

Listing 7.10 shows an example that illustrates differentiation of the following function: $f(x) = \cos(x) \cdot \ln(x)$.

LISTING 7.10. Numeric and Symbolic Differentiation

$$x := 0.01$$

$$\frac{d}{dx}\cos(x) \cdot \ln(x) = 100.041$$

$$\frac{d}{dx}\cos(x) \cdot \ln(x) \rightarrow -\sin\left(1. \cdot 10^{-2}\right) \cdot \ln\left(1. \cdot 10^{-2}\right) + 1. \cdot 10^{2} \cdot \cos\left(1. \cdot 10^{-2}\right)$$

NOTE

Don't forget to define the point where you need to perform numeric differentiation in advance (consider the first line of the Listing 7.10), otherwise you'll get an error message shown in Figure 7.4, informing you that the variable or function within your expression is not defined. Symbolic differentiation, in contrast to this situation, doesn't require you to explicitly specify the differentiation point. Thus, if you forget to specify the differentiation point, the symbolic processor will produce the analytical dependence (see the first line in Figure 7.4) instead of the derivative value (a number or numeric expression).

Of course, as is the case with any other operator, you can define the function in a separate expression in advance, and then calculate its derivative (see Listing 7.11); or you can apply the differentiation operator for defining custom functions (Listing 7.12).

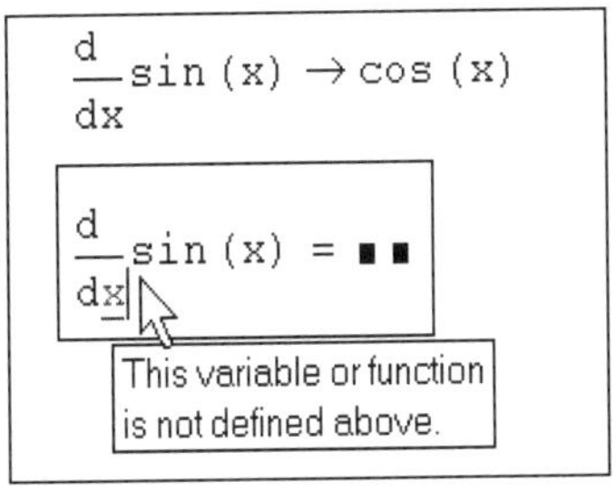

$$\frac{d}{dx} \sin(x) \rightarrow \cos(x)$$

$$\frac{d}{dx} \sin(x) \ = \ \blacksquare\,\blacksquare$$

This variable or function
is not defined above.

FIGURE 7.4. Error message displayed when using the differentiation operator incorrectly

LISTING 7.11. Symbolic and Numeric Differentiation of Custom Functions

$$f(x) := \frac{1}{x}$$

$$\frac{d}{dx} f(x) \ \rightarrow \ \frac{-1}{x^2}$$

$$x := 0.1$$

$$\frac{d}{dx} f(x) \ = -100$$

$$\frac{d}{dx} f(x) \ \rightarrow -\left(1. \cdot 10^2\right)$$

LISTING 7.12. Defining Custom Functions Using the Differentiation Operator

$$f(x) := \frac{1}{x}$$

$$g(x) := \frac{d}{dx} f(x)$$

$$g(0.1) \ = -100$$

$$g(0.1) \ \rightarrow -\left(1. \cdot 10^2\right)$$

In both listings, the first line defines the function $f(x) = 1/x$. The second line of Listing 7.11 uses the symbolic processor to find an analytical solution for its derivative, and the remaining lines, similar to those in Listing 7.10, calculate the values of this variable at the point $x = 0.1$. In Listing 7.12, the derivative of the $f(x)$ function is used to define another custom function $g(x)$, and then we find its value at the same point $x = 0.1$.

As you might have already noticed, the differentiation operator mainly corresponds to its commonly accepted math designation. Sometimes, however, you have to be very careful when inserting it. Listing 7.13 illustrates such a situation. The first two lines of this listing evaluate the derivative of $\sin(x)$ at the point $x = 0.5$. The last line demonstrates incorrect usage of the differentiation operator. Instead of calculating the derivative of $\sin(x)$ at the same point (as might be expected), the obtained result is equal to zero. This happened because the argument of the $\sin(x)$ function was introduced as a number rather than the variable (x). Thus, Mathcad interprets the last line as follows: first, it evaluates the sine value for $x = 0.5$, then it differentiates this value (a constant) at the same point $(x = 0.5)$ according to the first line of the listing. Thus, the result is incorrect, but this is not surprising, since the derivative of a constant is zero at any point.

LISTING 7.13. Example of Incorrect Usage of the Differentiation Operator

$$x := 0.5$$

$$\frac{d}{dx}\sin(x) = 0.878$$

$$\frac{d}{dx}\sin(0.5) = 0$$

For numerical differentiation, Mathcad uses a rather complicated algorithm that calculates the derivative with an impressive accuracy of about 7 to 8 digits after the decimal point. This algorithm (the Ridder method) is described in detail in the Mathcad on-line Help system, available via the Help menu. In contrast to most other numerical methods, a differentiation error doesn't depend on TOL or CTOL built-in constants, but rather it is defined by the algorithm itself.

The only exception is made up of the functions differentiated in the proximity of the singularity point. For example, for the $f(x) = 1/x$ function considered earlier, this will be the point $x = 0$. If you make an attempt to find the derivative of this function at $x = 0$, Mathcad will produce an error message informing you of a division by zero error such as "Can't divide by zero" or "Found a singularity while evaluating this expression. You may be dividing by zero." If you attempt to evaluate the derivative numerically for the value very close to the singularity point (for example, for $x = 10^{-100}$), the following error message might appear: "Can't converge to a solution." When you encounter such an error, study the function being differentiated very carefully and make sure that you are not dealing with the singularity point.

7.2.2. Higher Order Derivatives

Mathcad allows you to numerically evaluate derivatives of higher orders, from the 0-th to the 5-th inclusively. To calculate the derivative of the N-th order of the function f(x) at the x point, proceed in the same way you would as when you are evaluating the first derivative (*see Section 7.2.1*), except for the fact that instead of the first derivative operator, you just use the N-th derivative operator (**Nth Derivative**). You can insert this operator from the **Calculus** toolbar or from the keyboard by pressing <Ctrl>+<?> keys. The N-th derivative operator contains two additional placeholders, which you need to fill in with the N number. According to the mathematical sense of the operator, when you fill in one of the derivative placeholders, the same number immediately appears in another placeholder.

According to the definition, the "derivative of N-th order when N = 0" is equal to the function itself, and when N = 1, you'll get the first derivative. Listing 7.14 demonstrates a numerical and symbolic evaluation of the second derivative. Notice that as with the first derivative, you need to assign a value to the function argument to get the correct result.

LISTING 7.14. Numeric and Symbolic Evaluation of the Second Derivative

    x := 0.1

$$\frac{d^2}{dx^2} \cos(x) \cdot x^2 = 1.94$$

$$\frac{d^2}{dx^2} \cos(x) \cdot x^2 \rightarrow 1.99 \cdot \cos(.1) - .4 \cdot \sin(.1)$$

*To make sure that the Mathcad symbolic processor produces the same result (the last line of Listing 7.14) as a numeric processor, you can simplify the expression. To achieve this, select the resulting expression, then select the **Simplify** command from the **Symbolics** menu. After this, another line will appear, containing the numeric result of the selected expression.*

To calculate the derivative of an order higher than the 5-th derivative, apply the N-th derivative operator sequentially several times, just as you would for a multiple integral *(see Section 7.1.4)*. Take note of the fact, however, that you don't need to do this for symbolic calculations since the symbolic processor is capable of evaluating derivatives of the order higher than the 5-th derivative. This is illustrated by Listing 7.15, where the 7-th derivative is calculated first numerically and then analytically at the point $x = 0.1$.

LISTING 7.15. Numeric and Symbolic Evaluation of the 7-th Derivative

$x := 0.1$

$$\frac{d^5}{dx^5} \frac{d^2}{dx^2} \sin(x) = -0.995$$

$$\frac{d^7}{dx^7} \sin(x) \rightarrow -\cos(.1)$$

Calculation of higher-order derivatives is performed using the same Ridder method, as was used for first derivatives. Keep in mind that for the first derivative, this method provides an accuracy of 7 to 8 significant digits, and it decreases by approximately one digit with the increase of the derivative order.

Thus, it is obvious that a fall in precision when calculating higher-order derivatives might be significant. For example, if you attempt to evaluate the 9-th derivative of the sine function using the idea of Listing 7.15, the result will be zero, although the actual value of the 9-th variable is $\cos(0.1)$.

7.2.3. Partial Derivatives

Using both processors, Mathcad can calculate the derivatives of the functions of any number of arguments. In this case, the derivatives are known as partial derivatives. To calculate a partial derivative, insert the derivative operator from the Calculus toolbar and fill in the appropriate placeholder with the name of variable with which it is necessary to differentiate. Listing 7.16 exemplifies this concept. The first line of this listing defines the function of two variables, while the next two lines evaluate its partial derivatives by both variables — x and y — respectively. In order to determine a partial derivative using a numeric method, it is necessary to specify the values of all arguments in advance, as is done in the next two lines of the listing. The last expression in this listing once again (as in the third line) symbolically defines the partial derivative by y. However, because both x and y variables have already been assigned specific values, the result is a number rather than an analytical expression.

LISTING 7.16. Symbolic and Numeric Evaluation of Partial Derivatives

$$f(x, y) := x^{2 \cdot y} + \cos(x) \cdot y$$

$$\frac{\partial}{\partial x} f(x, y) \rightarrow 2 \cdot x^{(2 \cdot y)} \cdot \frac{y}{x} - \sin(x) \cdot y$$

$$\frac{\partial}{\partial y} f(x, y) \rightarrow 2 \cdot x^{(2 \cdot y)} \cdot \ln(x) + \cos(x)$$

$$x := 1 \qquad y := 0.1$$

$$\frac{\partial}{\partial y} f(x, y) = 0.54$$

$$\frac{\partial}{\partial y} f(x, y) \rightarrow \cos(1)$$

Partial derivatives of higher orders are evaluated in the same way that normal derivatives of higher orders are *(see Section. 7.2.2)*. Listing 7.17 outlines the procedure of calculating the second derivatives of the function from the previous example with x, y variables and a mixed derivative.

LISTING 7.17. Evaluating the Second Partial Derivative

$$\frac{\partial^2}{\partial x^2}\left(x^{2 \cdot y} + \cos(x) \cdot y\right) \rightarrow 4 \cdot x^{(2 \cdot y)} \cdot \frac{y^2}{x^2} - 2 \cdot x^{(2 \cdot y)} \cdot \frac{y}{x^2} - \cos(x) \cdot y$$

$$\frac{\partial^2}{\partial y^2}\left(x^{2 \cdot y} + \cos(x) \cdot y\right) \rightarrow 4 \cdot x^{(2 \cdot y)} \cdot \ln(x)^2$$

$$\frac{\partial}{\partial x}\frac{\partial}{\partial y}\left(x^{2 \cdot y} + \cos(x) \cdot y\right) \rightarrow 4 \cdot x^{(2 \cdot y)} \cdot \frac{y}{x} \cdot \ln(x) + 2 \cdot \frac{x^{(2 \cdot y)}}{x} - \sin(x)$$

You have probably noticed that in both Listings 7.16 and 7.17 the differentiation operator is written as a partial derivative. You can choose to write differentiation operators like a normal or partial variable, just like you can for an assignment operator. The form doesn't influence the calculations, but rather, serves as a form to which you are accustomed when representing the calculation results. To change the display of the differentiation operator to that of the partial derivative representation, proceed as follows:

1. Right-click the differentiation operator.
2. From the right-click menu, select the View Derivative As command.
3. Select the **Partial Derivative** command from the pop-up menu (Figure 7.5).

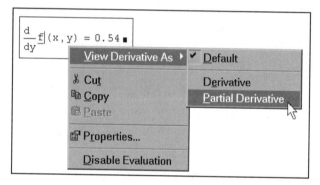

FIGURE 7.5. Changing the view of the differentiation operator

To reset the view of the differentiation operator to the one set by default, select the **Default** command from the pop-up menu or, to represent it as a normal derivative, — the **Derivative** command.

To complete our discussion of partial variables, let us consider two examples, which are common in the practice of numerical calculations. Programmatic implementation of the first example, which performs an evaluation of the gradient of the function of two variables, is shown in Listing 7.18. As an example, we have taken the f(x,y) function defined in the first line of this listing. The graph of this function is shown as level curves in Figure 7.6. According to the definition, the gradient of the f(x,y) function is the vector function of the same arguments determined via its partial variables (see the second line of the Listing 7.18). The remaining part of this listing defines range variables and matrices required for producing the plots of the function and its gradient.

LISTING 7.18. Calculating the Gradient

$$f(x, y) := x^2 + 0.1 \cdot y^3$$

$$grad(x, y) := \begin{pmatrix} \dfrac{\partial}{\partial x} f(x, y) \\[2ex] \dfrac{\partial}{\partial y} f(x, y) \end{pmatrix}$$

$$N := 5$$

$$i := 0 \ .. \ 2 \cdot N \qquad j := 0 \ .. \ 2 \cdot N$$

$$F_{i,j} := f(i - N, j - N)$$

$$V_{i,j} := grad(i - N, j - N)$$

$$X_{i,j} := (V_{i,j})_0 \qquad Y_{i,j} := (V_{i,j})_1$$

The vector field of the calculated gradient of the $f(x, y)$ function is shown in Figure 7.7. As you can see by comparing Figures 7.6 and 7.7, the gradient specifies the direction within the plane for each (x, y) point, in which the $f(x, y)$ function grows most rapidly.

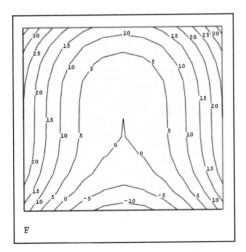

FIGURE 7.6. Level curves of the $f(x, y)$ function (Listing 7.18)

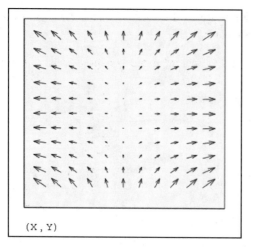

(X, Y)

FIGURE 7.7. Vector field of the gradient of the f(x,y) function (Listing 7.18)

Up to this point, we have considered scalar functions to which the differentiation operators are applicable. However, quite often, you must deal with vector functions. For example, in various fields of mathematics *(see Chapter 11)*, we encounter the problem of evaluating the Jacobian (or Jacobi matrix) — the matrix composed from the partial variables of a vector function with all its arguments. Listing 7.19 presents an example of evaluating the Jacobian of a vector function f(x) of vector argument x. In this example, to evaluate partial derivatives of the Jacobian, each i-th scalar argument f(x)$_i$ is differentiated by the Mathcad symbolic processor.

LISTING 7.19. Evaluating the Jacobian of a Vector Function of a Vector Argument

$$
f(x) := \begin{bmatrix} x_0 \cdot \sin(x_1) \\ -(x_1)^{x_0} + x_2 \\ 3 \cdot x_1 + x_2 \end{bmatrix}
$$

$$
\begin{bmatrix}
\frac{\partial}{\partial x}\left(\left(f\left(\begin{pmatrix} x \\ y \\ z \end{pmatrix}\right)\right)_0\right) & \frac{\partial}{\partial y}\left(\left(f\left(\begin{pmatrix} x \\ y \\ z \end{pmatrix}\right)\right)_0\right) & \frac{\partial}{\partial z}\left(\left(f\left(\begin{pmatrix} x \\ y \\ z \end{pmatrix}\right)\right)_0\right) \\
\frac{\partial}{\partial x}\left(\left(f\left(\begin{pmatrix} x \\ y \\ z \end{pmatrix}\right)\right)_1\right) & \frac{\partial}{\partial y}\left(\left(f\left(\begin{pmatrix} x \\ y \\ z \end{pmatrix}\right)\right)_1\right) & \frac{\partial}{\partial z}\left(\left(f\left(\begin{pmatrix} x \\ y \\ z \end{pmatrix}\right)\right)_1\right) \\
\frac{\partial}{\partial x}\left(\left(f\left(\begin{pmatrix} x \\ y \\ z \end{pmatrix}\right)\right)_2\right) & \frac{\partial}{\partial y}\left(\left(f\left(\begin{pmatrix} x \\ y \\ z \end{pmatrix}\right)\right)_2\right) & \frac{\partial}{\partial z}\left(\left(f\left(\begin{pmatrix} x \\ y \\ z \end{pmatrix}\right)\right)_2\right)
\end{bmatrix}
\rightarrow
\begin{pmatrix}
\sin(y) & x \cdot \cos(y) & 0 \\
-y^x \cdot \ln(y) & -y^x \cdot \frac{x}{y} & 1 \\
0 & 3 & 1
\end{pmatrix}
$$

The same Jacobian can be evaluated differently, if we define the function of three scalar arguments $f(x,y,z)$ (Listing 7.20).

LISTING 7.20. Evaluating the Jacobian of a Vector Function of Three Scalar Arguments

$$f(x,y,z) := \begin{pmatrix} x \cdot \sin(y) \\ -y^x + z \\ 3 \cdot y + z \end{pmatrix}$$

$$\begin{bmatrix} \dfrac{\partial}{\partial x}\big(f(x,y,z)_0\big) & \dfrac{\partial}{\partial y}\big(f(x,y,z)_0\big) & \dfrac{\partial}{\partial z}\big(f(x,y,z)_0\big) \\[6pt] \dfrac{\partial}{\partial x}\big(f(x,y,z)_1\big) & \dfrac{\partial}{\partial y}\big(f(x,y,z)_1\big) & \dfrac{\partial}{\partial z}\big(f(x,y,z)_1\big) \\[6pt] \dfrac{\partial}{\partial x}\big(f(x,y,z)_2\big) & \dfrac{\partial}{\partial y}\big(f(x,y,z)_2\big) & \dfrac{\partial}{\partial z}\big(f(x,y,z)_2\big) \end{bmatrix} \rightarrow \begin{pmatrix} \sin(y) & x \cdot \cos(y) & 0 \\[4pt] -y^x \cdot \ln(y) & -y^x \cdot \dfrac{x}{y} & 1 \\[4pt] 0 & 3 & 1 \end{pmatrix}$$

Don't forget that for a numeric evaluation of the Jacobian, it is necessary to define the point where it will be evaluated, for example, the x vector (Listing 7.19), or the three variables – x, y, z (Listing 7.20).

8 Algebraic Equations and Optimization

In this chapter we are going to consider the procedures involved in solving non-linear algebraic equations and systems of such equations.

The task is formulated as follows. Let us imagine that we have a single algebraic equation with an unknown value x

$$f(x) = 0,$$

or the following system of N algebraic equations

$$\begin{cases} f_1(x_1, \ldots, x_M) = 0, \\ \ldots \\ f_N(x_1, \ldots, x_M) = 0, \end{cases}$$

where $f(x)$ is some function. It is necessary to find all equation roots, in other words, all x values that turn the equation (or the system of equations) to the true equality (or set of true equalities).

Solving systems of linear equations where all functions have the following form: $f_i(x) = a_{i1} * x_1 + a_{i2} * x_2 + \ldots + a_{iN} * x_N$, is a separate problem of linear algebra. It will be covered in more detail in Chapter 9.

205

Usually, numeric methods of finding equation roots are related to solving the following problems:

- Investigating root existence, determining the number of existing roots and their approximate location
- Calculating the roots with the predefined TOL error

The latter problem makes it necessary to find such x_0 values whose $f(x_0)$ difference from zero doesn't exceed the TOL value. Nearly all built-in Mathcad functions intended for solving non-linear algebraic equations are aimed at solving the second task. This means that they suppose that equation roots are approximately localized. To solve the first task (root localization), one can use various methods, such as, graphical $f(x)$ representation *(see Section 8.1)*, or a sequential search for the root starting with a set of text points covering the scan area. Mathcad provides several built-in functions that can be used, depending on the specific properties of the equation to be solved (i.e., the $f(x)$ properties). The root function implementing the secant method is intended for solving a single equation with a single unknown *(see Section 8.1)*, while the Given/Find solve block, which combines various gradient methods *(see Sections 8.3 and 8.4)*, is used for solving a system of equations. If $f(x)$ is a polynomial, it is also possible to calculate all of its roots using the polyroots function *(see Section 8.2)*. Additionally, in some cases, one has to reduce the solution of equations to the task of finding the extreme *(see Section 8.5)*. Various methods of finding function extremes are implemented using the following built-in functions: Minerr, Maximize and Minimize *(see Sections 8.5 and 8.6)*. At the end of this chapter we will go over how to solve equations symbolically *(see Section 8.7)* and the possibilities of implementing an efficient method of programmatically solving a series of algebraic equations or a parametric optimization task *(see Section 8.8)*.

8.1. ONE EQUATION AND ONE UNKNOWN

Let us consider a single algebraic equation with the single unknown x:

$$f(x) = 0 \tag{1}$$

For example, $\sin(x) = 0$.

For solving such equations, Mathcad provides the built-in root function that, depending on the type of the task, can accept either two or four arguments, each of which work slightly differently.

- ▨ root(f(x),x)
- ▨ root(f(x),x,a,b)

Here:

- ● f(x) — Scalar function defining the equation (1)
- ● x — Scalar variable, with respect to which the root is found
- ● a,b — Boundaries of the interval, within which the root is found

The first type of root function requires that you additionally specify the initial value of (or guess a value for) the x variable. To do so, it is necessary to assign some value to the x variable in advance. Your search of the root will take place within close range of this number. Thus, assignment of a guess value requires some a priori information on the approximate location of the root.

To illustrate this idea, let us consider an example of a very simple equation, namely, sin(x) = 0, the roots of which are already known.

LISTING 8.1. Finding the Root of a Non-Linear Algebraic Equation

```
x := 0.5

f (x) := sin (x)

solution := root (f (x) , x)

solution = -6.2 × 10⁻⁷
```

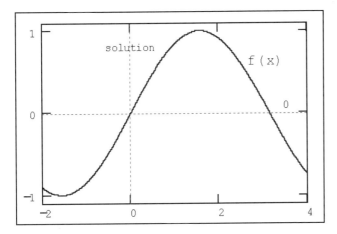

FIGURE 8.1. Graphic solution of the sin(x)=0 equation

The graph of the $f(x) = \sin(x)$ function and location of the root is shown in Figure 8.1. Notice that despite the fact that this equation has an infinite number of roots $x_N = N \cdot \pi$ ($N = 0, \pm 1, \pm 2, \ldots$), Mathcad finds (with the specified precision) only one of them, x_0, which is the closest to the x=0.5 value. If another guess value, say, x = 3, is specified, Mathcad will find another root of this equation, $x_1 = \pi$, and so on. Thus, to search for the root using Mathcad, it is necessary to have already localized it. This is related to the specific features of the selected numeric method, known as the *secant method*. The basic principle of this method is as follows (Figure 8.2):

1. The initial guess is taken for the 0-th approximation of the root: x0 = x.
2. Then, the step h = TOL · x is selected, and Mathcad finds the first approximation of the root x1 = x0 + h. If x = 0, then it is assumed that h = TOL.
3. The *secant* — a straight line connecting these two points — is drawn. The secant intersects the x-axis at the point x2. This point is taken as the second approximation.
4. Another secant is drawn, connecting the first and second points, which also intersects the x-axis at the point x3, thus defining the third approximation, and so on.
5. If at some iteration $|f(x)| < \text{TOL}$ is satisfied, the equation requirement has been met. The iterations halt, and the x value is returned as a result.

The result shown in Figure 8.2 is obtained for the calculation error, which was previously assigned the value of TOL = 0.5 to make the example more illustrative.

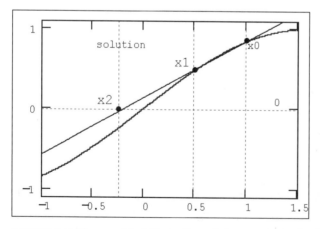

FIGURE 8.2. The graphical illustration of the secant method

Because of a low level of precision, a single iteration is sufficient to find the root. In the calculations provided in Listing 8.1, the calculation error TOL = 0.001 was set by default, and therefore, the solution found using the numeric method is much closer to the real value of the root (x = 0). In other words, the lower the value of the TOL constant, the closer to zero f(x) will be for the obtained root, and the longer it will take the Mathcad numeric processor to find the root.

In the Resource Center, you can find an illustrative example in the Solving Equations topic. It is titled Effects of TOL on Solving Equations.

If the equation has no solutions, an attempt to find its root will result in an error message. Furthermore, an attempt to use the secant method near the local maximum or minimum of the f(x) function will also result in either an error message or an incorrect solution. This happens because in such a situation, the secant might prove to be very close to a horizontal line, thus producing the point of the next approximation located rather far from the guessed location of the root. For solving such equations, it is preferable to use another built-in function, Minerr *(see Section 8.5).* Similar problems may arise, if the initial guess selected is too far from the actual root or if f(x) has the singularities of the infinity type.

Gradient methods, which in Mathcad are related to the equation systems, are also applicable for solving equations with a single unknown (see Section 8.3).

Sometimes it is more convenient to specify the [a,b] interval within which the root is definitely located rather than the initial guess. In this case, it is necessary to use the root function with four arguments, as shown in Listing 8.2. Also, you don't need to assign the initial guess value. The function will scan for the root within the interval between a and b using an alternative numeric method (Ridder or Brent).

LISTING 8.2. Finding the Root of an Algebraic Equation within the Specified Interval

```
solution := root (sin (x) , x, -1, 1)

solution = 0
```

Pay special attention to the fact that the f(x) function can be explicitly defined directly in the body of the root function.

When using the root function with four arguments, remember the following two specific features of this function:

- No more than one root must be located within the [a,b] interval, otherwise the function will find only one of them, and you won't be able to guess which one.
- The values of f(a) and f(b) must have opposite signs, otherwise the error message will appear.

Even if the equation has imaginary roots rather than real ones, such roots can also be found. Listing 8.3 provides an example where the $x^2 + 1 = 0$ equation that has two purely imaginary roots is solved two times using different initial guess values. When an initial guess is set to 0.5 (the first line), the numeric method finds the first root (negative imaginary unit -i), and when the initial guess is set to -0.5 (the third line), the numeric method finds the second root (i).

LISTING 8.3. Finding Imaginary Roots

x := 0.5

$$\text{root}\left(x^2 + 1 , x\right) = -i$$

x := -0.5

$$\text{root}\left(x^2 + 1 , x\right) = i$$

The second form of the root function (accepting four arguments rather than two arguments) is not applicable for solving this equation. This is due to the fact that because the f(x) function is always positive. Therefore, it is impossible to find an interval at the boundaries of which it will take values of opposite signs.

Now the only thing left to mention here is that f(x) may be the function of any number of arguments, not only that of x. As a result, it is necessary to specify the argument in which to solve the equation within the root function itself. This capability is illustrated by Listing 8.4 with an example of the function of two arguments: $f(x,y) = x^2 - y^2 + 3$. In this listing, we first solve the $f(x,0) = 0$ equation in relation to the x variable, and then we solve another equation — $f(1,y) = 0$ — in relation to the y variable.

LISTING 8.4. Finding the Root of an Equation Specified by the Function of Two Variables

$$f(x,y) := x^2 - y^2 + 3$$

$$x := 1$$

$$y := 0$$

$$root(f(x,y),x) = -1.732i$$

$$root(f(x,y),y) = 2$$

The first line of this listing defines the $f(x,y)$ function, while the second and the third lines define the values for which the equation will be solved in y and x, respectively. The fourth line of this listing solves the $f(x,0) = 0$ equation, while the last one solves $f(1,y) = 0$. When finding a numeric solution of an equation in one of its variables, never forget to define the values of other variables first, otherwise you'll get the following error message: "This variable or function is not defined above", which, in this case, informs you that another variable was not defined previously. Of course, you can specify the values of other variables directly within the root function. If you do, the second and third lines of Listing 8.4 are not required, and the last lines of this listing will be represented as $root(f(x,0),x) =$ and $root(f(1,y),y) =$, respectively.

There are efficient special-purpose algorithms that allow you to find the dependence between the equation roots in one of the remaining variables (see Section 8.8).

8.2. THE ROOTS OF A POLYNOMIAL

If the $f(x)$ function is a polynomial, then all its roots can be found using the following built-in function:

```
polyroots(v)
```

where v is the vector composed of polynomial coefficients.

Since a polynomial of the N-th power has exactly N roots (some of which might be repeated), the v vector must comprise $N + 1$ elements. The result of applying the `polyroots` function is a vector composed from N roots of the polynomial under consideration. Notice that in this case it is unnecessary to specify an initial guess, just like with the root function. Listing 8.5 gives an example of how to find a polynomial of the 4-th power's roots.

LISTING 8.5. Finding the Polynomial's Roots

$$v := (\begin{array}{ccccc} 3 & -10 & 12 & -6 & 1 \end{array})^T$$

$$\text{polyroots }(v) = \begin{pmatrix} 1 \\ 1 - 5.113i \times 10^{-6} \\ 1 + 5.113i \times 10^{-6} \\ 3 \end{pmatrix}$$

The coefficients of the polynomial considered in this example

$$f(x) = (x-3) \cdot (x-1)^3 = x^4 - 6x^3 + 12x^2 - 10x + 3 \tag{1}$$

are written as a vector in the first line of the listing. The first element of the vector must represent the free member of the polynomial, the second, the coefficient of x^1, and so on. The final $N + 1$ element of the vector must represent the coefficient of the highest-order term x^N.

TIP

Sometimes the initial polynomial is represented as a product of several other polynomials rather than in canonic form. In this case, you can determine all its coefficients by selecting the polynomial and selecting the Expand command from the Symbolics menu. As a result, Mathcad's symbolic processor will convert the polynomial into the canonic form. You'll only need to pass the coefficients of the resulting polynomial as the arguments to the polyroots function.

The second line of Listing 8.5 shows the result of applying the polyroots function. Notice that the numeric processor produces two imaginary numbers instead of the two of three real roots equal to 1 (in other words, instead of the multiple 1 root). However, the imaginary part of these roots fits within the error limits defined by the TOL constant and, thus, it shouldn't confuse users. You simply have to remember that the roots of the polynomial can be complex, and the calculation error may influence both the real and imaginary parts of the root.

When using the polyroots function, you have an option of selecting one of the following numeric methods — the Laguerre polynomials method (used by default) or the companion matrix method.

To change the method, proceed as follows:

1. Right-click the `polyroots` keyword.
2. From the right-click menu select either the **LaGuerre** or the **Companion Matrix** command.
3. Click somewhere outside the `polyroots` function. If the automatic calculation mode is enabled, the roots of the polynomials will be recalculated according to the newly selected method.

If you want to rely on Mathcad when selecting a solution method, set the AutoSelect option in the same right-click menu.

8.3. SYSTEMS OF EQUATIONS

Let us consider the procedure of solving a system of N nonlinear equations in M variables.

$$
\begin{cases}
f_1(x_1, \dots, x_M) = 0 \\
\dots \\
f_N(x_1, \dots, x_M) = 0
\end{cases}
\tag{1}
$$

Here $f_1(x_1, \dots, x_M), \dots, f_N(x_1, \dots, x_M)$ are some scalar functions of scalar variables $x_1, x_2, \dots, x_M$, which possibly can depend on some other variables. The number of equations can be either greater or less than the number of variables. Notice that the system (1) can be formally rewritten as follows:

$$
\mathbf{f}(\mathbf{x}) = 0 \tag{2}
$$

where $\mathbf{x}$ is the vector composed of variables $x_1, x_2, \dots, x_M$, and $\mathbf{f}(\mathbf{x})$ is the respective vector function.

For solving systems of equations, Mathcad provides a special solve block, consisting of three parts directly following each other:

- ■ `Given` — The keyword
- ■ System of equations written using logical operators as equations and, possibly, inequalities
- ■ `Find(x_1, ..., x_M)` — The built-in function intended for solving the system in variables $x_1, \dots, x_M$

To insert logical operators, use the **Boolean** toolbar. If you prefer to type the operators from the keyboard, remember that a logical equal sign is entered by pressing the <Ctrl>+<=> keyboard shortcut. The Given/Find solve block uses iterative methods for searching the solution. Consequently, as is the case with the root function, you need to specify the initial guess values for all unknowns $x_1, \ldots, x_M$. Notice that this has to be done before entering the Given keyword. The value of the Find function is the vector composed from solutions in each variable. Thus, the number of vector members is equal to the number of arguments of the Find function.

Listing 8.6 illustrates the solution of a system of two equations.

LISTING 8.6. Solution of a System of Two Equations

$$f(x, y) := x^4 + y^2 - 3$$

$$g(x, y) := x + 2 \cdot y$$

$$x := 1 \qquad y := 1$$

Given

$$f(x, y) = 0$$

$$g(x, y) = 0$$

$$v := \text{Find}(x, y)$$

$$v = \begin{pmatrix} 1.269 \\ -0.635 \end{pmatrix}$$

$$f(v_0, v_1) = -1.954 \times 10^{-7}$$

$$g(v_0, v_1) = 0$$

The first two lines of this listing define the functions that determine the system of equations. Next, initial values are assigned to the x and y variables in which the system will be solved. After this, the Given keyword is followed by two logical operators expressing the system of equations under consideration. The solve block is closed by the Find function, whose value is assigned to the v vector. The next line displays the contents of the v vector representing the solution of the system of equations. The first element of the vector is the first argument of the Find function, and the second element is its second argument. The last two lines check the accuracy of the obtained solution.

TIP

Quite often, it proves useful to check if the equations have been solved correctly. This can be performed by evaluating the values of functions in roots found by the numeric processor, as was done in the final lines of Listing 8.6.

Take note of the fact that the equations can be defined directly within the solve block. Thus, you don't need to have defined the f(x,y) and g(x,y) functions previously, as shown in the first two lines of Listing 8.6. Rather, you can formulate the solve block as follows:

```
Given

x⁴ + y² = 3

x + 2·y = 0
```

This form represents the equations in a more common and illustrative form, which is especially convenient for documenting your work.

Figure 8.3 provides a graphic representation of the equation system under consideration. Each equation is represented by its graph (the first equation is shown

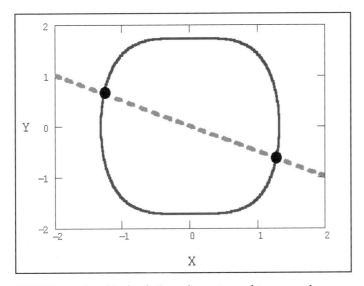

FIGURE 8.3. Graphical solution of a system of two equations

by a thick black line, the second appears as a gray dashed line). Since the second equation is linear, it is represented by a straight line on the XY plane. The two points of intersection of the two graphs correspond to the points where the conditions specified by both equations are satisfied (i.e., to the real roots of the system). As can be clearly seen from the listing, the process has found only one of the two possible solutions — namely, the one located in the lower-right part of the graph. To find the second solution, it is necessary to repeat the calculations with other initial guess values (change them to the ones closer to the other intersection point — for example $x = -1$, $y = -1$).

For the moment, we are considering an example of a system that has two equations and two unknowns, which is the most frequent case. However, the numbers of equations and unknowns can be different. Furthermore, the solve block can contain additional conditions specified in the form of inequalities. For example, you can introduce a restriction by instructing Mathcad to find only negative values of x. For example, introducing this restriction into Listing 8.6 will result in finding another solution, as shown in Listing 8.7.

LISTING 8.7. Solving a System of Equations and Inequalities

```
x := 1      y := 1

Given

 4    2
x  + y  = 3

x + 2·y = 0

x < 0
```

$$\text{Find}(x, y) = \begin{pmatrix} -1.269 \\ 0.635 \end{pmatrix}$$

Notice that despite the fact that the same guess values are specified here in Listing 8.7 as were specified in Listing 8.6, we have obtained another root. This is because of an additional inequality defined in the Given block in the next to last line of Listing 8.7.

If you attempt to solve a system that has no solutions, Mathcad will display an error message, informing you that no solution has been found, and it will then prompt you to change initial values or the calculation error value.

Solve block uses the CTOL constant as the calculation error for the equations en-tered after the Given keyword. For example, if CTOL = 0.001, then the x = 10 equa-tion will be satisfied at x = 10.001, and at x = 9.999. Another constant, TOL, specifies the condition that, being satisfied, halts the iterations done by a numeric algorithm (see Section. 8.4). The value of CTOL can be changed by the user in just the same way as that for the TOL value, for example, CTOL := 0.01. By default, CTOL = TOL = 0.001, but you can redefine these constants as desired.

When solving the systems with more unknowns than equations, be very careful. For example, you can delete one of the equations from Listing 8.6 and try to solve the remaining equation g(x, y) = 0 using the two variables, x and y. A task formulated in such a way has an infinite number of roots: for each x and, respec-tively, y = –x/2, the condition defining a single equation is satisfied. However, even if the number of roots is infinite, the numeric processor will perform calculations only until the logical expressions in the solve block are satisfied within the limits determined by the calculation error. After that, the iterations will cease, and the numeric processor will display the solution. As a result, only one pair of values (x, y) will be produced, and this will be the pair found first.

In Section 8.7 we will consider a method of finding all the solutions of this task.

The solve block with the Find function can also be used to find the root of the single equation in the single unknown. In this case, the Find function will work the same way as it has in the examples considered in this section. In this case, finding the root is considered the solution to a system comprising only one equa-tion. The only difference is that here the Find function will return a scalar value rather than a vector value. Listing 8.8 puts forth the solution of the equation from the previous section.

LISTING 8.8. Finding the Root of an Equation with One Unknown Using the Find Function

```
x := 0.5

Given

sin (x) = 0

Find (x) = −3.814 × 10⁻⁷
```

What is the difference between this solution and the one provided in Listing 8.1 using the `root` function? Basically, the same task has been solved using different numeric methods. In this case, the selection of the numeric method has no influence on the final result. However, there might be such situations when the usage of a specific method is of critical importance.

8.4. ON THE NUMERIC METHODS OF SOLVING SYSTEMS OF EQUATIONS

If you are solving "good" equations similar to the ones provided in the previous sections, you don't need to figure out how Mathcad searches for their roots. However, even though this may be the case, it would be rather useful to have a sound understanding of the processes that take place "under the hood" — for instance, what actions are made in the interval between entering the required conditions after the `Given` keyword and obtaining the result of applying the `Find` function. For example, this understanding is important for the correct selection of initial guess values before the solve block. In this section, we will concentrate on some specific features of numeric methods and the possibilities for customizing their parameters provided by Mathcad.

The `Find` function implements gradient numeric methods. Let us consider their main idea in the example of one equation and one unknown — $f(x) = 0$ for the function $f(x) = x^2 + 5x + 2$, the graph of which is shown in Figure 8.4. The main concept of the gradient methods lies in sequential approximations of the actual solution of the equation, which are calculated using the derivative of the $f(x)$ function. Let us discuss the simplest form of this algorithm, also known as the Newton method.

1. The initial guess value provided by the user is taken as the 0-th approximation — $x_0 = x$.
2. Using the finite difference method, the derivative in the point x_0 is calculated: $f'(x_0)$.
3. Using the Taylor expansion, we can replace $f(x)$ in close range of x_0 by the tangent — a straight line specified by the following equation: $f(x) \approx f(x_0) + f'(x_0) \cdot (x - x_0)$.
4. Next, the point x_1 where the tangent intersects the x-axis (Figure 8.4) is found.
5. If $f(x_1) < \text{TOL}$, the iterations halt, and the x_1 value is returned as the root. Otherwise x_1 is taken for the next iterations and the process continues repeatedly: the tangent of $f(x)$ in x_1 is drawn, and the x_2 — the point where that tangent intersects with the x-axis is determined, and so on.

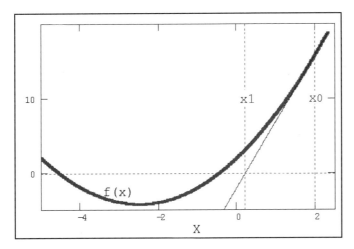

FIGURE 8.4. Graphical illustration of the Newton method

Modified Newton algorithm for solving a system of multiple equations implies that appropriate functions of multiple variables are linearized, — approximated by their linear dependence using partial derivatives. For example, the following expressions are used for 0-th iterations when solving a system of two equations:

$$f_1(x, y) \approx f_1(x0, y0) + \frac{\partial f_1(x0, y0)}{\partial x}(x - x0) + \frac{\partial f_1(x0, y0)}{\partial y}(y - y0),$$

$$f_2(x, y) \approx f_2(x0, y0) + \frac{\partial f_2(x0, y0)}{\partial x}(x - x0) + \frac{\partial f_2(x0, y0)}{\partial y}(y - y0).$$

To find a point corresponding to each new iteration, it is necessary to equate both expressions to zero, i.e., at each iteration one needs to solve a system of linear equations.

Mathcad provides three various types of gradient methods. To change the numeric method, proceed as follows:

1. Right-click the `Find` keyword.
2. Select the **Nonlinear** command from the right-click menu.
3. The submenu will appear (Figure 8.5). Select one of the available methods from this submenu: **Conjugate Gradient**, **Levenberg-Marquardt**, or **Quasi-Newton**.

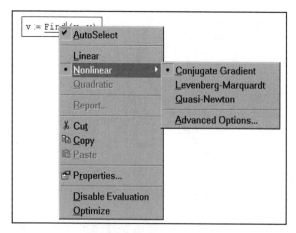

FIGURE 8.5. Selecting the numeric method

If you want to rely on Mathcad when selecting the numeric method, set the AutoSelect option in the right-click menu. If the automatic selection option is enabled (i.e., if the appropriate command in the right-click menu is checked), you can view the currently selected numeric method by opening the same submenu and viewing which of the available methods is marked with a point. The last two methods are quasi-Newtonian, and their main principle was considered above. The first one, the conjugate gradients method, is a two-stage procedure — to find the next iteration, it uses both the current and previous iterations. The Levenberg method is described in detail in the Mathcad on-line Help system. You can find detailed information about the conjugate gradients method and modifications of the Newton method, in most modern books on numeric methods.

Besides selecting the numeric method, you can also set some parameters of the selected method. To accomplish this, use the right-click menu to open the **Advanced Options** window by selecting the **Nonlinear | Advanced options** commands from this menu. This window (Figure 8.6) has three groups of radio buttons.

The first group, **Derivative estimation**, allows you to specify one of two methods to be used for calculating the derivative: **Forward** or **Central**. These options correspond to different methods of derivative approximation — either by right finite difference (two-point forward scheme) or by central finite difference (three-point symmetric scheme).

Notice that derivative calculation in gradient numeric methods of solving equations is performed by a more economic method than what is used when performing numeric differentiation (see Chapter 7).

NOTE

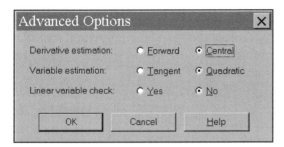

FIGURE 8.6. The **Advanced Options** window

The second group, Variable estimation, allows you to specify the type of approximation with the Taylor series. For the case considered in this section (approximation by the tangent), select the **Tangent** option. For a more precise approximation (by parabola), select the **Quadratic** option.

Finally, the last group of radio buttons, Linear variable check, enables you to reduce the calculation time in some specific cases. For example, if you are absolutely sure that the non-linear characteristics of all functions included in the equation have little or no effect on the values of their partial derivatives, you can set the Yes option. In this case, the derivatives will be considered constant, and they won't be calculated at each step under this method.

Be very careful when changing parameters of numeric methods. Use these options when the solution can't be found with the default parameters or when the calculations are time-consuming.

TIP

As is obvious from the materials covered in this section, all gradient methods implemented in the `Find` function require repeated calculation of derivatives. If the functions you are dealing with are rather smooth, the gradient methods find the root quickly and reliably. On the other hand, if you need to find the roots of the functions that are not sufficiently smooth, it is better to use the secant method (the `root` function). Thus, if you need to solve non-trivial tasks, it's rather important to make the proper selection of the numeric method to be used.

8.5. FINDING APPROXIMATE SOLUTIONS OF EQUATIONS

Sometimes it is necessary to replace the task of isolating the roots of the system of equations with that of finding exteremes of the function of multiple variables. For example, when it is impossible to obtain a solution using the `Find` function, you might attempt to minimize the error of closure rather than obtaining the precise solution. To achieve this, replace the `Find` function with the `Minerr` function, which has the same set of arguments in the solve block. This function must also be located within the solve block, where:

- $x_1 := c_1$ ... $x_M := c_M$ — Initial guess values for all unknowns
- `Given` — The keyword
- System of algebraic equations and inequalities, written using logical operators.
- `Minerr(x_1, ..., x_M)` — Approximate solution of the system in variables $x_1, ..., x_M$, minimizing the closure error of the system of equations

The `Minerr` function implements the same algorithms as the `Find` function. The only difference is that another condition of termination is used. Thus, the user can select the numeric algorithm of the approximate solution for the `Minerr` function via the same right-click menu (see Section 8.4).

Listing 8.9 presents an example of how to use the `Minerr` function. As you can see, it is sufficient to change the function name to `Minerr` to obtain an approximate solution of the equation specified after the `Given` keyword rather than a precise one (with `TOL` precision).

LISTING 8.9. An Approximate Solution of the Equation Having the Root $(x=0, y=0)$

$$x := 1 \qquad y := 1$$

$$k := 10^6$$

Given

$$k \cdot x^2 + y^2 = 0$$

$$v := \text{Minerr}(x, y)$$

$$v = \begin{pmatrix} 0 \\ 0 \end{pmatrix}$$

Listing 8.9 gives an approximate solution of the $k \cdot x^2 + y^2 = 0$ equation, which, at any value of the k coefficient, has the only precise root $(x = 0, y = 0)$. Still, any attempts to solve this equation for large values of k using the `Find` function result in the following error message: "No solution was found." This is because $f(x,y) = k \cdot x^2 + y^2$ behaves in a different way than the functions considered earlier in this chapter (see Figures 8.1 and 8.2) while close to its root. In contrast to these functions, $f(x,y)$ doesn't intersect the $f(x,y) = 0$ plane, but rather, it has only one common point $(x = 0, y = 0)$ with it (Figure 8.7). Accordingly, it is more problematic to find the root using the gradient methods described in the previous section, since the derivatives of the $f(x,y)$ function are close to zero around the root and iterations may lead the guess solution far from the actual root.

The situation gets even worse if, besides the tangency root (Figure 8.7), there exist (probably, rather distant) several intersection roots. If these roots do exist, the attempt at solving the equation or the system of equations using the Find function might result in finding the root of the second type, even if the initial guess value was taken very close to the first root. Consequently, if you suppose that the system of equations has the tangency root, it is preferable to use the Minerr function, especially if one always has the possibility to check the accuracy of the solution by simply substituting the obtained value in the equations (see Listing 8.6).

In Listing 8.9, we considered the example of finding the existing solution of the equation. To conclude our discussion of this topic, let us provide an example in which the Minerr function is used to find an approximate solution of an insoluble system of equations and inequalities (Listing 8.10). The solution produced by the Minerr function minimizes the closure error of this system.

According to its mathematical principles, the Minerr function can be used for creating a regression series according to the rule specified by the user (see Chapter 14).

NOTE

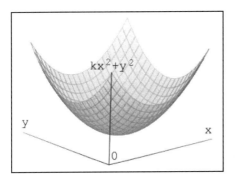

FIGURE 8.7. The graph of the $k \cdot x^2 + y^2$ function

LISTING 8.10. An Approximate Solution of an Insoluble System of Equations and Inequalities

```
x := 1        y := 1
```

```
Given
```

$$x^2 + y^2 = -1$$

$$x > 0.1$$

$$y \le -0.2$$

$$\text{Minerr}(x, y) = \begin{pmatrix} -0.042 \\ -0.085 \end{pmatrix}$$

As follows from the listing, the function returns the values of the variables that satisfy the equation and inequalities specified within the solve block. A conscientious reader will also be able to notice that the solution returned by the Minerr function in this example is not the only one possible since there is a range of (x, y) pairs that minimize the closure error equally. Hence, a different solution will be obtained for different initial guess values, similar to the Find function, which produces different values for the case of an infinite number of roots *(see Listing 8.6 in Section 8.3)*. There is an even more dangerous case when there are several local minimums of the closure error function. In this case, an incorrect selection of the initial guess will produce a local minimum, despite the fact that another (global) minimum of the closure error function, which fits the system much better, might exist.

8.6. SEARCHING THE FUNCTION EXTREME

The tasks of finding the function extreme require that one find its maximum or minimum value within a specific domain of its arguments' definition. Limitations of the argument values specifying this domain of definition, along with other additional conditions, must be specified in the form of a system of inequalities and/or equations. If this is the case, we are dealing with the conditional extreme task.

For solving the tasks involved in searching for the maximum and minimum values Mathcad provides the following built-in functions: Minerr, Minimize and Maximize. All these functions use the same gradient numeric methods as the Find function. Thus, you are capable of selecting the numeric algorithm of minimization or maximization from the number of available numeric methods, which we have already covered *(see Section 8.4)*.

8.6.1. Extreme of the Function of One Variable

Searching for a function's extreme includes the tasks of finding local and global extremes. The latter tasks are also known as optimization tasks. Let us consider a specific example of the $f(x)$ function, shown by the graph presented in Figure 8.8, within the interval of $(-2, 5)$. This function has a global maximum value at the left limit of the interval, a global minimum and local maximum within the interval, and a local maximum at the right boundary of the interval.

Using Mathcad built-in functions, it is only possible to find the local extreme of the function. To find the global maximum or minimum value, one needs to either first calculate all local extreme values and then select the lowest (or the highest) value, or scan the area with the predefined step to locate the sub-area of the highest (lowest) values and then find the global extreme in its vicinity. The last way is potentially risky, since it might lead you to the zone of another local extreme. However, quite often it is preferable since it is faster.

Mathcad provides two built-in functions for finding local extremes, which can be used both within the solve block or as stand-alone functions.

- $\text{Minimize}(f, x_1, \ldots, x_M)$ — The vector of argument values, at which the function reaches its minimum
- $\text{Maximize}(f, x_1, \ldots, x_M)$ — The vector of argument values, at which the function reaches its maximum
- $f(x_1, \ldots, x_M, \ldots)$ — The function under consideration
- $x_1, \ldots, x_M$ — Arguments, by which the maximization (minimization) is done

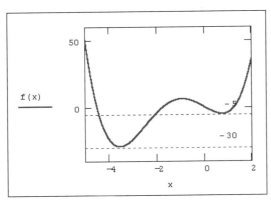

FIGURE 8.8. The graph of the $f(x) = x^4 + 5 \cdot x^3 - 10 \cdot x$ function

All arguments of the f function must be assigned some values first. Keep in mind that for variables with which minimization takes place, these values will be interpreted as initial guess values. Examples illustrating calculations of the extreme of a function of one variable (Figure 8.8) without additional limitations are shown in Listings 8.11–8.12. Since no additional conditions are formulated, the search of extremes is done for all values of x from $-\infty$ to ∞.

LISTING 8.11. Minimum Value of the Function of One Variable

$f(x) := x^4 + 5 \cdot x^3 - 10 \cdot x$

$x := -1$

$\text{Minimize}(f, x) = -3.552$

$x := 1$

$\text{Minimize}(f, x) = 0.746$

LISTING 8.12. Maximum Value of the Function of One Variable

$f(x) := x^4 + 5 \cdot x^3 - 10 \cdot x$

$x := 1$

$\text{Maximize}(f, x) = -0.944$

$x := -10$

$\text{Maximize}(f, x) = \blacksquare$

As follows from the listings, the selection of the initial guess has a significant effect on the result. Namely, depending on the value selected as the initial guess, different local extremes will be returned as the result. In the latter case, the numeric method was unable to solve the problem since the initial guess of $x = -10$ was selected very far from the area of the local maximum value, and since the search leads to the direction in which $f(x)$ grows (i.e., diverges to $x \rightarrow \infty$).

8.6.2. Conditional Extreme

In tasks that deal with finding the conditional extreme, the minimization and maximization functions must be included within the solve block, i.e., they must be preceded by the Given keyword. Within the interval between the Given keyword and the function under consideration, logical expressions (inequalities and equations) specifying limitations of the argument values must be written using Boolean operators. Listing 8.13 demonstrates examples of searching the conditional extreme within different intervals defined by inequalities. Compare the results of this listing with the two previous ones.

LISTING 8.13. Three Examples of Finding a Conditional Extreme

$f(x) := x^4 + 5 \cdot x^3 - 10 \cdot x$

$x := 1$

Given

$-5 < x < -2$

$\text{Minimize}(f, x) = -3.552$

$x := 1$

Given

$x > 0$

$\text{Minimize}(f, x) = 0.746$

$x := -10$

Given

$-3 < x < 0$

$\text{Maximize}(f, x) = -0.944$

Don't forget that the correct selection of an initial guess is important even for the tasks involving the conditional extreme. For example, in the last example, if instead of the $-3 < x < 0$ condition, we were to specify that $-5 < x < 0$, the same initial guess (x = -10) would produce the maximum value `Maximize(f,x) = -0.944`, which is incorrect, since the maximum value is achieved by the `f(x)` function at the left boundary of the interval (x = -5). If the initial guess is set to x = -4 the task will be solved correctly: `Maximize(f,x) = -5`.

8.6.3. Extreme of a Function of Multiple Variables

Finding an extreme of a function of multiple variables is not a significantly different operation from that performed with functions of one variable. Therefore, let us limit our discussion to providing an example (Listing 8.14) of finding maximum and minimum values of the function shown in the form of the graphs (3D surface and level curves) in Figure 8.9. We'd like to draw your attention to the method of specifying the domain of definition at the (X, Y) plane using inequalities written with logical operators.

LISTING 8.14. Extreme of a Function of Two Variables

$$f(x,y) := 2 \cdot (x - 5.07)^2 + (y - 10.03)^2 - 0.2 \cdot (x - 5.07)^3$$

$$x := 3 \qquad y := 3$$

Given

$$0 < x < 15$$

$$0 < y < 20$$

$$\text{Minimize}(f, x, y) = \begin{pmatrix} 5.07 \\ 10.03 \end{pmatrix}$$

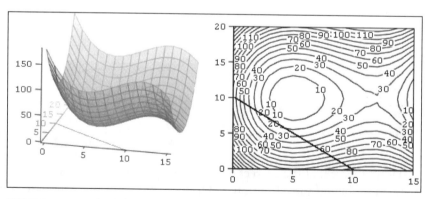

FIGURE 8.9. Graph of the `f(x,y)` function and section of the line $x + y = 10$

Additional conditions can be specified by equations. For example, if the Given keyword is followed by the $x + y = 10$ equation, the following solution of the conditional extreme task will be provided:

$$\text{Minimize } (f, x, y) = \begin{pmatrix} 3.589 \\ 6.411 \end{pmatrix}$$

Obviously, any additional limitation will result in a search for the minimum value of the $f(x, y)$ function along the section of a straight line shown in Figure 8.9.

You can also organize the minimum value's search by using the Minerr function. To do so, change the Minimize function to Minerr in Listing 8.14, and then add the expression equating the $f(x, y)$ function to the value, which is definitely smaller than the minimal one, for example $f(x, y) = 0$, after the Given keyword.

8.6.4. Linear Programming

The tasks of finding conditional exteremes for functions of multiple variablesare rather common in economic calculations that are used to try to minimize expenses or financial risks, maximize income, etc. There is a special class of tasks in economics that can be described by systems of linear equations and inequalities. They are also known as linear programming tasks. Let us provide a typical example of the so-called transportation problem, which solves one of the issues of organizing the optimal delivery of goods to customers economically (Listing 8.15).

LISTING 8.15. Solution of the Linear Programming Problem

$$a := \begin{pmatrix} 145 \\ 210 \\ 160 \end{pmatrix}$$

$$b := \begin{pmatrix} 237 \\ 278 \end{pmatrix}$$

$$\sum a = 515 \qquad\qquad \sum b = 515$$

$$M := \text{rows}(a) \qquad\qquad N := \text{rows}(b)$$

$$M = 3 \qquad\qquad N = 2$$

$$c := \begin{pmatrix} 11.5 & 7 & 12 \\ 6.2 & 10 & 9.0 \end{pmatrix}$$

$$f(x) := \sum_{i=0}^{N-1} \sum_{j=0}^{M-1} c_{i,j} \cdot x_{i,j}$$

$CTOL := 0.5$

$x_{N-1,M-1} := 0$

Given

$x_{0,0} + x_{1,0} = a_0$	$x_{0,0} \geq 0$	$x_{1,0} \geq 0$
$x_{0,1} + x_{1,1} = a_1$	$x_{0,1} \geq 0$	$x_{1,1} \geq 0$
$x_{0,2} + x_{1,2} = a_2$	$x_{0,2} \geq 0$	$x_{1,2} \geq 0$

$x_{0,0} + x_{0,1} + x_{0,2} = b_0$

$x_{1,0} + x_{1,1} + x_{1,2} = b_1$

$sol := \text{Minimize}(f, x)$

$$sol = \begin{pmatrix} 0 & 210 & 27 \\ 145 & 0 & 133 \end{pmatrix} \qquad f(sol) = 3.89 \times 10^3$$

The model of a typical transportation task is as follows. Let us imagine that there are N manufacturers that have produced $b_0, \dots, b_{N-1}$ tons of their products. These products must be delivered to M customers in the quantities of $a_0, \dots, a_{M-1}$ tons each. Numeric definitions of the a and b vectors are on the first line of this listing. The sum of all customer orders a_i is equal to the sum of all manufactured products, (i.e., it is equal to the sum of all b_i values — the second line checks this condition). If the price of transporting a ton of one product from the i-th manufacturer to j-th customer is c_{ij}, the solution of this task specifies the optimal distribution of the x_{ij} traffic from the point of view of minimization of transportation expenses. The c matrix and the $f(x)$ function being minimized in the matrix argument x are specified in the middle of the Listing 8.15.

The conditions delineating the positive goods traffic and equations specifying the total quantity of the goods manufactured by each vendor and the sums of the orders from each customer are placed after the Given keyword. The solution assigned to the `sol` matrix variable is displayed in the last line of the listing along with the corresponding sum of expenses. Notice that in order to obtain the result, it is necessary to increase the value of the `CTOL` error specifying the maximum allowable closure error of additional conditions. The line preceding the `Given` keyword defines initial values for x by simply creating the 0-th element of the $x_{N-1,M-1}$ matrix.

If other initial values for x are defined, you'll probably get another solution! Perhaps you'll be able to find another local minimum, which will further optimize the transportation expenses. This can be considered additional proof that global minimum tasks, the class that includes linear programming tasks, require careful and cautious selection of the initial values. Quite often you won't have any choice other than to scan the whole range of initial values in order to select the deepest minimum value from the set of existing local minimum values.

8.7. SOLVING EQUATIONS SYMBOLICALLY

Some equations can be solved analytically using the Mathcad symbolic processor. This is accomplished in a way very similar to numeric solution of equations using the numeric processor. However, in this case you are not required to assign initial values to the variables. Listings 8.16 and 8.17 demonstrate the symbolic solution of the equation with one unknown and the system of two equations, respectively.

LISTING 8.16. Symbolic Solution of the Algebraic Equation with One Unknown

```
Given
```

$$x^2 + 2 \cdot x - 4 = 0$$

$$\text{Find (x)} \rightarrow \left(\sqrt{5} - 1 \quad -1 - \sqrt{5} \right)$$

LISTING 8.17. Symbolic Solution of the System of Two Equations

```
Given
```

$$x^4 + y^2 - 3 = 0$$

$$x + 2 \cdot y = 0$$

$$\text{Find}(x,y) \rightarrow \begin{bmatrix} \frac{-1}{4}\cdot\left(-2+2\cdot\sqrt{193}\right)^{\left(\frac{1}{2}\right)} & \frac{1}{4}\cdot\left(-2+2\cdot\sqrt{193}\right)^{\left(\frac{1}{2}\right)} \\ \frac{1}{8}\cdot\left(-2+2\cdot\sqrt{193}\right)^{\left(\frac{1}{2}\right)} & \frac{-1}{8}\cdot\left(-2+2\cdot\sqrt{193}\right)^{\left(\frac{1}{2}\right)} \end{bmatrix}$$

$$\begin{bmatrix} \frac{-1}{4}\cdot\left(-2-2\cdot\sqrt{193}\right)^{\left(\frac{1}{2}\right)} & \frac{1}{4}\cdot\left(-2-2\cdot\sqrt{193}\right)^{\left(\frac{1}{2}\right)} \\ \frac{1}{8}\cdot\left(-2-2\cdot\sqrt{193}\right)^{\left(\frac{1}{2}\right)} & \frac{-1}{8}\cdot\left(-2-2\cdot\sqrt{193}\right)^{\left(\frac{1}{2}\right)} \end{bmatrix}$$

As you can see, in these listings the Find function is followed by the symbolic evaluation sign rather than by an equal sign. To insert the symbolic evaluation sign, click an appropriate button in the Symbolic toolbar or press the <Ctrl>+<.> keys on the keyboard. Don't forget that the equations themselves must be in the form of logical expressions, which means that the equal signs must be entered from the Boolean toolbar. Notice that Listing 8.17 calculates both the first two real roots, which were already found using the numeric method *(see Section 8.3)*, and the next two imaginary roots. The last two roots are purely imaginary, since the multiplier is equal to $(-2-2\cdot\sqrt{193})^{\frac{1}{2}} \approx 5.46\cdot i$.

The symbolic processor allows you to solve equations with one unknown using another method:

1. Enter the equation using the **Boolean** toolbar or by pressing the <Ctrl>+<.> keyboard combination to insert the logical equality sign. For example, $x^2 + 2\cdot x - 4 = 0$.
2. Click the variable for which you are going to solve the equation.
3. From the Symbolics menu select the **Variable | Solve** commands.

The new line will appear next to the line containing the equation, and it will contain either a solution or an error message informing you that this equation can't be solved symbolically.

In our example, after you have accomplished the instructions provided above, the vector containing two roots of the equation will appear:

$$\begin{pmatrix} \sqrt{5}-1 \\ -1-\sqrt{5} \end{pmatrix}$$

Symbolic calculations can also be performed over equations that contain various parameters besides unknowns. Listing 8.18 provides an example of a solution to an equation of the 4-th order with the `a` parameter. As you can see, Mathcad allows you to obtain the result in analytical form.

LISTING 8.18. Symbolic Solution of the Parametric Equation

```
Given

 4    4
x  - a  = 0

Find (x) → ( a   -a   i·a   -i·a )
```

In the next section we will concentrate on the methods provided by Mathcad for solving similar tasks numerically.

8.8. METHOD OF CONTINUATION BY PARAMETER

Solving "good" nonlinear equations and systems of equations similar to the ones that were considered in the previous sections of this chapter is not a complicated task in terms of calculation. In real-world engineering and scientific calculation, a more complicated problem arises: solving a sequence of equations depending on the specific parameter (or set of parameters) rather than solving a single equation or system. There are rather efficient methods for solving such tasks known of as continuation methods. These methods are not directly built into Mathcad, but they can be easily programmed using the tools that we have already considered. In this discussion, we are going to consider a single equation, bearing in mind that the task can be generalized to the case of a system of equations.

Let us suppose that we have the `f(a,x) = 0` equation depending both on the unknown x and on the parameter a. It is necessary to determine the dependence of the x root on the a parameter, i.e., `x(a)`. Listing 8.18 presented the simplest example of such a task. In that case we were lucky enough to obtain the general form of a solution using symbolic calculations. Now, let us consider another, more complicated example, the analytical solution of which is also well known, but which still can't be solved using the symbolic processor. Let us generalize the task of solving one algebraic equation with one variable by adding the following dependence on the parameter a:

$$\sin(a \cdot x) = 0 \tag{1}$$

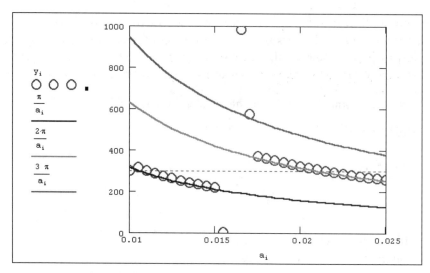

FIGURE 8.10. The solution of the $\sin(a \cdot x) = 0$ equation (see Listing 8.21)

An analytical solution to this equation can be found from the $a \cdot x = N \cdot \pi$ relation where $N = 0, \pm 1, \pm 2, \ldots$, which means that there is an infinite number (for each N) of solution families $x_N = N \cdot \pi / a$. Several continuous curves from the families for $N = 1, 2, 3$ are shown in Figure 8.10. Furthermore, bear in mind that for $N = 0$ — i.e., for $x = 0$ — a straight line coinciding with the x-axis forms the solution. Notice (Listing 8.19) that the Mathcad symbolic processor output produces only this $x = 0$ series as the solution.

LISTING 8.19. Symbolic Solution of the $\sin(a \cdot x) = 0$ Equation

```
root (sin (a·x) , x) → 0
```

For the moment, let us forget that the analytical solution is known and approach the equation (1) as we would any other new task. Let us attempt to solve it using the secant method with the built-in `root` function. Listing 8.20 illustrates this method, which though the simplest, is not the best. The root of the equation (1) must be evaluated numerically for each value of the a parameter. To achieve this, the first two lines of this listing create a range variable i, which is used for defining the vector of parameter values a_i at which the calculations will be performed. The elements of this vector take the values from 0.010 to 0.025 in increments of 0.0005 (these values have been chosen just for illustrative purposes, you can experiment with other values). The last line of the listing assigns the elements of another vector y the values of the equation (1) roots calculated using the root function for each a_i. However, to make the root function work, it is necessary to have previously specified the initial guess for the solution, which is done in the third line of the listing. The key idea of the method applied in Listing 8.20 is that the same initial value $x = 300$ is used for all a_i.

LISTING 8.20. Solving the $\sin(a \cdot x) = 0$ Equation Numerically

```
i := 0 ..  30

a_i := 0.01 + i · 0.0005

x := 300

y_{i+1} := root (sin(a_i · x) , x)
```

The calculation results y_i are denoted in Figure 8.10 as points, where the leftmost point corresponds to the initial guess $y_0 = 300$. Notice that as a increases, the curve of the equation roots first goes along the solution family $y = \pi/a$, and then (in the area of $a \approx 0.17$) "jumps" to the next family $y = 2 \cdot \pi/a$. This situation is the most unfavorable one in terms of calculation, since we want to find a continuous solution family. The jumps of the $y(a)$ dependence might confuse the user and hide the existence of the solution family at $a > 0.17$.

Why does this happen? This behavior is caused by the initial guess selection for calculating each of the roots. The line of initial values, $y = 300$, is designated in Figure 8.7 as a dashed horizontal line. For $a_0 = 0.01$, and generally for several first values of a_i, the initial value of $y = 300$ is the closest to the lower family of solutions $y = \pi/a$. Thus, it is not surprising that these roots are the ones found by the numeric method. In the right part of the graph shown in Figure 8.10 the second family of solutions $y = 2 \cdot \pi/a$ is closer to the line of initial values, and therefore, it is the one produced by the numeric method as a solution.

The considerations provided above lead to a very simple method of eliminating the jumps and finding one of the continuous solution families. To achieve this, when searching each $(i + 1)$-th root, it is necessary to select an initial guess as close to the family being searched as possible. Selecting the previous i-th root found for the previous value of the a_i parameter as an approximation would be a good idea. One of the possible implementations of this method, known as continuation by parameter, is presented in Listing 8.21. Here the root function is applied within the custom function $f(x0,a)$ defined at the beginning of this listing using programming tools. The main aim of the f(x0,a) function is to return the root value for the specified value of the a parameter and initial guess x0. Concerning other aspects, Listing 8.21 is very similar to the previous one, except for the fact that its next to last line explicitly specifies the initial guess $y_0 = 300$ only to find y_1. For all subsequent points, the previous y_i value is taken as an initial guess (the last line of the listing).

LISTING 8.21. Solving the $\sin(a \cdot x) = 0$ Equation Using Continuation by the Parameter Method

$$f(x0, a) := \begin{vmatrix} x \leftarrow x0 \\ \text{root}(\sin(a \cdot x), x) \end{vmatrix}$$

$$i := 0 .. \ 30$$

$$a_i := 0.01 + i \cdot 0.0005$$

$$y_0 := 300$$

$$y_{i+1} := f(y_i, a_i)$$

The result of the calculation provided in Figure 8.11 is strikingly different from the previous one. As you can see, a rather small change in the usage ideology of the numerical method has resulted in a continuous family of roots being defined. Notice that it is possible to obtain the result shown in Figure 8.10 (without continuation by parameter) in the terms of the f(x0,a) function by replacing its first argument in the last line of Listing 8.21 with a constant, for example, f(300,a_i).

To find another family of solutions, select the appropriate y_0 initial guess, for example, $y_0 = 600$. The result of the calculations that will be obtained using Listing 8.21 for this case is presented in Figure 8.12. If you select y_0 closer to the third family of solutions (for example, $y_0 = 900$), then it is this solution that will be found by the Mathcad numeric processor, and so on.

The continuation method is also applicable for solving appropriate parametric optimization tasks. The main idea of this method in this case remains the same, however; instead of the root or Find functions, you need to apply one of the functions intended for finding an extreme – Minerr, Maximize, or Minimize.

Here we have described the basic idea and one of the possible implementations of the continuation with the parameter method. Readers interested in this topic can certainly suggest other solutions for this problem that are both mathematical and programmatic. In particular, you can use the result of extrapolating the $x(a)$ dependence as an initial approximation, you can implement more sophisticated algorithms for branching families of solutions, and so on.

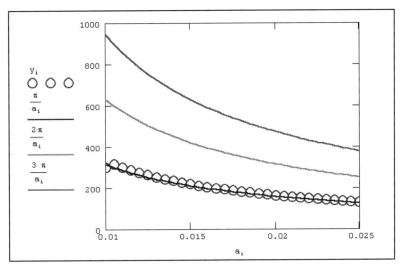

FIGURE 8.11. Solution of the $sin(a \cdot x) = 0$ equation obtained using the continuation by the parameter method for $y_0 = 300$ (Listing 8.21)

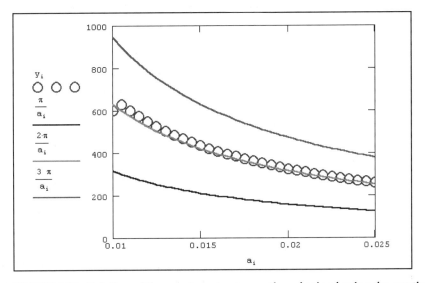

FIGURE 8.12. Solution of the $sin(a \cdot x) = 0$ equation obtained using the continuation by the parameter method for $y_0 = 600$

9 | Matrix Calculations

Matrix calculations can be classified into several types. The first type represents the simplest actions implemented by operators *(see Section 9.1)* and several functions intended for creating, joining, and sorting matrices, and obtaining their basic properties *(see Section 9.2)*. The second type comprises more complicated functions implementing numerical algorithms of linear algebra, such as solving systems of linear equations *(see Section 9.3)*, calculating eigenvalues and eigenvectors *(see Section 9.4)*, and various matrix decompositions *(see Section 9.5)*.

9.1. THE SIMPLEST MATRIX OPERATIONS

Mathcad implements the simplest operations of matrix algebra as operators. The syntax of the operators is very close to the mathematical principle of their respective operations. Each operator is designated by an appropriate symbol. In this section, we will consider matrix and vector operations available in Mathcad 2001i. A *vector* is a particular case in which the matrix contains one column (i.e., one with the dimension of $N \times 1$). Thus, all operations that are valid for matrices are also applicable for vectors, except when restrictions are specially mentioned (for example, some operations are only applicable to square matrices like N×N). Some operations are valid for vectors only (such as a vector inner product, also known as a "dot product"), while other operations have a different effect on vectors and matrices, despite the fact that they are written identically.

Vector operations aren't directly applicable to strings (1 ×N matrices). Before applying such an operation, you need to first convert it into a vector by transposing it.

9.1.1. Transpose Operation

Transposition is the matrix operation that converts a M×N matrix to an N×M matrix by interchanging its rows and columns. Listing 9.1 presents an example illustrating this operation. To insert the transpose operator, use the **Matrix** toolbar or press the <Ctrl>+<1> keyboard combination. Don't forget that to insert the transpose operator you need to place the matrix between the insertion lines. A warning about the position of insertion lines in relation to the matrix is provided in Section 9.1.4.

LISTING 9.1. Transposition of Vectors and Matrices

$$\begin{pmatrix} 1 \\ 2 \\ 3 \end{pmatrix}^{T} = \begin{pmatrix} 1 & 2 & 3 \end{pmatrix}$$

$$\begin{pmatrix} 1 & 2 & 3 \\ 4 & 5 & 6 \end{pmatrix}^{T} = \begin{pmatrix} 1 & 4 \\ 2 & 5 \\ 3 & 6 \end{pmatrix}$$

FIGURE 9.1. The Matrix toolbar

9.1.2. Addition

Mathcad supports both addition and subtraction operations of matrices. These operators are designated by the <+> and <−> symbols, respectively. Matrices must have the same dimension, otherwise an error message will be displayed. Each element of the sum of two matrices is equal to the sum of the respective elements of the summand matrices (Listing 9.2).

LISTING 9.2. Adding and Subtracting Matrices

$$A := \begin{pmatrix} 1 & 2 & 3 \\ 4 & 5 & 6 \end{pmatrix} \qquad B := \begin{pmatrix} 1 & 0 & 0 \\ -1 & -3 & -4 \end{pmatrix}$$

$$A + B = \begin{pmatrix} 2 & 2 & 3 \\ 3 & 2 & 2 \end{pmatrix}$$

$$A - B = \begin{pmatrix} 0 & 2 & 3 \\ 5 & 8 & 10 \end{pmatrix}$$

Besides matrix addition, Mathcad supports the operation of adding a scalar value to a matrix (Listing 9.3). Each element of the resulting matrix is equal to the sum of the respective element of the source matrix and a scalar value.

LISTING 9.3. Adding Matrices and Scalar Values

$$A := \begin{pmatrix} 1 & 2 & 3 \\ 4 & 5 & 6 \end{pmatrix}$$

$$x := 1$$

$$A + x = \begin{pmatrix} 2 & 3 & 4 \\ 5 & 6 & 7 \end{pmatrix}$$

$$A - x = \begin{pmatrix} 0 & 1 & 2 \\ 3 & 4 & 5 \end{pmatrix}$$

The result of matrix negation (changing the sign) is equivalent to the negation of all of its elements. To change the sign of the matrix, it must be preceded by a minus sign, like any number (Listing 9.4).

LISTING 9.4. Matrix Negation

$$A := \begin{pmatrix} 1 & 2 & 3 \\ 4 & 5 & 6 \end{pmatrix} \qquad -A = \begin{pmatrix} -1 & -2 & -3 \\ -4 & -5 & -6 \end{pmatrix}$$

9.1.3. Multiplication

When performing multiplication, remember that the M×N matrix can only be multiplied by the N × P matrix (P may take any value). As a result of this operation, you'll get an M × P matrix.

To insert the multiplication operator, press the asterisk key on the keyboard <*> or use the **Matrix** toolbar by clicking the **Dot Product** button (Figure 9.1). Matrix multiplication is by default designated by a dot, as shown in Listing 9.5. You can select the symbol of matrix multiplication in the same way you did for the scalar expressions *(see Chapter 2)*.

LISTING 9.5. Multiplying Matrices

$$A := \begin{pmatrix} 1 & 2 & 3 \\ 4 & 5 & 6 \end{pmatrix} \qquad B := \begin{pmatrix} 1 & 0 & 0 \\ -1 & -3 & -4 \end{pmatrix}$$

$$\mathbf{A \cdot B} = \blacksquare$$

$$C := B^T$$

$$C = \begin{pmatrix} 1 & -1 \\ 0 & -3 \\ 0 & -4 \end{pmatrix}$$

$$A \cdot C = \begin{pmatrix} 1 & -19 \\ 4 & -43 \end{pmatrix}$$

Notice that an attempt to multiply the matrices A and B of an inappropriate dimension (both 2 × 3) will fail: Mathcad displays an empty placeholder after the equal sign. Also notice that Mathcad editor highlights such an expression in red (in Listing 9.5 this expression is shown in bold). When you position the cursor over this expression, an error message will appear, informing you that the number of rows in the first matrix doesn't match the number of columns in the second matrix.

Listing 9.6 presents just another example that illustrates the multiplication of a vector by a row matrix and, inversely, a row by a vector. The second line of this listing displays how the formula looks when the No Space option is selected for displaying the multiplication operator.

LISTING 9.6. Multiplication of Vectors and Row Matrices

$$(\ 3 \quad 4 \ ) \cdot \begin{pmatrix} 1 \\ 2 \end{pmatrix} = (\ 11 \ )$$

$$\begin{pmatrix} 1 \\ 2 \end{pmatrix} (\ 3 \quad 4 \ ) = \begin{pmatrix} 3 & 4 \\ 6 & 8 \end{pmatrix}$$

The same multiplication operator has different effects when two vectors are multiplied (see Section 9.1.6).

Matrix multiplication and division by a scalar value are defined in a similar way to the matrix addition to a scalar value (Listing 9.7). A multiplication operator is inserted with the same procedure used for the multiplication of two matrices. Any MxN matrix can be multiplied by a scalar value.

LISTING 9.7. Matrix Multiplication by a Scalar Value

$$A \cdot 2 = \begin{pmatrix} 2 & 4 & 6 \\ 8 & 10 & 12 \end{pmatrix}$$

$$\frac{A}{2} = \begin{pmatrix} 0.5 & 1 & 1.5 \\ 2 & 2.5 & 3 \end{pmatrix}$$

9.1.4. Determinant of a Square Matrix

The *determinant* of the matrix is designated by a standard math character. To insert the determinant operator, click the Determinant button on the Matrix toolbar (Figure 9.2) or enter the <|> character from the keyboard (by pressing <Shift>+<\>). As a result, the placeholder will appear, which you must fill in with

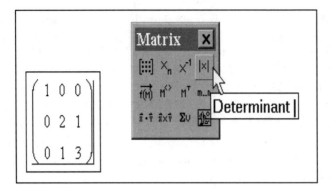

FIGURE 9.2. Inserting the determinant operator

the matrix. To calculate the determinant of the existing matrix (see Figure 9.2), proceed as follows:

1. Position the cursor in such a way as to place the matrix between the insertion lines (insertion lines are blue sections of vertical and horizontal lines specifying the current editing area).
2. Insert the matrix determinant operator.
3. To calculate the determinant, enter the equal sign.

The result of determinant evaluation is shown in Listing 9.8.

LISTING 9.8. Evaluating the Determinant of the Square Matrix

$$
\begin{vmatrix} 1 & 0 & 0 \\ 0 & 2 & 1 \\ 0 & 1 & 3 \end{vmatrix} = 5
$$

9.1.5. Vector Magnitude

Vector magnitude is designated with the same character as the matrix determinant. By definition, vector magnitude is equal to the value of the square root of the sum of the squares of all its elements (Listing 9.9).

LISTING 9.9. Finding Vector Magnitude

$$
\begin{Vmatrix} 1 \\ 2 \\ 3 \end{Vmatrix} = 3.742
$$

9.1.6. Vector Inner Product

The *vector inner product* is the scalar value equal to the sum of the products of the corresponding elements of the vectors being multiplied. The vectors must have the same dimension, and the resulting inner product will have the same dimension. The inner product of the two vectors u and v is equal to $u \cdot v = |u| \cdot |v| \cdot \cos\theta$, where θ is the angle between the two vectors. If the vectors are orthogonal, their inner product is equal to zero. The inner product is designated by the multiplication character (Listing 9.10). The user is able to select the representation of the multiplication character for designating the inner product.

Don't use the $\times$ character to represent the vector inner product since this symbol is commonly accepted for designating the vector cross product (see Section 9.1.7).

LISTING 9.10. Vector Inner Product

$$\begin{pmatrix} 1 \\ 2 \\ 3 \end{pmatrix} \cdot \begin{pmatrix} 4 \\ 5 \\ 6 \end{pmatrix} = 32$$

Be very careful when multiplying several (more than two) vectors. The order of multiplication specified by the brackets influences the result of the operation. Listing 9.11 presents examples illustrating this influence.

LISTING 9.11. Vector Inner Product Multiplied by the Third Vector

$$\begin{pmatrix} 1 \\ 2 \\ 3 \end{pmatrix} \cdot \begin{pmatrix} 4 \\ 5 \\ 6 \end{pmatrix} \cdot \begin{pmatrix} 7 \\ 8 \\ 9 \end{pmatrix} = \begin{pmatrix} 224 \\ 256 \\ 288 \end{pmatrix}$$

$$\left[\begin{pmatrix} 1 \\ 2 \\ 3 \end{pmatrix} \cdot \begin{pmatrix} 4 \\ 5 \\ 6 \end{pmatrix} \right] \cdot \begin{pmatrix} 7 \\ 8 \\ 9 \end{pmatrix} = \begin{pmatrix} 224 \\ 256 \\ 288 \end{pmatrix} \qquad \begin{pmatrix} 1 \\ 2 \\ 3 \end{pmatrix} \cdot \left[\begin{pmatrix} 4 \\ 5 \\ 6 \end{pmatrix} \cdot \begin{pmatrix} 7 \\ 8 \\ 9 \end{pmatrix} \right] = \begin{pmatrix} 122 \\ 244 \\ 366 \end{pmatrix}$$

9.1.7. Cross Product

The *cross product* of the two vectors u and v with the angle θ between them is equal to another vector with a magnitude of $|u| \cdot |v| \cdot \sin\theta$, perpendicular to u and v and pointing in the direction determined by the right hand rule (the right-hand rule states that the orientation of the vectors' cross product is determined by placing u and v tail-to-tail, flattening the right hand, extending it in the direction of **u**, and then curling the fingers in the direction that the angle that v makes with u. The thumb then points in the direction of the cross product). The vector cross product is designated by the $\times$ character, which can be inserted either by clicking the Cross Product button in the **Matrix** toolbar or by pressing the <Ctrl>+<8> keys on the keyboard. An example is provided in Listing 9.12.

LISTING 9.12. Vector Cross Product

$$
\begin{pmatrix} 1 \\ 2 \\ 3 \end{pmatrix} \times \begin{pmatrix} 4 \\ 5 \\ 6 \end{pmatrix} = \begin{pmatrix} -3 \\ 6 \\ -3 \end{pmatrix}
$$

9.1.8. Sum of All Elements of the Vector and Trace of the Matrix

Sometimes it might be necessary to calculate the sum of all elements of the vector. Expressly for this purpose, there is an auxiliary operator (see the first line of Listing 9.13), which can be inserted by clicking the Vector Sum button on the **Matrix** toolbar or by pressing the <Ctrl>+<4> keys. Most frequently, this operator proves to be useful when organizing loops with indexed variables rather than in vector algebra.

The lower part of Listing 9.13 shows the usage of the summation of elements along the diagonal of the square matrix. This sum is known as the *trace of the matrix*. This operation is implemented by the built-in tr function:

■ tr(A) — The trace of the square matrix A

LISTING 9.13. The Sum of All Elements of the Vector and Trace of the Matrix

$$
\Sigma \begin{pmatrix} 1 \\ 2 \\ 3 \end{pmatrix} = 6
$$

$$
A := \begin{pmatrix} 3 & 4 \\ 7 & 2 \end{pmatrix} \qquad \qquad \text{tr}(A) = 5
$$

9.1.9. Matrix Inverse

The matrix can be inverted if it is square and its determinant is not equal to zero (Listing 9.14). According to the definition, the product of the source matrix and its inverse matrix is the identity matrix (also known as the unit matrix). To insert the matrix inverse operator, click the Inverse button on the **Matrix** toolbar.

LISTING 9.14. Finding the Inverse Matrix

$$\begin{pmatrix} 1 & 0 & 0 \\ 0 & 2 & 0 \\ 0 & 0 & 3 \end{pmatrix}^{-1} = \begin{pmatrix} 1 & 0 & 0 \\ 0 & 0.5 & 0 \\ 0 & 0 & 0.333 \end{pmatrix}$$

$$\begin{pmatrix} 1 & 0 & 0 \\ 0 & 2 & 0 \\ 0 & 0 & 3 \end{pmatrix} \cdot \begin{pmatrix} 1 & 0 & 0 \\ 0 & 0.5 & 0 \\ 0 & 0 & 0.333 \end{pmatrix} = \begin{pmatrix} 1 & 0 & 0 \\ 0 & 1 & 0 \\ 0 & 0 & 0.999 \end{pmatrix}$$

$$\begin{pmatrix} 1 & 0 & 0 \\ 0 & 0.5 & 0 \\ 0 & 0 & 0.333 \end{pmatrix} \cdot \begin{pmatrix} 1 & 0 & 0 \\ 0 & 2 & 0 \\ 0 & 0 & 3 \end{pmatrix} = \begin{pmatrix} 1 & 0 & 0 \\ 0 & 1 & 0 \\ 0 & 0 & 0.999 \end{pmatrix}$$

9.1.10. Matrix Exponentiation

Formally, square matrices could be raised to the n power. Notice that n must be an integer. The results of this operation are listed in Table. 9.1. To insert the matrix exponentiation operator (to raise the matrix M to the power of n), proceed in the same way as you would with a scalar value: click the **Raise to Power** button on the Calculator toolbar or press the <^> key. When the placeholder appears, fill it in with the value of the power n.

TABLE 9.1. The results of matrix exponentiation

n	M^n
0	Unit matrix of dimension M
1	The matrix M itself
-1	M^{-1} — the inverse matrix
2, 3, ...	$M \cdot M$, $(M \cdot M) \cdot M$, ...
-2, -3, ...	$M^{-1} \cdot M^{-1}$, $(M^{-1} \cdot M^{-1}) \cdot M^{-1}$, ...

Several examples of matrix exponentiation are shown in Listing 9.15.

LISTING 9.15. Matrix Exponentiation Examples

$$
\begin{pmatrix} 1 & 0 & 0 \\ 0 & 0.5 & 0 \\ 0 & 0 & 0.333 \end{pmatrix}^{-1} = \begin{pmatrix} 1 & 0 & 0 \\ 0 & 2 & 0 \\ 0 & 0 & 3.003 \end{pmatrix}
$$

$$
\begin{pmatrix} 1 & 0 & 0 \\ 0 & 2 & 0 \\ 0 & 0 & 3 \end{pmatrix}^{-2} = \begin{pmatrix} 1 & 0 & 0 \\ 0 & 0.25 & 0 \\ 0 & 0 & 0.111 \end{pmatrix}
$$

$$
\begin{pmatrix} 1 & 0 & 0 \\ 0 & 2 & 0 \\ 0 & 0 & 3 \end{pmatrix}^{2} = \begin{pmatrix} 1 & 0 & 0 \\ 0 & 4 & 0 \\ 0 & 0 & 9 \end{pmatrix}
$$

9.1.11. Array Vectorization

Mathcad implementation of vector algebra includes a specific and somewhat un-usual operator known of as the vectorize operator. This operator is intended for working with arrays. It changes the meaning of the operator or function to make them applicable to all elements of the array (matrix or vector), thus simplifying the task of loop programming and enabling you to perform calculations in paral-lel. For example, sometimes it might be necessary to multiply each element of one vector by the respective element of another vector. Mathcad doesn't have a built-in operation for performing this task, but this aim can be easily achieved using vectorization (Listing 9.16). To do this, proceed as follows:

1. Enter the vector expression as shown in the second line of the listing (notice that this representation of the multiplication character implies a vector inner product).
2. Position the cursor so that you can make the editing lines frame the whole expression that needs to be vectorized (Figure 9.3).
3. Insert the vectorize operator by clicking the Vectorize button at the **Matrix** toolbar (Figure 9.3), or press <Ctrl>+<-> keys.
4. To evaluate the result, insert <=>.

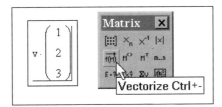

FIGURE 9.3. The vectorize operator

LISTING 9.16. Using the Vectorize Operator for Multiplying the Vector's Elements

$$v := \begin{pmatrix} 1 \\ 2 \\ 3 \end{pmatrix}$$

$$v \cdot \begin{pmatrix} 1 \\ 2 \\ 3 \end{pmatrix} = 14$$

$$\overrightarrow{\left[v \cdot \begin{pmatrix} 1 \\ 2 \\ 3 \end{pmatrix} \right]^2} = \begin{pmatrix} 1 \\ 4 \\ 9 \end{pmatrix}$$

The vectorize operator can only be used with vectors and matrices of the same dimension.

Most non-specific functions in Mathcad don't require the vectorize operator to perform the same operation over all elements of the vector. For example, according to its definition, an argument of trigonometric functions is scalar. If you attempt to evaluate the sine of a vector, Mathcad will perform vectorization by default by calculating the sine of each element, and it will produce the result as a vector. Listing 9.17 shows an example of such an operation.

LISTING 9.17. Examples of Vectorization Performed by Default

$$\sin(v) = \begin{pmatrix} 0.841 \\ 0.909 \\ 0.141 \end{pmatrix} \qquad \sin(\overrightarrow{v}) = \begin{pmatrix} 0.841 \\ 0.909 \\ 0.141 \end{pmatrix}$$

9.1.12. Symbolic Operations over Matrices

All matrix and vector operators discussed earlier in this chapter can be used in symbolic calculations. The power of symbolic operators lies in the fact that besides specific numbers, they can also be applied to variables. Several examples are shown in Listing 9.18.

LISTING 9.18. Examples of Symbolic Operations over Vectors and Matrices

$$
\begin{pmatrix} a & b \\ c & d \\ f & g \end{pmatrix} \cdot \begin{pmatrix} o & p & q \\ r & s & t \end{pmatrix} \rightarrow \begin{pmatrix} a \cdot o + b \cdot r & a \cdot p + b \cdot s & a \cdot q + b \cdot t \\ c \cdot o + d \cdot r & c \cdot p + d \cdot s & c \cdot q + d \cdot t \\ f \cdot o + g \cdot r & f \cdot p + g \cdot s & f \cdot q + g \cdot t \end{pmatrix}
$$

$$
\begin{pmatrix} a & b \\ c & d \end{pmatrix}^{-1} \rightarrow \begin{bmatrix} \dfrac{d}{(a \cdot d - b \cdot c)} & \dfrac{-b}{(a \cdot d - b \cdot c)} \\ \dfrac{-c}{(a \cdot d - b \cdot c)} & \dfrac{a}{(a \cdot d - b \cdot c)} \end{bmatrix}
\qquad
\begin{vmatrix} a & 0 & 0 \\ 0 & b & 1 \\ 0 & 1 & c \end{vmatrix} \rightarrow a \cdot (b \cdot c - 1)
$$

$$
\begin{pmatrix} a \\ b \\ c \end{pmatrix} \times \begin{pmatrix} r \\ s \\ t \end{pmatrix} \rightarrow \begin{pmatrix} b \cdot t - c \cdot s \\ c \cdot r - a \cdot t \\ a \cdot s - b \cdot r \end{pmatrix}
$$

$$
\sum \begin{pmatrix} a \\ b \\ c \end{pmatrix} \rightarrow a + b + c
$$

TIP

Don't hesitate to use the Mathcad symbolic processor as a powerful and comprehensive reference on all aspects of math. For example, it might help you to remember some definitions of linear algebra (for example, matrix multiplication and inversion, as shown in Listing 9.18).

9.2. MATRIX FUNCTIONS

In this section, we are going to consider the main built-in functions intended to simplify the process of working with vectors and matrices. These functions are needed for creating and joining matrices, selecting specific parts of matrices, and for obtaining their basic properties.

9.2.1. Functions for Creating Matrices

The simplest and most illustrative method of creating a matrix or a vector involves using the first button of the **Matrix** toolbar *(see Section 4.3, "Arrays," in Chapter 4)*. However, when programming complex projects, you will frequently find it more convenient to create arrays using various built-in functions.

Defining Matrix Elements Using the Function

The `matrix(M,N,f)` function creates the M×N matrix, each `(i,j)` element of which is `f(i,j)` (Listing 9.19), where:

- `M` — Number of rows
- `N` — Number of columns
- `f(i,j)` — Some function

LISTING 9.19. Creating a Matrix

```
f(i,j) := i + 0.5·j

A := matrix(2,3,f)
```

$$A = \begin{pmatrix} 0 & 0.5 & 1 \\ 1 & 1.5 & 2 \end{pmatrix}$$

Mathcad provides two other specific functions for creating matrices: the `CreateSpace` function and the `CreateMesh` function. These are mainly used for quick and efficient representations of specific dependencies as 3D graphs (surfaces or 3D curves). Except for the first argument (a function), all other arguments of these functions are optional.

Let us consider the `CreateSpace` function.

The `CreateSpace(F(or f1,f2,f3),t0,t1,tgrid,fmap)` function creates a nested array, representing x-, y-, and z-coordinates by a parametric 3D curve specified by the F function, where:

- `F(t)` — Vector function comprising three elements, specified by parametric dependence on its only argument `t`
- `f1(t),f2(t),f3(t)` — Scalar functions
- `t0` — Lower limit of `t` (by default set to `-5`)
- `t1` — Upper limit of `t` (by default set to `5`)

- `tgrid` — Grid granularity (integer number of grid points) by variable t (by default set to 20)
- `fmap` — Vector function of three arguments specifying coordinate mapping

Nested arrays were covered in the section entitled "Creating a Tensor" in Chapter 4.

An example illustrating the usage of the `CreateSpace` function is shown in Figure 9.4. Notice that we didn't require any additional code for creating the spiral graph, except for needing to specify the parametric dependence in the F vector function F!

The `CreateMesh` function for creating the matrix for a 3D surface graph works similarly, except that we need two variables to define a surface. An example of how to use the `CreateMesh` function is illustrated in Figure 9.5.

The `CreateMesh(F (or g, or f1,f2,f3),s0,s1,t0,t1,sgrid,tgrid,fmap)` function creates a nested array representing x-, y-, and z-coordinates of a parametric surface specified by the F function, where:

- `F(s,t)` — Vector function comprising three elements and specified parametrically in relation to s and t arguments

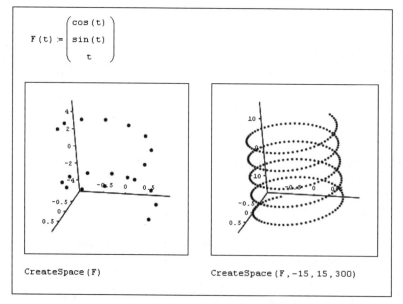

FIGURE 9.4. Using the `CreateSpace` function with different sets of parameters

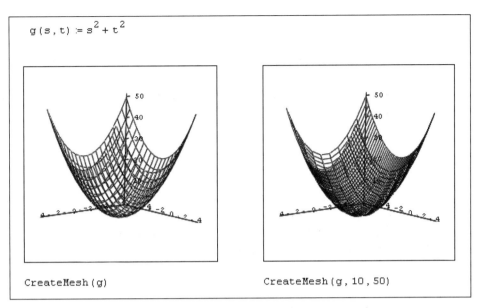

FIGURE 9.5. Usage of the `CreateMesh` function with different sets of parameters

- `g(s,t)` — Scalar function
- `f1(s,t),f2(s,t),f3(s,t)` — Scalar functions
- `s0,t0` — Lower limits of the arguments `s,t` (by default set to `-5`)
- `s1,t1` — Upper limits of the arguments `s,t` (by default set to `5`)
- `sgrid,tgrid` — Numbers of grid points by `s` and `t` variables (by default set to `20`)
- `fmap` — Vector function of three arguments specifying coordinate mapping

Examples of nested arrays created by `CreateMesh` and `CreateSpace` functions are shown in Listing 9.20. Each of the three nested matrices that make up the array defines `x`-, `y`-, and `z`-coordinates of the points belonging to the 3D curve or surface, respectively.

LISTING 9.20. The Results of Applying the `CreateMesh` and `CreateSpace` Functions (Figures 9.4 and 9.5)

$$
\text{CreateMesh}(g,2,3) = \left[\left[\begin{pmatrix} -5 & -5 & -5 \\ 5 & 5 & 5 \end{pmatrix}\right]\right.
\begin{pmatrix} -5 & 0 & 5 \\ -5 & 0 & 5 \end{pmatrix}
\left.\begin{pmatrix} 50 & 25 & 50 \\ 50 & 25 & 50 \end{pmatrix}\right]
$$

$$\text{CreateSpace}(F, 3) = \left[\left[\begin{pmatrix} 0.284 \\ 1 \\ 0.284 \end{pmatrix} \begin{pmatrix} 0.959 \\ 0 \\ -0.959 \end{pmatrix} \begin{pmatrix} -5 \\ 0 \\ 5 \end{pmatrix} \right] \right]$$

Creating Matrices of a Specific Type

Mathcad enables you to easily create matrices of a specific type using built-in functions. Some examples of how to use these functions are given in Listing 9.21.

- `identity(N)` — An identity $N \times N$ matrix (a matrix of 0's with 1's along the diagonal)
- `diag(v)` — Diagonal matrix that contains the elements of vector v on its diagonal
- `geninv(A)` — Left inverse function. Returns the left inverse matrix of A
- `rref(A)` — Returns row-reduced echelon form of matrix or vector A

Here:

- N — Integer value
- v — Vector
- A — Matrix composed of real numbers

Dimension $N \times M$ of the A matrix for the `geninv` function must satisfy the condition $N \geq M$.

NOTE

LISTING 9.21. Creating Matrices of Special Types

$$\text{identity}(2) = \begin{pmatrix} 1 & 0 \\ 0 & 1 \end{pmatrix}$$

$$\text{diag}\left(\begin{pmatrix} 1 \\ 2 \\ 3 \end{pmatrix}\right) = \begin{pmatrix} 1 & 0 & 0 \\ 0 & 2 & 0 \\ 0 & 0 & 3 \end{pmatrix}$$

$$A := \begin{pmatrix} 1 & 2 \\ 3 & 4 \\ 5 & 6 \end{pmatrix}$$

$$\text{geninv}(A) = \begin{pmatrix} -1.333 & -0.333 & 0.667 \\ 1.083 & 0.333 & -0.417 \end{pmatrix} \qquad \text{geninv}(A) \cdot A = \begin{pmatrix} 1 & 0 \\ 0 & 1 \end{pmatrix}$$

$$\text{rref}(A) = \begin{pmatrix} 1 & 0 \\ 0 & 1 \\ 0 & 0 \end{pmatrix}$$

$$\text{rref}\left(\begin{pmatrix} 1 & 2 & 3 \\ 4 & 5 & 6 \end{pmatrix}\right) = \begin{pmatrix} 1 & 0 & -1 \\ 0 & 1 & 2 \end{pmatrix}$$

9.2.2. Joining and Unjoining Matrices

You can easily select a submatrix, vector, or specific element from a matrix or vector. And, conversely, it is possible to join several matrices to form another one.

Selecting Part of a Matrix

To select a specific part of the matrix, use one of the following methods:

- To select a single element of a matrix, use the subscript operator (more detailed information on this operator is provided in *Section 4.3.1, "Accessing Array Elements," in Chapter 4*). This operator is inserted by clicking the **Sub-script** button labeled with the x_n icon at the **Matrix** toolbar or by pressing the <[> key (Listing 9.22, the second line from the top).
- To select a specific column from the matrix, use the superscript operator by clicking the **Matrix Column** button labeled by angle brackets <> on the **Matrix** toolbar. Alternately, you can press the <Ctrl>+<6> keys on the keyboard (Listing 9.22).
- To select an entire row from the matrix, transpose the matrix and then apply the superscript operator (Listing 9.22, last line).

■ To select a smaller submatrix, use the built-in `submatrix(A,ir,jr,ic,jc)` function, returning the part of the A matrix located between the specified `ir,jr` rows and `ic,jc` columns inclusively (Listing 9.23).

The `submatrix` function can also be used to select a single column or row from the matrix.

NOTE

LISTING 9.22. Accessing Matrix Elements, Columns and Rows

$$A := \begin{pmatrix} 1 & 2 & 3 \\ 4 & 5 & 6 \end{pmatrix}$$

$A_{0,1} = 2$ $\qquad\qquad A_{1,1} = 5$

$$A^{\langle 2 \rangle} = \begin{pmatrix} 3 \\ 6 \end{pmatrix} \qquad A^{\langle 0 \rangle} = \begin{pmatrix} 1 \\ 4 \end{pmatrix} \qquad A^{\langle 1 \rangle} = \begin{pmatrix} 2 \\ 5 \end{pmatrix}$$

$$\left(A^T\right)^{\langle 0 \rangle} = \begin{pmatrix} 1 \\ 2 \\ 3 \end{pmatrix} \qquad \left(A^T\right)^{\langle 0 \rangle T} = \begin{pmatrix} 1 & 2 & 3 \end{pmatrix}$$

LISTING 9.23. Selecting a Submatrix

$$\text{submatrix}\left[\begin{pmatrix} 1 & 2 & 3 \\ 4 & 5 & 6 \end{pmatrix}, 0, 1, 0, 1\right] = \begin{pmatrix} 1 & 2 \\ 4 & 5 \end{pmatrix}$$

$$\text{submatrix}\left[\begin{pmatrix} 1 & 0 & 0 \\ -1 & -3 & -4 \end{pmatrix}, 0, 0, 0, 1\right] = \begin{pmatrix} 1 & 0 \end{pmatrix}$$

The same operations are applicable to vector and row matrices. You only need to remember that their dimensions are N×1 and 1×N, respectively (Listing 9.24).

LISTING 9.24. Selecting Parts from Vectors and Rows

$$\begin{pmatrix} 1 & 2 & 3 \end{pmatrix}^{\langle 0 \rangle} = \begin{pmatrix} 1 \end{pmatrix} \qquad \left(\begin{pmatrix} 1 \\ 2 \\ 3 \end{pmatrix}^T\right)^{\langle 1 \rangle} = \begin{pmatrix} 2 \end{pmatrix}$$

$$\text{submatrix}\left[\begin{pmatrix} 1 \\ 2 \\ 3 \end{pmatrix}, 0, 1, 0, 0\right] = \begin{pmatrix} 1 \\ 2 \end{pmatrix}$$

Joining Matrices

To combine two or more matrices into a new one, Mathcad provides two matrix functions (Listing 9.25):

■ augment (A,B,C,...) — The matrix created by joining argument matrices A, B, C left to right

■ stack (A,B,C,...) — The matrix created by joining the argument matrices A, B, C top to bottom

Here:

● A,B,C,... — Vectors or matrices of an appropriate size

LISTING 9.25. Examples of Matrix Joining

$$A := \begin{pmatrix} 1 & 2 & 3 \\ 4 & 5 & 6 \end{pmatrix} \qquad B := \begin{pmatrix} 0 & 0 & 0 \\ 0 & 0 & 0 \end{pmatrix}$$

$$\text{stack} (A, B) = \begin{pmatrix} 1 & 2 & 3 \\ 4 & 5 & 6 \\ 0 & 0 & 0 \\ 0 & 0 & 0 \end{pmatrix}$$

$$\text{augment} (A, B) = \begin{pmatrix} 1 & 2 & 3 & 0 & 0 & 0 \\ 4 & 5 & 6 & 0 & 0 & 0 \end{pmatrix}$$

$$\text{augment}\left[\begin{pmatrix} 1 \\ 1 \end{pmatrix}, A, B, \begin{pmatrix} 3 \\ 3 \end{pmatrix}\right] = \begin{pmatrix} 1 & 1 & 2 & 3 & 0 & 0 & 0 & 3 \\ 1 & 4 & 5 & 6 & 0 & 0 & 0 & 3 \end{pmatrix}$$

9.2.3. Displaying the Matrix Size

To obtain information on matrix and vector properties, Mathcad provides several built-in functions (Listing 9.26):

- ▪ rows(A) — Number of rows
- ▪ cols(A) — Number of columns
- ▪ length(v) — Number of vector elements
- ▪ last(v) — Index of the last vector element

Here:

- ● A — Matrix or vector
- ● v — Vector

The number of vector elements and the index of its last element are the same when indexing starts from 1, for example, the ORIGIN system constant is set to 1 (see Chapter 4).

LISTING 9.26. Sizes of Matrices and Vectors

$$w := (1 \quad 2 \quad 3)$$

$$A := \begin{pmatrix} 1 & 2 \\ 3 & 4 \\ 5 & 6 \end{pmatrix} \qquad v := \begin{pmatrix} 1 \\ 2 \\ 3 \end{pmatrix}$$

rows (A) = 3   rows (v) = 3   rows (w) = 1

cols (A) = 2   cols (v) = 1   cols (w) = 3

length (v) = 3

last (v) = 2

9.2.4. Sorting Matrices

Quite often it might be necessary to re-order matrix or vector elements by placing them in ascending or descending order for a specific row or column. For this purpose, there are several built-in functions that provide flexible sorting capabilities for matrices:

- ■ sort (v) — Sorts vector elements in ascending order (Listing 9.27)
- ■ csort (A, i) — Rearranges matrix rows by sorting the elements of the i-th column in ascending order (Listing 9.28)
- ■ rsort (A, i) — Rearranges matrix columns by sorting the elements of the i-th row in ascending order (Listing 9.29)
- ■ reverse (v) — Rearranges the vector elements by reversing their order (Listing 9.27)

Here:

- ● v — Vector
- ● A — Matrix
- ● i — Row or column index

If a matrix or vector contains complex elements, sorting is performed by the real part of the number. The imaginary part is ignored.

LISTING 9.27. Sorting Vectors

$$v := \begin{pmatrix} 3 \\ 1 \\ 4 \\ 2 \end{pmatrix} \qquad \text{sort (v)} = \begin{pmatrix} 1 \\ 2 \\ 3 \\ 4 \end{pmatrix} \qquad \text{reverse (v)} = \begin{pmatrix} 2 \\ 4 \\ 1 \\ 3 \end{pmatrix}$$

LISTING 9.28. Sorting Matrices by Column

$$A := \begin{pmatrix} 1 & 9 \\ 3 & 0 \\ 2 & 8 \end{pmatrix}$$

$$\text{csort (A, 1)} = \begin{pmatrix} 3 & 0 \\ 2 & 8 \\ 1 & 9 \end{pmatrix} \qquad \text{csort (A, 0)} = \begin{pmatrix} 1 & 9 \\ 2 & 8 \\ 3 & 0 \end{pmatrix}$$

LISTING 9.29. Sorting Matrices by Row *(See Matrix A in Listing 9.28)*

$$\text{rsort}(A, 1) = \begin{pmatrix} 9 & 1 \\ 0 & 3 \\ 8 & 2 \end{pmatrix} \qquad \text{rsort}(A, 2) = \begin{pmatrix} 1 & 9 \\ 3 & 0 \\ 2 & 8 \end{pmatrix}$$

9.2.5. Norm of a Square Matrix

In linear algebra, it is common to use various norms of a matrix. Matrix norms map some scalar numeric value to the matrix. A matrix norm reflects the order of the matrix elements' values. Specific problems of linear algebra operate with different types of matrix norms. Mathcad provides four built-in functions for calculating various types of norms for square matrices:

- norm1(A) — The norm of the matrix in L1 space
- norm2(A) — The norm of the matrix in L2 space
- norme(A) — The Euclidean norm
- normi(A) — The max-norm, or ∞-norm (infinity norm)

Here:

- A — A square matrix

Listing 9.30 illustrates the calculation of various norms of the two matrices, A and B, where elements of the second matrix are several orders of magnitude greater than those of the first. The last line of this listing explains the Euclidean norm, the definition of which is similar to the definition of vector length.

For most tasks, it is not principally important which norm is used. As you can see from the examples provided in Listing 9.30, different norms usually produce the same values that reflect the order of magnitude of the matrix elements quite adequately. If you are interested in this topic, you can easily find definitions of other norms in reference literature on linear algebra or in the Mathcad Resource Center.

LISTING 9.30. Matrix Norms

$$A := \begin{pmatrix} 1 & 2 \\ 3 & 4 \end{pmatrix} \qquad B := \begin{pmatrix} 100 & 200 \\ 300 & 400 \end{pmatrix}$$

norm1 (A) = 6   norm1 (B) = 600

norm2 (A) = 5.465   norm2 (B) = 546.499

normi (A) = 7   normi (B) = 700

norme (A) = 5.477   norme (B) = 547.723

$$\sqrt{1^2 + 2^2 + 3^2 + 4^2} = 5.477$$

9.2.6. Condition Number of the Square Matrix

The *condition number* is just another important characteristic of a matrix. The condition number is the measure of sensitivity of the system of linear equations $A \cdot x = b$, defined by the matrix A, to the errors of specifying vector b errors in the right parts of equations. The higher the condition number, the stronger the influence of these errors, and the less stable the process of solving this system. The condition number is related to the matrix norm and is calculated differently for each norm:

- cond1 (A) — Condition number for the L1 norm
- cond2 (A) — Condition number for the L2 norm
- conde (A) — Condition number for the Euclidean norm
- condi (A) — Condition number for the ∞-norm

Here:

- A — A square matrix

The calculation of the condition numbers for A and B matrices is shown in Listing 9.31. Notice that the first matrix is well-conditioned, while the second one is ill-conditioned (its two rows define very close equations, with the precision up to the multiplier having the value of 3). The second line of the listing provides a formal definition of the condition number as the product of the norms of initial and inverse matrices. For other norms, this definition is the same.

As you can easily notice, the matrices A and B considered in Listing 9.30 have the same condition numbers since B = 100 · A, and, consequently, both matrices define the same system of equations.

NOTE

LISTING 9.31. Condition Numbers of Square Matrices

$$A := \begin{pmatrix} 1 & 2 \\ 3 & 4 \end{pmatrix} \qquad\qquad B := \begin{pmatrix} 1 & 2 \\ 3 & 6.01 \end{pmatrix}$$

$$\text{norme}(A) \cdot \text{norme}\left(A^{-1}\right) = 15$$

$$\text{conde}(A) = 15 \qquad\qquad \text{conde}(B) = 5.012 \times 10^3$$

$$\text{cond1}(A) = 21 \qquad\qquad \text{cond1}(B) = 7.217 \times 10^3$$

$$\text{cond2}(A) = 14.933 \qquad\qquad \text{cond2}(B) = 5.012 \times 10^3$$

$$\text{condi}(A) = 21 \qquad\qquad \text{condi}(B) = 7.217 \times 10^3$$

9.2.7. Rank of the Matrix

Matrix rank is the largest natural value k, for which there exists a nonzero determinant of the k-th order for a submatrix composed from any intersection of k columns and k rows of the matrix.

To calculate matrix rank, Mathcad provides the rank function.

◼ rank(A) — Matrix rank

Here:

● A — The matrix

LISTING 9.32. Rank of the Matrix

$$\text{rank}\begin{pmatrix} 1 & 2 \\ 3 & 6 \end{pmatrix} = 1 \qquad\qquad \text{rank}\begin{pmatrix} 1 & 2 \\ 3 & 4 \end{pmatrix} = 2$$

$$\text{rank}\begin{pmatrix} 1 & 2 \\ 3 & 6 \\ 1 & 2 \end{pmatrix} = 1 \qquad\qquad \text{rank}\begin{pmatrix} 1 & 2 \\ 3 & 4 \\ 5 & 6 \end{pmatrix} = 2$$

9.3. SYSTEMS OF LINEAR ALGEBRAIC EQUATIONS

The central problem of the numerical methods of linear algebra is the solution of the systems of linear algebraic equations, i.e., systems of equations of the following type:

$$a_{i1} \cdot x_1 + a_{i2} \cdot x_2 + \ldots + a_{iN} \cdot x_N = b_i. \tag{1}$$

In matrix form, a system of linear algebraic equations can be written using an equivalent representation:

$$A \cdot x = b, \tag{2}$$

where A is the matrix of coefficients of a system of linear algebraic equations with the dimension N×N, x is a vector of unknowns, and b is a vector of the right parts of equations.

NOTE

There is a wide range of tasks of numerical mathematics (probably most tasks) that can be reduced to solving systems of linear algebraic equations. One such example will be presented in Section 12.3, "Difference Schemes for Ordinary Differential Equations," in Chapter 12.

A system of linear algebraic equations has a single solution if the A matrix is not singular, i.e., its determinant is not equal to zero. From the point of view of numeric calculations, the solution of the system of linear algebraic equations doesn't present any difficulties if the A matrix is not large. On the other hand, a large matrix also doesn't present any problems, if it is well-conditioned. In Mathcad, you can solve systems of linear algebraic equations both in a more illustrative form (1) and in a more convenient form (2). When using the first form of notation, use the Given/Find solve block *(see Chapter 8)*, and for the second form, use the lsolve built-in function.

■ lsolve(A,b) — Solution of the system of linear algebraic equations

Here:

- A — Matrix of the coefficients of the system
- b — Vector of right parts

The usage of the lsolve function is illustrated by Listing 9.33. Here the matrix A can be defined using any of the available methods *(see Section 4.3, "Arrays,"* in Chapter 4), not necessarily in the explicit form, as in all the examples from this section. The `lsolve` built-in function can be used when solving systems of linear algebraic equations symbolically (Listing 9.34).

In Listing 9.35, the system of equations corresponding to matrix A and vector b is written explicitly.

LISTING 9.33. Solving a System of Linear Algebraic Equations

$$A := \begin{pmatrix} 1 & 5 & 2 \\ 0.7 & 12 & 5 \\ 3 & 0 & 4 \end{pmatrix} \qquad b := \begin{pmatrix} 1 \\ 2.9 \\ 3.1 \end{pmatrix}$$

$$\text{lsolve} (A, b) = \begin{pmatrix} -0.186 \\ -0.129 \\ 0.915 \end{pmatrix}$$

LISTING 9.34. Symbolic Solution of a System of Linear Algebraic Equations (*Listing 9.33, Continued*)

$$\text{lsolve} (A, b) \rightarrow \begin{pmatrix} -.18648648648648648649 \\ -.12864864864864864865 \\ .91486486486486486486 \end{pmatrix}$$

In some cases, to represent a system of linear algebraic equations more illustratively, it can be solved in just the same way as a system of non-linear equations can *(see Chapter 8)*. Listing 9.35 provides an example of a numeric solution of a system of linear algebraic equations from the previous listings. Don't forget that when solving the system numerically, you need to assign initial guess values to all unknowns (the first line of Listing 9.35). The initial guess values can be set arbitrarily, since the system of linear algebraic equations with a non-singular matrix has only one solution.

When solving a system of linear algebraic equations using the Find function, Mathcad automatically selects a linear numeric algorithm, which you can check by right-clicking the Find keyword and viewing the right-click menu.

LISTING 9.35. Solving the System of Linear Algebraic Equations Using the Solve Block

$x := 0 \qquad y := 0 \qquad z := 0$

Given

$1x + 5y + 2z = 1$

$0.7x + 12y + 5z = 2.9$

$3x + 0y + 4z = 3.1$

$$find(x, y, z) = \begin{pmatrix} -0.186 \\ -0.129 \\ 0.915 \end{pmatrix}$$

9.4. EIGENVECTORS AND EIGENVALUES

Another frequent task of linear algebra is that of finding eigenvectors x and eigenvalues λ of the matrix A, i.e., the solution of the matrix equation $A \cdot x = \lambda \cdot x$. Such an equation has solutions in the form of eigenvalues $\lambda_1, \lambda_2, \ldots$ and corresponding eigenvectors $x_1, x_2, \ldots$. Mathcad provides several built-in functions that implement rather complicated numerical algorithms for solving the tasks of finding eigenvectors and eigenvalues:

- ◼ `eigenvals(A)` — Calculates the vector composed from eigenvalues of the A matrix.
- ◼ `eigenvecs(A)` — Calculates the matrix containing normalized eigenvectors corresponding to the eigenvalues of the matrix A. The n-th column of the matrix being calculated corresponds to the eigenvector of the n-th eigenvalue calculated by `eigenvals`.
- ◼ `eigenvec(A, λ)` — Returns an eigenvector for matrix A and a specified eigenvalue λ.

Here:

- ● A — Square matrix

Listing 9.36 maps out the usage of these functions. A check for accuracy of the values of eigenvectors and eigenvalues is provided in Listing 9.37. Notice that an accuracy check for the $A \cdot x = \lambda \cdot x$ expression is performed twice — first for numeric values of x and λ, then by multiplying the appropriate matrix components.

LISTING 9.36. Finding Eigenvectors and Eigenvalues

$$A := \begin{pmatrix} 1 & 5 & 2 \\ 0.7 & 12 & 5 \\ 3 & 0 & 4 \end{pmatrix}$$

$$\text{eigenvals (A)} = \begin{pmatrix} 0.938 \\ 3.024 \\ 13.037 \end{pmatrix}$$

$$\text{eigenvecs (A)} = \begin{pmatrix} 0.68936 & -0.27605 & 0.3989 \\ 0.26171 & -0.45121 & 0.90738 \\ -0.6755 & 0.84865 & 0.13242 \end{pmatrix}$$

$$\text{eigenvec (A, 0.938)} = \begin{pmatrix} -0.68936 \\ -0.26171 \\ 0.6755 \end{pmatrix}$$

$$\text{eigenvec (A, 3.024)} = \begin{pmatrix} -0.27605 \\ -0.45121 \\ 0.84865 \end{pmatrix}$$

LISTING 9.37. Accuracy Check for Eigenvalues and Eigenvectors (*Listing 9.36, Continued*)

$$\begin{pmatrix} 1 & 5 & 2 \\ 0.7 & 12 & 5 \\ 3 & 0 & 4 \end{pmatrix} \cdot \begin{pmatrix} 0.68936 \\ 0.26171 \\ -0.6755 \end{pmatrix} = \begin{pmatrix} 0.647 \\ 0.246 \\ -0.634 \end{pmatrix} \qquad 0.938 \cdot \begin{pmatrix} 0.68936 \\ 0.26171 \\ -0.6755 \end{pmatrix} = \begin{pmatrix} 0.647 \\ 0.245 \\ -0.634 \end{pmatrix}$$

$$A \cdot \text{eigenvecs (A)}^{\langle 0 \rangle} = \begin{pmatrix} 0.647 \\ 0.246 \\ -0.634 \end{pmatrix}$$

$$\text{eigenvals (A)}_0 \cdot \text{eigenvecs (A)}^{\langle 0 \rangle} = \begin{pmatrix} 0.647 \\ 0.246 \\ -0.634 \end{pmatrix}$$

In addition to needing to find eigenvalues and eigenvectors, sometimes one needs to solve a more general task, that is, by finding generalized eigenvalues $A \cdot x = \lambda \cdot B \cdot x$. The formulation of this task, besides the A matrix, contains another square matrix — B. For solving the task of finding generalized eigenvalues and eigenvectors, Mathcad also provides two built-in functions, similar to the two functions that we have just considered (Listings 9.38 and 9.39):

- genvals(A,B) — Calculates the vector of eigenvalues, each of which satisfies the generalized eigenvalues problem.
- genvecs(A,B) — Calculates the matrix composed from normalized eigenvectors corresponding to the eigenvalues in the v vector calculated using the genvals function. In this matrix, the i-th column is the x eigenvector satisfying the generalized eigenvalues problem.

Here:

- A,B — Square matrices

LISTING 9.38. Finding Generalized Eigenvalues and Eigenvectors

$$A := \begin{pmatrix} 1 & 5 & 2 \\ 0.7 & 12 & 5 \\ 3 & 0 & 4 \end{pmatrix} \qquad B := \begin{pmatrix} 1 & 2 & 3 \\ 4 & 5 & 6 \\ 7 & 8 & 9 \end{pmatrix}$$

$$\text{genvals}(A, 1B) = \begin{pmatrix} 0.5 \\ -0.674 \end{pmatrix}$$

$$\text{genvecs}(A, B) = \begin{pmatrix} -0.3067 & -0.70544 \\ 0.15571 & -0.23715 \\ -0.93898 & 0.66792 \end{pmatrix}$$

LISTING 9.39. Accuracy Check for Calculated Values of Generalized Eigenvectors and Eigenvalues *(Listing 9.39, Continued)*

$$A \cdot \text{genvecs}(A, B)^{\langle 0 \rangle} = \begin{pmatrix} -1.406 \\ -3.041 \\ -4.676 \end{pmatrix}$$

$$\text{genvals}(A, B)_0 \cdot B \cdot \text{genvecs}(A, B)^{\langle 0 \rangle} = \begin{pmatrix} -1.406 \\ -3.041 \\ -4.676 \end{pmatrix}$$

$$A \cdot \text{genvecs} (A, B)^{\langle 1 \rangle} = \begin{pmatrix} -0.555 \\ 0 \\ 0.555 \end{pmatrix}$$

$$\text{genvals} (A, B)_1 \cdot B \cdot \text{genvecs} (A, B)^{\langle 1 \rangle} = \begin{pmatrix} -0.555 \\ 0 \\ 0.555 \end{pmatrix}$$

9.5. MATRIX DECOMPOSITIONS

Modern numeric algorithms of linear algebra represent a scientific area that is constantly developing and growing at a rapid rate. The most important problem considered in this area is that of solving systems of linear equations. For the moment, there are lots of methods that simplify this task. These methods depend on the matrix of a system of linear algebraic equations. Most methods are based on the representation of the matrix as a product of other special-type matrices, and of matrix decompositions. Usually, after the matrix has been decomposed, the main task of linear algebra is significantly simplified. Mathcad provides a range of built-in functions for implementing the most popular matrix decomposition algorithms.

9.5.1. Cholesky Decomposition

Cholesky decomposition of the symmetric matrix A is a representation of the following form: A=L·L^T, where L is a triangular matrix (i.e., a matrix filled with zero values at one side of its diagonal). The Cholesky algorithm is implemented with the built-in cholesky function.

■ cholesky(A) — Cholesky decomposition

Here:

● A — Square, positively defined matrix

An example of Cholesky decomposition is presented in Listing 9.40. Notice that the function returns the lower triangular matrix (zero elements are above the diagonal), and the transposed matrix is the upper triangular matrix. The last line of this listing contains the accuracy check for the decomposition found.

LISTING 9.40. Cholesky Decomposition

$$A := \begin{pmatrix} 13 & 7 & 4 \\ 7 & 9 & -3 \\ 4 & -3 & 9 \end{pmatrix}$$

$L := cholesky\ (A)$

$$L = \begin{pmatrix} 3.606 & 0 & 0 \\ 1.941 & 2.287 & 0 \\ 1.109 & -2.253 & 1.64 \end{pmatrix}$$

$$L \cdot L^{T} = \begin{pmatrix} 13 & 7 & 4 \\ 7 & 9 & -3 \\ 4 & -3 & 9 \end{pmatrix}$$

9.5.2. QR-Decomposition

QR-decomposition of the A matrix is decomposition that can be represented as $A = Q \cdot R$, where Q is the orthogonal matrix and R is the upper triangular matrix.

■ qr (A) — Returns QR-decomposition

Here:

● A — Vector or matrix of any dimension

The result of applying the qr (A) function is the L matrix, which is composed from the Q and R matrices, respectively. To extract the matrices of the QR-decomposition, it is necessary to use the submatrix function (Listing 9.41).

LISTING 9.41. QR-Decomposition

$$A := \begin{pmatrix} 13 & 7 & 4 \\ 7 & 9 & -3 \\ 4 & -3 & 9 \end{pmatrix}$$

$L := qr\ (A)$

$$L = \begin{pmatrix} 0.85 & 0.122 & 0.513 & 15.297 & 9.283 & 4.38 \\ 0.458 & -0.654 & -0.603 & 0 & -7.268 & 9.171 \\ 0.261 & 0.747 & -0.612 & 0 & 0 & -1.646 \end{pmatrix}$$

$$Q := submatrix\left(L, 0, rows\ (L) - 1, 0, \frac{cols\ (L) - 1}{2} \right)$$

$$Q = \begin{pmatrix} 0.85 & 0.122 & 0.513 \\ 0.458 & -0.654 & -0.603 \\ 0.261 & 0.747 & -0.612 \end{pmatrix}$$

$$R := \text{submatrix}\left(L, 0, \text{rows}(L) - 1, \frac{\text{cols}(L) + 1}{2}, \text{cols}(L) - 1\right)$$

$$R = \begin{pmatrix} 15.297 & 9.283 & 4.38 \\ 0 & -7.268 & 9.171 \\ 0 & 0 & -1.646 \end{pmatrix}$$

$$Q \cdot R = \begin{pmatrix} 13 & 7 & 4 \\ 7 & 9 & -3 \\ 4 & -3 & 9 \end{pmatrix}$$

9.5.3. LU-Decomposition

LU-decomposition of the A matrix, or *triangular decomposition* is matrix decomposition that can be represented as follows: $P \cdot A = L \cdot U$, where L and U are lower and upper triangular matrices of the same order, respectively. P, A, L, and U are square matrices of the same size.

■ lu(A) — LU-decomposition of the matrix

Here:

● A — square matrix

Actually, you perform triangular decomposition of the system of linear equations when solving it using the Gauss method.

The LU-decomposition function, similar to the previous QR-decomposition function, returns the composite matrix B (Listing 9.42). Using the built-in submatrix function, it is not difficult to extract the P, L, and U matrices.

LISTING 9.42. LU-Decomposition

$$A := \begin{pmatrix} 13 & 7 & 4 \\ 7 & 9 & -3 \\ 4 & -3 & 9 \end{pmatrix}$$

$$B := \text{lu}(A)$$

$$B = \begin{pmatrix} 1 & 0 & 0 & 1 & 0 & 0 & 13 & 7 & 4 \\ 0 & 1 & 0 & 0.538 & 1 & 0 & 0 & 5.231 & -5.154 \\ 0 & 0 & 1 & 0.308 & -0.985 & 1 & 0 & 0 & 2.691 \end{pmatrix}$$

$$P := \mathrm{submatrix}\left(B, 0, \mathrm{rows}(B) - 1, 0, \frac{\mathrm{cols}(B) - 1}{3}\right)$$

$$P = \begin{pmatrix} 1 & 0 & 0 \\ 0 & 1 & 0 \\ 0 & 0 & 1 \end{pmatrix}$$

$$L := \mathrm{submatrix}\left(B, 0, \mathrm{rows}(B) - 1, \frac{\mathrm{cols}(B) + 1}{3}, \frac{\mathrm{cols}(B) - 1}{3} \cdot 2\right)$$

$$L = \begin{pmatrix} 1 & 0 & 0 \\ 0.538 & 1 & 0 \\ 0.308 & -0.985 & 1 \end{pmatrix}$$

$$U := \mathrm{submatrix}\left(B, 0, \mathrm{rows}(B) - 1, \frac{\mathrm{cols}(B) + 1}{3} \cdot 2, \mathrm{cols}(B) - 1\right)$$

$$U = \begin{pmatrix} 13 & 7 & 4 \\ 0 & 5.231 & -5.154 \\ 0 & 0 & 2.691 \end{pmatrix}$$

$$P \cdot A - L \cdot U = \begin{pmatrix} 0 & 0 & 0 \\ 0 & 0 & 0 \\ 0 & 0 & 0 \end{pmatrix}$$

9.5.4. Singular Value Decomposition

Singular value decomposition of the A matrix with an N×M dimension (where N ≥ M) is the decomposition of the following form: $A = U \cdot s \cdot V^T$, where U and V are orthogonal matrices of the size $N \times N$ and $M \times M$ respectively, and s is a diagonal matrix with singular numbers of the A matrix at its diagonal.

■ svds (A) — Vector composed of the singular numbers
■ svd (A) — Singular decomposition

Here:

● A — Real matrix

Respectively, Listings 9.43 and 9.44 present examples illustrating the calculation of the singular numbers of nonsingular and singular matrices. Listing 9.45 presents the confirmation of the accuracy of the singular value decomposition. The calculated singular numbers are placed at the main diagonal of the s matrix (its remaining elements, according to the definition, are equal to zero). By comparing the matrices from Listings 9.44 and 9.45, you'll easily understand how to extract the singular value decomposition matrices from the result returned by the svd function.

LISTING 9.43. Singular Numbers and Eigenvalues of a Nonsingular Matrix

$$A := \begin{pmatrix} 13 & 7 & 4 \\ 7 & 9 & -3 \\ 4 & -3 & 9 \end{pmatrix}$$

$$\text{eigenvals (A)} = \begin{pmatrix} 18.522 \\ 11.629 \\ 0.85 \end{pmatrix} \qquad \text{svds (A)} = \begin{pmatrix} 18.522 \\ 11.629 \\ 0.85 \end{pmatrix}$$

LISTING 9.44. Singular Value Decomposition of the Singular Matrix

$$A := \begin{pmatrix} 1 & 2 & 3 \\ 3 & 6 & 9 \\ 0 & 0 & 0 \end{pmatrix} \qquad \text{svds (A)} = \begin{pmatrix} 11.832 \\ 0 \\ 0 \end{pmatrix}$$

$$\text{svd (A)} = \begin{pmatrix} -0.316 & -0.949 & 0 \\ -0.949 & 0.316 & 0 \\ 0 & 0 & 1 \\ -0.267 & 0.964 & 0 \\ -0.535 & -0.148 & 0.832 \\ -0.802 & -0.222 & -0.555 \end{pmatrix}$$

LISTING 9.45. Accuracy Check of the Singular Value Decomposition
(Listing 9.44, Continued)

$$\begin{pmatrix} -0.316 & -0.949 & 0 \\ -0.949 & 0.316 & 0 \\ 0 & 0 & 1 \end{pmatrix} \cdot \begin{pmatrix} 11.832 & 0 & 0 \\ 0 & 0 & 0 \\ 0 & 0 & 0 \end{pmatrix} \cdot \begin{pmatrix} -0.267 & 0.964 & 0 \\ -0.535 & -0.148 & 0.832 \\ -0.802 & -0.222 & -0.555 \end{pmatrix}^T = \begin{pmatrix} 1 & 2 & 3 \\ 3 & 6 & 9 \\ 0 & 0 & 0 \end{pmatrix}$$

10 Special Functions

This chapter is dedicated to a discussion of the various mathematical built-in functions provided by Mathcad. We'll cover both special functions (Bessel, Airy, and so on), and elementary functions (sine, exponential, hyperbolic), and those functions designed for certain, more specialized areas (for example, financial functions). Furthermore, we'll cover some other functions, rather simple yet quite useful in terms of programmatic implementation, such as the step function, delta functions, and so on.

Although most of the functions that will be covered in this chapter are calculated without using any special-purpose numerical methods, we have joined them together in a single chapter. This has been done in order to simplify the task of finding the description of the required function, if necessary. The list of special functions is classified into sections based on the mathematical principle of the function being discussed and/or its area of usage.

The simplest way to insert a new special function into the document is by using the Insert Function dialog, which you can open by clicking the f(x) button on the standard toolbar (see the "First Acquaintance with Mathcad" section in Chapter 1). In this dialog the functions are classified into several groups, thus, it is not difficult

to open the required one. When you select a specific group from the left list, the list of functions belonging to this group appears in the right part of the window. Names of functions appearing in the left list of the Insert Function window are listed in parentheses after the name of each section of this chapter.

10.1. BESSEL FUNCTIONS (BESSEL)

By definition, Bessel functions are solutions of various boundary problems for some ordinary differential equations.

10.1.1. Ordinary Bessel Functions

Bessel functions of the first and second kind are normally generated as solutions of wave equations with cylindrical boundary conditions.

Specific differential equations can be easily found in special function reference literature or in the Mathcad on-line Help system.

Provided below is a list of built-in Bessel functions available in Mathcad:

- J0(x) — 0-th order Bessel function of the first kind
- J1(x) — First order Bessel function of the first kind
- Jn(m,x) — m-th order Bessel function of the first kind

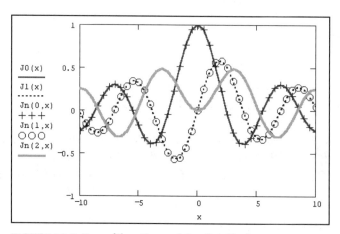

FIGURE 10.1. Bessel functions of the first kind

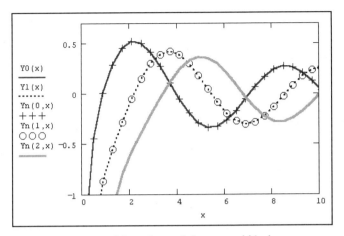

FIGURE 10.2. Bessel functions of the second kind

■   Y0(x) — 0-th order Bessel function of the second kind, x>0
■   Y1(x) — First order Bessel function of the second kind, x>0
■   Yn(m,x) — m-th order Bessel function of the second kind, x>0

Here:

●   x — Real dimensionless scalar value
●   m — Order, integer value where 0<m<100

Several graphs of the first and second kind of Bessel functions are shown in Figures 10.1 and 10.2 respectively.

10.1.2. Modified Bessel Functions

Modified Bessel functions are listed below:

■   I0(x) — Modified Bessel function of the first kind, 0-th order
■   I1(x) — Modified Bessel function of the first kind, first order
■   In(m,x) — Modified Bessel function of the first kind, m-th order
■   K0(x) — Modified Bessel function of the second kind, 0-th order, x>0
■   K1(x) — Modified Bessel function of the second kind, first order, x>0
■   Kn(m,x) — Modified Bessel function of the second kind, m-th order, x>0

Here:

●   x — Real dimensionless scalar value
●   m — Order, integer value where 0<m<100

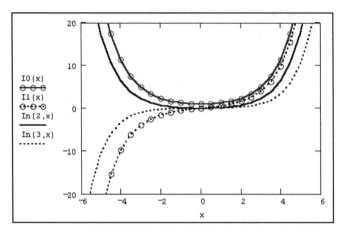

FIGURE 10.3. Modified Bessel function of the first kind

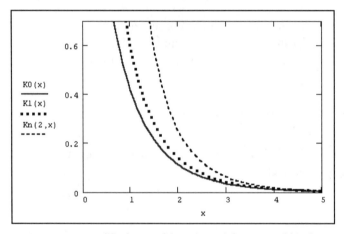

FIGURE 10.4. Modified Bessel function of the second kind

Several examples of the modified Bessel functions are shown in Figures 10.3 and 10.4.

10.1.3. Airy Functions

Airy functions are independent solutions of the following ordinary differential equation: y''=xy. Their graphs are shown in Figure 10.5, where the following are applied:

■ Ai(x) — Airy function of the first kind
■ Bi(x) — Airy function of the second kind

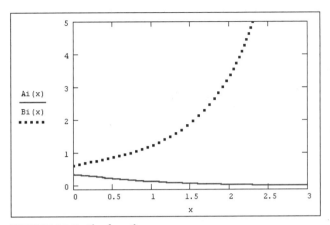

FIGURE 10.5. Airy functions

Here:

- x — Real dimensionless scalar value, x<103.892

10.1.4. Bessel Kelvin Functions

The complex combination of the Bessel Kelvin functions, such as ber(n,x) + i · bei(n,x), represents the solution of the respective standard differential equation depending on the n parameter. Graphs of the bei functions for n = 1 and 2 are shown in Figure 10.6.

- bei(n,x) — Imaginary part of Bessel Kelvin function of n-th order
- ber(n,x) — Real part of Bessel Kelvin function of n-th order

Here:

- n — Order (dimensionless positive integer number)
- x — Real dimensionless scalar value

10.1.5. Spherical Bessel Functions

Graphs of the spherical Bessel functions of the first order are shown in Figure 10.7.

- js(n,x) — Spherical Bessel function of the first kind, n-th order, x > 0
- ys(n,x) — Spherical Bessel function of the second kind, n-th order, x > 0

Here:

- n — Order (integer number), n ≥ 200
- x — Real dimensionless scalar value, x > 0

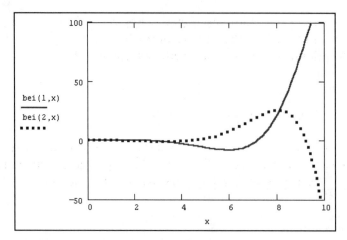

FIGURE 10.6. Bessel Kelvin function

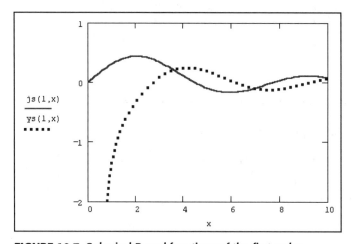

FIGURE 10.7. Spherical Bessel functions of the first order

10.2. FUNCTIONS FOR WORKING WITH COMPLEX NUMBERS

Mathcad provides several functions that simplify work with complex numbers:

- `Re(z)` — Real part of the complex number `z`
- `Im(z)` — Imaginary part of the complex number `z`
- `arg(z)` — The argument of the complex number `z`, $-\pi < \arg(z) \le \pi$

▪ csgn(z) — Sign of the complex number. Returns 0 if z = 0; 1 if Re(z) > 0 or Re(z) = 0 and Im(z)>0; otherwise, returns −1)
▪ signum(z) — Returns 1 if z = 0; otherwise, returns z/|z|

Here:

- z — Real, imaginary or complex number

The complex number can be entered as usual, in the form of the sum of real and imaginary parts, or as a result of any complex expression. Listings 10.1–10.3 provide several examples illustrating the usage of functions for complex numbers.

LISTING 10.1. Basic Functions for Working with Complex Numbers

$$\text{Re}(3.9 + 2.4i) = 3.9 \qquad \text{Im}(3.9 + 2.4i) = 2.4$$

$$\left|1.7 \cdot e^{0.1i}\right| = 1.7 \qquad \arg\left(1.7 \cdot e^{0.1i}\right) = 0.1$$

LISTING 10.2. Example of the Usage of the csgn Function

$$\text{csgn}(0) = 0 \qquad\qquad \text{csgn}(0 - i) = -1$$

$$\text{csgn}(i) = 1 \qquad\qquad \text{csgn}(0 + i) = 1$$

$$\text{csgn}(0.1) = 1 \qquad\qquad \text{csgn}(0.1 + 2i) = 1$$

$$\text{csgn}(-0.1) = -1 \qquad\qquad \text{csgn}(-0.1 - 3i) = -1$$

LISTING 10.3. Example of the Usage of the signum Function

$$\text{signum}(0) = 1 \qquad\qquad \text{signum}(0 - i) = -i$$

$$\text{signum}(i) = i \qquad\qquad \text{signum}(0 + i) = i$$

$$\text{signum}(0.1) = 1 \qquad\qquad \text{signum}(0.1 + 2i) = 0.05 + 0.999i$$

$$\text{signum}(-0.1) = -1 \qquad\qquad \text{signum}(-0.1 - 3i) = -0.033 - 0.999i$$

10.3. LOGARITHMS AND THE EXPONENTIAL FUNCTION

In this section, we will list, without any comments, some well-known logarithmic (Figure 10.8) and exponential functions:

- `exp(z)` — e (base of the natural logarithm) raised to the power of z
- `ln(z)` — The natural logarithm
- `log(z)` — The decimal logarithm
- `log(z,b)` — The base b logarithm of z

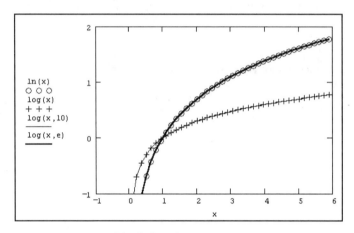

FIGURE 10.8. Logarithmic functions

10.4. TRIGONOMETRIC FUNCTIONS

The list of built-in trigonometric functions available in Mathcad is provided below:

- `acos(z)` — Arccosine
- `acot(z)` — Arccotangent
- `acsc(z)` — Arccosecant (Listing 10.4)
- `angle(x,y)` — Angle from the line containing the origin (0,0) and the (x,y) point to the OX axis
- `asec(z)` — Arcsecant
- `asin(z)` — Arcsine (Listing 10.4)
- `atan(z)` — Arctangent

- `atan2(x,y)` — Angle from the ox axis to the line containing the origin $(0,0)$ and the (x,y) point (Listing 10.5)
- `cos(z)` — Cosine
- `cot(z)` — Cotangent
- `csc(z)` — Cosecant (Listing 10.4)
- `sec(z)` — Secant
- `sin(z)` — Sine (Listing 10.4)
- `tan(z)` — Tangent

Here:

- `z` — Dimensionless scalar

An argument of trigonometric and inverse trigonometric functions is expressed in radians. When angle values are specified in degrees, first convert them to radians (Listing 10.6). Trigonometric functions can accept a complex argument.

LISTING 10.4. Examples of Trigonometric Functions

$$\sin(0.5) = 0.479$$

$$\frac{1}{\csc(0.5)} = 0.479$$

$$\text{asin}(0.479) = 0.5$$

$$\text{acsc}\left(\frac{1}{0.479}\right) = 0.5$$

LISTING 10.5. Examples of Calculations of the Angle between a Straight Line and the ox Axis

$$\text{atan2}(1,1) = 0.785 \qquad \text{atan2}(-1,-1) = -2.356$$

$$\text{angle}(1,1) = 0.785 \qquad \text{angle}(-1,-1) = 3.927$$

LISTING 10.6. Calculation of Trigonometric Functions in Degrees

$$z := 47$$

$$\cos\left(\frac{\pi \cdot z}{180}\right) = 0.682$$

$$\text{acos}(0.682) \cdot \frac{180}{\pi} = 47$$

10.5. HYPERBOLIC FUNCTIONS

Hyperbolic functions, according to their definitions, are expressed via various combinations of e^z and e^{-z} (an example is provided in Listing 10.7). Hyperbolic functions can accept complex arguments. Graphs of three main hyperbolic functions are shown in Figure 10.9.

- ▪ `acosh(z)` — Hyperbolic arccosine
- ▪ `acoth(z)` — Hyperbolic cotangent
- ▪ `asinh(z)` — Hyperbolic arcsine
- ▪ `acsch(z)` — Hyperbolic arccosecant
- ▪ `atanh(z)` — Inverse hyperbolic tangent
- ▪ `asech(z)` — Inverse hyperbolic secant
- ▪ `cosh(z)` — Hyperbolic cosine
- ▪ `coth(z)` — Hyperbolic cotangent
- ▪ `sinh(z)` — Hyperbolic sine
- ▪ `csch(z)` — Hyperbolic cosecant
- ▪ `tanh(z)` — Hyperbolic tangent
- ▪ `sech(z)` — Hyperbolic secant

Here:

- • `z` — Dimensionless scalar

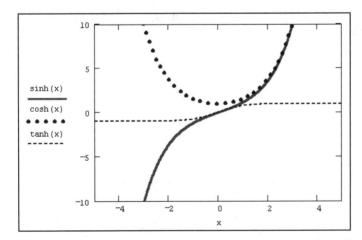

FIGURE 10.9. Main hyperbolic functions

LISTING 10.7. Examples of Hyperbolic Functions

$z := 1.27$

$$\frac{e^z + e^{-z}}{2} = 1.921$$

$\cosh(z) = 1.921$

$\mathrm{acosh}(1.921) = 1.27$

10.6. OTHER SPECIAL FUNCTIONS

Let us provide a list of some other special functions that can be calculated using Mathcad's built-in functions. Some of these are illustrated by Listing 10.8, while several polynomials are shown in the graphs in Figures 10.11–10.13.

- ■ $\mathrm{erf}(x)$ — Error function (*see Section 13.1.1, "Normal (Gauss) Distribution," in Chapter 13*)
- ■ $\mathrm{erfc}(x) \equiv 1 - \mathrm{erf}(x)$
- ■ $\mathrm{fhyper}(a, b, c, x)$ — Gauss hypergeometric function
- ■ $\mathrm{mhyper}(a, b, x)$ — Confluent hypergeometric function

Here:

- ● a, b, c — Parameters
- ● x — Real scalar value, $-1 < x < 1$

- ■ $\mathrm{Gamma}(z)$ — Euler's gamma function

Here:

- ● z — Scalar, $|z| < 1$

- ■ $\mathrm{Gamma}(a, x)$ — Incomplete gamma function of a order

Here:

- ● x — Real positive scalar value

Gamma functions in Mathcad documents are designated by the Greek letter Γ (Listing 10.8).

NOTE

■ Her(n,x) — Hermite polynomial of order n with argument x

Here:

- n — Order (non-negative integer)
- x — Scalar

■ ibeta(a,x,y) — Incomplete beta function of x and y with the parameter a

Here:

- a — Real scalar, $0 \le a \le 1$
- x,y — Real scalars, $x > 0, \ y > 0$

■ Jac(n,a,b,x) — Jacobi polynomial of n degree in x point with the parameters a and b

■ Lag(n,x) — Laguerre polynomial of degree n in point x (Figure 10.10)

■ Leg(n,x) — Legendre polynomial of degree n in point x (Figure 10.11)

Here:

- n — Order (non-negative integer)
- x — Real scalar
- a,b — Real scalars, $a > -1, \ b > -1$

■ Tcheb(n,x) — Chebyshev polynomial of the first kind of order n in point x (Figure 10.12)

■ Ucheb(n,x) — Chebyshev polynomial of the second kind of degree n in point x (Figure 10.12)

Here:

- n — Order (non-negative integer)
- x — Real scalar

LISTING 10.8. Examples of Calculations for Some Special Functions

fhyper(1,2,3,0.34) = 1.306

Γ(0.7i) = −0.29 − 0.961i

Γ(1.3,7.7) = 8.655×10^{-4}

Jac(1,2,1,−0.13) = 0.175

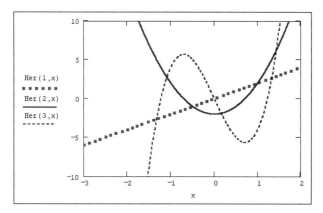

FIGURE 10.10. Hermite polynomials

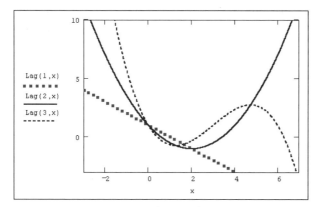

FIGURE 10.11. Laguerre polynomials

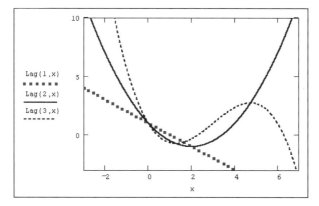

FIGURE 10.12. Legendre polynomials

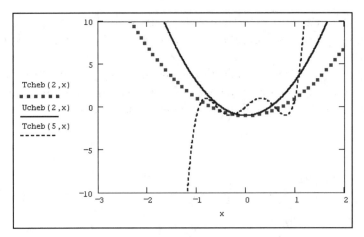

FIGURE 10.13. Chebyshev polynomials

10.7. STRING FUNCTIONS

Let us provide a brief list of functions that you can use in operations with string variables in a way similar to the way you'd use numeric variables:

- `concat(S1,S2,...)` — Returns a string variable obtained by concatenating string variables or constants S1, S2, ... (Listing 10.9)
- `error(S)` — Returns string S as an error message (Figure 10.14)
- `IsString(x)` — Returns 1 if x is a string variable; otherwise, returns 0 (Listing 10.10)
- `num2str(z)` — Returns the string obtained from converting the decimal value of z into the string (Listing 10.10)

This function is used when you need to manipulate a number as a string rather than as a math variable.

- `search(S,Subs,m)` — Returns the starting position of the Subs substring within the S string, starts the search from position m, and in case of failure, returns −1 (Listing 10.9)
- `str2num(S)` — Converts a string representation of the number S (in any form) to number (Listing 10.10)

- `str2vec(S)` — Converts the string s to a vector of ASCII codes (Listing 10.10)
- `strlen(S)` — Returns a number of characters in the string s (Listings 10.9 and 10.10)
- `substr(S,m,n)` — Returns a substring obtained from the s string by extracting n characters starting from the position m within the initial string s (Listing 10.9)
- `vec2str(v)` — Returns a string representation of the elements of the vector v of ASCII codes

Here:

- s — String
- v — Vector of ASCII codes (integer numbers, $0 < v < 255$)

LISTING 10.9. Examples of String Functions Usage

```
concat ("Hello," , " " , "World" , "!" ) = "Hello, World!"

substr ("Hello, World!" , 4 , 8) = "o, World"

substr ("Hello, World!" , 0 , 5) = "Hello"

search ("Hello, World!" , "Wo" , 1) = 7

search ("Hello, World!" , "wo" , 1) = −1

strlen ("hello" ) = 5
```

LISTING 10.10. Mutual Conversions of Strings and Numbers

```
IsString (1) = 0

IsString ("!" ) = 1

strlen ("Hello, World!" ) = 13

num2str (579 + 3i) = "579 + 3i"

num2str (12.345) = "12.345"

str2num ("123.4567" ) = 123.457
```

$$\text{str2vec ("17")} = \begin{pmatrix} 49 \\ 55 \end{pmatrix}$$

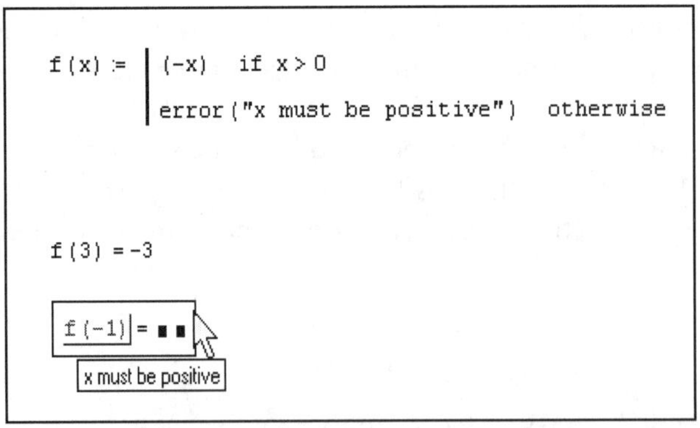

```
f(x) :=  │ (-x)   if x > 0
         │ error("x must be positive")   otherwise


f(3) = -3


 f(-1) = ■ ■
 ┌──────────────────┐
 │ x must be positive │
 └──────────────────┘
```

FIGURE 10.14. Example of creating custom error message

10.8. TRUNCATION AND ROUND-OFF FUNCTIONS

The list of built-in truncation and round-off functions available in Mathcad is provided below:

■ ceil(x) — Returns the smallest integer no less than x (Listing 10.11)
■ floor(x) — Returns the greatest integer less than or equal to x (Listing 10.11)
■ round(x,n) — When n > 0, returns the x value rounded with precision to n decimal positions, when n < 0, returns x rounded to n decimal positions to the left of the decimal point, when n = 0, returns x rounded to the nearest integer (Listing 10.12)
■ trunc(x) — Returns the integer part of a number by removing its fractional part (Listing 10.11)

Here:

● x — Real scalar or integer

LISTING 10.11. Truncation and Round-off Functions

```
ceil(3.7) = 4      floor(3.7) = 3      trunc(3.7) = 3

ceil(-3.7) = -3    floor(-3.7) = -4    trunc(-3.7) = -3
```

LISTING 10.12. Rounding Numbers

```
round(1.23456789, 0) = 1       round(12.3456789, 0) = 12

round(12.3456789, 1) = 12.3    round(12.3456789, -1) = 10
```

round (12.3456789 , 2) = 12.35   round (12.3456789 , −2) = 0

round (12.3456789 , 5) = 12.34568

 *When performing the rounding operation, don't forget the principles of number representation in Mathcad. To display the required number of digits after the decimal point, use the **Result Format** dialog (see Chapter 4).*

10.9. PIECEWISE CONTINUOUS FUNCTIONS

The list of piecewise continuous functions available in Mathcad is provided below:

- `heaviside step(x)` — The Heaviside function returns 1 if $x \geq 0$; otherwise, it returns 0 (Figure 10.15)

Here:

- x — Real scalar

- `if(cond,x,y)` — Returns x if the logical condition cond is true (non-zero); otherwise, returns y (Listing 10.13)
- `Kronecker delta(x,y)` — The Kronecker delta function returns 1 if $x = y$; otherwise, it returns 0 (Figure 10.15)
- `sign(x)` — Returns 0 if $x = 0$ and 1 if $x > 0$; otherwise, returns −1 (Listing 10.13)

Here:

- x — Real number

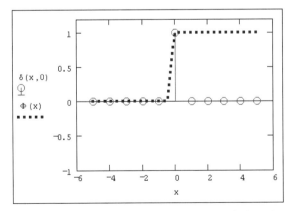

FIGURE 10.15. The Kronecker and Heaviside functions

In Mathcad documents, the Kronecker function is designated by the Greek letter δ, and the Heaviside function by the letter Φ.

LISTING 10.13. The Sign and Condition Functions

```
sign (-4) = -1

sign (1.3) = 1

if (1 > 3 , 1 , 3) = 3

if (1 > 3 , "Yes" , "No") = "No"

if (2 + 5 ≥ 0 , "Yes" , "No") = "Yes"
```

10.10. MAPPING FUNCTIONS (VECTOR AND MATRIX)

Mathcad 2001i introduces a family of new functions that enable you to perform coordinate mapping, both for 2D and 3D coordinate systems:

- xy2pol(x,y) — Converts rectangular coordinates of a point in 2D space to polar coordinates
- pol2xy(r,θ) — Converts polar coordinates of a point in 2D space to rectangular coordinates
- angle(x,y) — Returns an angle from the ox axis to the point (x,y) *(see Section 10.4)*
- atan2(x,y) — Returns an angle from the positive ox axis to the line containing the origin (0,0) and the point (x,y) *(see Section 10.4)*
- xyz2cyl(x,y,z) — Converts rectangular coordinates of a point in 3D space to cylindrical coordinates
- cyl2xyz(r,θ,z) — Converts cylindrical coordinates of a point in 3D space to rectangular coordinates
- xyz2sph(x,y,z) — Converts rectangular coordinates of a point in 3D space to spherical coordinates
- sph2xyz(r,θ,ϕ) — Converts spherical coordinates of a point in 3D space to rectangular coordinates

Here:

- x,y — Rectangular coordinates of a point in 2D space
- x,y,z — Rectangular coordinates of a point in 3D space
- r,θ — Polar coordinates of a point in 2D space
- r,θ,z — Cylindrical coordinates
- r,θ,ϕ — Spherical coordinates

Several examples of performing coordinate mapping are provided in Listings 10.14 and 10.15. Notice that you can enter function arguments both as a list and in the form of a vector.

LISTING 10.14. Coordinate Mapping in 2D Space

$$\text{xy2pol}\begin{pmatrix}1\\7\end{pmatrix}=\begin{pmatrix}7.071\\1.429\end{pmatrix}\qquad \text{xy2pol}\,(1,7)=\begin{pmatrix}7.071\\1.429\end{pmatrix}$$

$$\text{pol2xy}\begin{pmatrix}7.071\\1.429\end{pmatrix}=\begin{pmatrix}0.999\\7\end{pmatrix}\qquad \text{pol2xy}\,(7.071,1.429)=\begin{pmatrix}0.999\\7\end{pmatrix}$$

LISTING 10.15. Coordinate Mapping in 3D Space

$$\text{xyz2cyl}\,(1,1,1)=\begin{pmatrix}1.414\\0.785\\1\end{pmatrix}\qquad \text{cyl2xyz}\begin{pmatrix}1,\pi,3.93\end{pmatrix}=\begin{pmatrix}-1\\0\\3.93\end{pmatrix}$$

$$\text{xyz2sph}\,(1,1,1)=\begin{pmatrix}1.732\\0.785\\0.955\end{pmatrix}\qquad \text{sph2xyz}\begin{pmatrix}\sqrt{2},\dfrac{\pi}{4},\dfrac{\pi}{2}\end{pmatrix}=\begin{pmatrix}1\\1\\0\end{pmatrix}$$

10.11. FINANCIAL FUNCTIONS

Starting with version 2000, Mathcad provides functions intended to simplify financial analysis. In this section, we will provide a brief list of these functions. Detailed descriptions of these functions along with examples of their practical usage can be found in the Mathcad on-line Help system.

■ cnper(rate,pv,fv) — Calculates the number of compound periods required to yield the future value of investment based on the specified initial value and rate of interest

Here:

- rate — Fixed interest rate, must be real scalar, rate > -1
- pv — Current value of the investment (present value, pv > 0)
- fv — Future value of investment, fv > 0

■ crate(nper,pv,fv) — Calculates the fixed interest rate per period required for the current value of investment to yield the specified future value based on the specified number of compound periods

Here:

- nper — Number of compound periods, must be an integer, nper≥1
- pv — Present value of investment, pv > 0
- fv — Future value of investment, fv > 0

■ cumint(rate,nper,pv,start,end,[type]) — Calculates the cumulative interest paid on a loan between initial and final periods based on the fixed rate of interest, the specified number of compound periods, and the present value of the loan

■ cumprn(rate,nper,pv,start,end,[type]) — Calculates the cumulative sum to be paid (principal) between the starting and ending periods based on the fixed rate of interest, the specified number of compound periods, and the current value of the loan

Here:

- rate — Fixed rate of interest, must be real scalar, rate ≥ 0
- nper — Total number of compound periods, must be a positive integer
- pv — Current (present) value of the loan, pv > 0
- start — Starting period of accumulation, must be a positive integer
- end — Ending period of accumulation, must be a positive integer, start≤ end
- type = 0, if the payment was made at the end of the period, or type = 1, if the payment takes place at the beginning of the period

■ eff(rate,nper) — Calculates the effective annual interest rate based on the specified nominal rate per period and the number of compound periods per year

Here:

- rate — Nominal interest rate, must be real scalar
- nper — Total number of compound periods per year, nper > 0

- `fv(rate,nper,pmt,[[pv],[type]])` — Calculates the future value of the investment or loan based on the specified number of compound periods given constant payment and a fixed rate of interest

Here:

- `rate` — Fixed rate of interest per period, must be real scalar, `rate > 0`
- `nper` — Total number of compound periods per year, `nper > 0`
- `pv` — Present value of investment or loan
- `type=0` for payment made at the end of the period or `1` for the payment made at the start of the period

- `fvadj(prin,v)` — Calculates future value of the initial principal, after applying a series of compound interest rates

Here:

- `prin` — The initial principal
- `v` — Vector of interest rates, each of which is applied to the same initial principal over the same amount of time

- `fvc(rate,v)` — Calculates the future value of the series of cash flow taking place at regular intervals and earning the specified interest rate

Here:

- `rate` — Fixed interest rate per period, must be real scalar
- `v` — Vector of cash flows

- `ipmt(rate,per,nper,pv,[[fv],type])` — Calculates the interest payment on the investment or loan for a specified period, based on constant, periodic payments over a given number of compound periods using a fixed interest rate and a specified future value

Here:

- `rate` — Fixed interest rate per period, $rate \geq 0$
- `per` — Period for which it is necessary to calculate the interest, must be a positive integer
- `nper` — Total number of compound periods, $per \leq nper$
- `pv` — Present value
- `fv` — Future value
- `type = 0` — For payment made at the end of the period; or `type = 1` — For the payment made at the start of the period

- ■ irr(v,[guess]) — Calculates the internal rate of return for a series of cash flows that occur at regular intervals

Here:

- ● v — Vector of cash flows occurring at regular intervals, must contain at least one positive and one negative value
- ● guess — Numerical value that you guess to approximate the answer. If omitted, guess = 0.1(10%)

- ■ mirr(v,fin_rate,rein_rate) — Calculates the modified internal rate of return for a series of cash flows occurring at regular intervals, provided that the financial rate is payable according to the loaned sum, and the reinvestment rate earns the income from the reinvested sum

Here:

- ● v — Vector of cash flow occurring at regular intervals, must contain at least one positive and one negative value
- ● fin_rate — Financial rate payable on the borrowed cash flow
- ● rein_rate — Reinvestment rate

- ■ nom(rate,nper) — Calculates the nominal annual interest rate, including the effective annual rate of interest and the number of compound periods per year

Here:

- ● rate — Effective annual interest rate, must be real scalar, rate > -1
- ● nper — Total number of compound periods per year, nper > 0

- ■ npv(rate,v) — Calculates the net present value of investment with an account at a discount rate and with a regular cash flow

Here:

- ● rate — Fixed rate of interest at which investment earns interest per period, must be real scalar
- ● v — Vector of regular cash flows

- ■ nper(rate,pmt,pv,[[fv],[type]]) — Calculates the number of periods for an investment or loan based on constant, periodic payments using the fixed interest rate and the specified present value
- ■ pmt(rate,nper,pv,[[fv],[type]]) — Calculates payment per period for investment or loan based on constant, periodic payments over a specified number of compound periods using the fixed rate of interest and the specified present value

- `ppmt(rate,per,nper,pv,[[fv],[type]])` — Calculates the payment on the principal of the investment or loan for a given period based on constant, periodic payments over a given number of compounding periods using a fixed interest rate and a specified future value
- `pv(rate,nper,pmt,[[fv],[type]])` — Calculates the present value of an investment or loan based on constant, periodic payments over a specified number of compound periods using a fixed interest rate and a specified payment
- `rate(nper,pmt,pv,[[fv],[type],[guess]])` — Calculates the interest rate for a period of investment or loan based on a specified number of compound periods, constant, periodic payment, and a specified present value

Here:

- `rate` — Fixed interest rate
- `per` — Period
- `nper` — Total number of compound periods per year, must be a positive integer
- `pmt` — Payment made each period
- `pv` — Present value of investment
- `fv` — Future value of the investment
- `type = 0` — For payment made at the end of the period; or `type = 1` — For the payment made at the start of the period
- `guess` — Numerical value that you guess to approximate the answer. If omitted, `guess = 0.1(10%)`

11 Ordinary Differential Equations

Differential equations are equations in which unknowns are functions of one or more variables rather than variables themselves (i.e., numbers). Such equations (or systems) include relationships between unknown functions and their derivatives. If equations include derivatives by only one variable, they are called ordinary differential equations; otherwise we are dealing with partial differential equations *(see Chapter 12)*. Thus, solving (or integrating) a differential equation involves finding an unknown function within the specified interval of values taken by its variables.

It is a well-known fact that one ordinary differential equation *(see Sections 11.1 and 11.2)* or system of ordinary differential equations *(see Section 11.3)* has a single solution provided that besides the equation itself, we have appropriately specified the initial or boundary conditions. In higher mathematics, the existence and uniqueness of solution theorems have been proven in accordance with specific conditions. Mathcad 2001i provides the tools to solve the following two types of tasks:

- *Cauchy problems* — These are problems for which initial conditions for unknown functions are specified, i.e., these functions' values are set in the initial point of the equation's integration interval.

■ *Boundary value problems* — These are problems for which specific conditions on both boundaries of the interval are specified (these problems will be *considered in Chapter 12*).

As a rule, solving Cauchy problems for ordinary differential equations and systems of such equations is thoroughly studied, and in terms of calculation, this task doesn't present serious difficulties. The most important aspects of this task are representation of the results and analysis of dependence between the solution and various parameters of the system *(see Section 11.4)*. Still, there is an entire class of ordinary differential equations, known as *stiff*, which can't be solved using standard methods, such as the Runge-Kutta method. Mathcad provides special capabilities for solving such equations *(see Section 11.5)*.

11.1. ORDINARY DIFFERENTIAL EQUATIONS OF THE FIRST ORDER

By definition, a differential equation of the first order can contain only its first derivative, `y'(t)`, in addition to the unknown function `y(t)`. In most cases, a differential equation can be written in the *standard form (Cauchy form)*:

$$y'(t) = f(y(t), t) \tag{1}$$

The Mathcad numeric processor can only work with this form. Mathematically speaking, accurate formulation of an appropriate Cauchy problem for an ordinary differential equation of the first order must, besides the equation itself, contain one initial condition — the value of the function `y(t₀)` at the specific point t_0. Given this, it is necessary to explicitly define the `y(t)` function within the interval from t_0 to t_1. In terms of formulation type, Cauchy problems are also known as initial value problems, as opposed to boundary value problems.

For numeric integration of one ordinary differential equation, Mathcad 2001i (to be more precise, starting from Mathcad 2000 Pro) provides the following alternatives: you can either use the `Given/Odesolve` solve block, or you can rely on the built-in functions, just like in all earlier versions of Mathcad. The first method is preferable, since it represents the problem and its solution more illustratively. On the other hand, the second method provides the user with more capabilities to influence the parameters of the numeric method. Let us consider both solution methods sequentially.

11.1.1. The *Given/Odesolve* Solve Block

The solve block for solving one ordinary differential equation when implementing the Runge-Kutta numeric method comprises three parts:

- Given — The keyword
- An ordinary differential equation and an initial condition written using logical operators, where the initial condition must be written in the following form: y(t0) = b·
- Odesolve(t,t1) — The built-in function for solving an ordinary differential equation using the t variable within the (t0,t1) interval

Specifying the Odesolve(t,t1,step) *function with three parameters is allowed, and sometimes it is even preferable. Here,* step *is the internal parameter of the numerical method, specifying the number of steps used by the Runge-Kutta method in the process of calculating the solution of a differential equation. The higher the* step *value, the more precise the final result will be, but, on the other hand, the longer it will take to calculate the solution. Remember that by correctly selecting this value, you can significantly speed up calculation (by several times) without noticeably impeding precision.*

An example illustrating the solution of a Cauchy problem for an ordinary differential equation of the first order $y' = y - y^2$ using the solve block is provided in Listing 11.1.

LISTING 11.1. Solution of a Cauchy Problem for an Ordinary Differential Equation of the First Order

```
Given

 d
 ──y(t) = y(t) − y(t)²
 dt

y(0) = 0.1

y := Odesolve (t, 10)
```

Don't forget that logical operators must be inserted using the buttons available on the **Boolean** toolbar. If you prefer keyboard input, remember that a logical equal sign must be entered using the <Ctrl>+<=> keyboard combination. The derivative symbol can be inserted both using the **Calculus** toolbar, as in Listing 11.1, and as an accent (prime) character by pressing the <Ctrl>+<F7> keys (an appropriate example will be presented in Listing 11.3). Select a method of representing the derivative based on the reasons for a more illustrative representation of the results. Naturally, the selected representation of the derivative won't influence the calculation results.

Mathcad requires that the final point of integration of an ordinary differential equation be positioned to the right of the initial point: t0 < t1 (in Listing 11.1, t0 = 0, t1 = 10); otherwise, the error message will be displayed. As you can see, the application of the Given/Odesolve solve block results in the y(t) function defined within the interval (t0,t1). To build its graph or calculate the function value in a specific point, use standard Mathcad tools, for example: y(3) = 0.691.

The user can choose between two modifications of the numeric Runge-Kutta method. To change the method, right-click anywhere within the area of the Odesolve function, and select one of the following two methods from the right-click menu: **Fixed** or **Adaptive**. By default, the fixed step Runge-Kutta method is used. More detailed information on the differences between these two methods will be provided in Section 11.3.

11.1.2. The *rkfixed, Rkadapt, Bulstoer* Built-in Functions

An alternative method of solving ordinary differential equations has been passed down from earlier versions of Mathcad. Based on the use of one of the following built-in functions: rkfixed, Rkadapt, or Bulstoer, this method is both less illustrative and more difficult to use. This is the main reason we advise you to use the Given/Odesolve solve block whenever possible. However, sometimes it is necessary to use the second method for solving ordinary differential equations of the first order. Generally, this is because of the following reasons:

- You are working with earlier versions of Mathcad software simultaneously, for example, Mathcad 5.0–8.0 or a less powerful version — Mathcad 2001i Standard — and your documents must be correctly interpreted in all these versions.
- One ordinary differential equation is being solved in the context of more complicated tasks, including systems of differential equations (for which the solve block is not applicable). In this case you might need to use a unified programming style.
- When it is preferable to obtain the answer in vector representation rather than as a function.

■ You are accustomed to writing ordinary differential equations using notation passed down from earlier versions of Mathcad, or you have lots of documents which were created with earlier Mathcad versions, and so on.

Since the solution of one ordinary differential equation using the second method is hardly or not at all different from a solution of systems of ordinary differential equations *(see Section 11.3)*, here we will only provide an example illustrating its usage for the task in Listing 11.1 (see Listing 11.2). In this example, a Cauchy problem is solved with the help of the rkfixed built-in function. Notice that it is necessary to explicitly specify the number of integration points (notice the M = 100 statement on the third line of this listing). Also, pay special attention to the fact that in contrast to the solve block, this built-in function provides the result as a matrix of the dimension M×2, rather than as a function. The matrix comprises two columns, the first of which contains the values of the t argument (from t0 to t1 inclusively), while the second contains the respective values of the unknown function y(t).

LISTING 11.2. Solution of a Cauchy Problem for an Ordinary Differential Equation of the First Order Using the Built-in Function

$$y := 0.1$$

$$D(t, y) := y - y^2$$

$$M := 100$$

$$y := rkfixed(y, 0, 10, M, D)$$

Mathcad's programming style, as presented in Listing 11.2, can't be considered to be a good one. First, the y variable is assigned the scalar value $y = 0.1$, then the same variable is assigned a matrix value (the result of solving an ordinary differential equation). Always try to avoid any style that impedes the readability of your program and in some cases might result in errors caused by barely recognizable reasons. It would be much better if you assigned another name for the result, for example, u.

TIP

Figure 11.1 shows a graph of the obtained solution of the equation under consideration. Notice that this graph corresponds to the solution obtained in the matrix form (Listing 11.2), therefore the columns extracted from the y matrix using the<> operator are plotted along the axes.

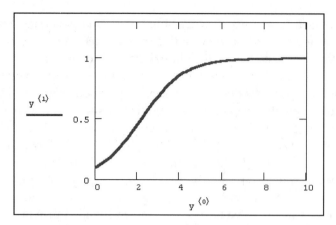

FIGURE 11.1. Solution of the $y' = y - y^2$ equation (Listing 11.2)

The example solved in Listings 11.1 and 11.2 relates to the field of mathematical ecology and describes population dynamics with intraspecific competition. At first, the population grows exponentially, then it stabilizes at a stationary rate.

11.2. ORDINARY DIFFERENTIAL EQUATIONS OF A HIGHER ORDER

An ordinary differential equation with the unknown function $y(t)$, including derivatives of this function up to $y^{(N)}(t)$, is known as an ordinary differential equation of the N-th order. Having such an equation, it is imperative that you specify N initial conditions for the $y(t)$ function itself, and its derivatives up to the (N - 1)-th order inclusively, in order to produce a correct formulation of a Cauchy problem. Mathcad 2001i is capable of solving ordinary differential equations of higher orders both using the Given/Odesolve solve block and by means of reducing them to the systems of differential equations of the first order.

The following requirements must be satisfied within the solve block:

- The ordinary differential equation must be linear in relation to the highest-order derivative, i.e., it must be represented in canonic form.
- Initial conditions must be formulated as $y(t) = b$ or $y^{(N)}(t) = b$, rather than by using a more complicated form (for example, for some math applications the $y(t) + y'(t) = b$ form is common).

In all other respects, there are no significant differences between solving ordinary differential equations of higher orders and solving ordinary differential equations of the first order *(see Section 11.1)*, a fact that is illustrated by List-ing 11.3. As you will recall, the derivative can be represented both by using a dif-ferential symbol (the equation itself is represented this way in Listing 11.3), and as a prime character (such as isshown in the initial condition for the first derivative). Don't forget to use Boolean operators when entering the equations and initial conditions. The resulting solution y(t) is shown in Figure 11.2.

LISTING 11.3. Solution of a Cauchy Task for an Ordinary Differential Equation of the Second Order

```
Given
```

$$\frac{d^2}{dt^2} y(t) + 0.1 \cdot \frac{d}{dt} y(t) + 1 \cdot y(t) = 0$$

```
y(0) = 0.1

y'(0) = 0

y := Odesolve (t, 50)
```

Listing 11.3 illustrates the procedure involved in solving the equation of damped harmonic oscillations, describing, for example, pendulum oscillations. For the pendulum model, y(t) describes the angular deviation from the vertical, y'(t) is the angular velocity of the pendulum, and y''(t) is acceleration. The initial de-viation of the pendulum is specified as y(0) = 0.1, and initial velocity as y'(0) = 0.

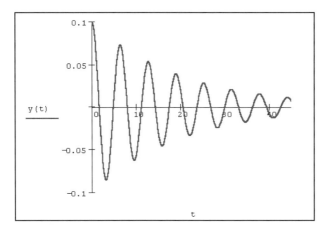

FIGURE 11.2. Solution of the oscillator equation (Listing 11.3)

The second method of solving ordinary differential equations of a higher order entails reducing it to the equivalent system of ordinary differential equations of the first order. Let us demonstrate this procedure outlined in Listing 11.3. If we formally designate $y_0(t) \equiv y(t)$, and $y_1(t) \equiv y'(t) = y_0'(t)$, then the initial equation can be re-written via $y_0(t)$ and $y_1(t)$ functions as follows:

$$\begin{cases} y_0' = y_1, \\ y_1' + 0.1 \cdot y_1 + 1 \cdot y_0 = 0. \end{cases} \tag{1}$$

In *Section 11.3* we are going to solve this system as an example. Thus, any ordinary differential equation of the N-th order, linear in relation to the highest-order derivative, can be reduced to a system of an equivalent system of N differential equations.

11.3. SYSTEMS OF DIFFERENTIAL EQUATIONS OF THE FIRST ORDER

Mathcad requires that a system of differential equations be represented in the standard form:

$$y_0'(t) = f_0(y_0(t), y_1(t), \ldots, y_{N-1}(t), t),$$
$$y_1'(t) = f_1(y_0(t), y_1(t), \ldots, y_{N-1}(t), t),$$
$$\ldots \tag{1}$$
$$y_{N-1}'(t) = f_{N-1}(y_0(t), y_1(t), \ldots, y_{N-1}(t), t)$$

The system specified in (1) is equivalent to the following vector representation:

$$Y'(t) = F(Y(t), t) \tag{2}$$

Here Y and Y' are the vector function and its derivative in variable t (dimension is $N \times 1$), while F is a vector function of the same size in $(N + 1)$ number of variables (N vector components and, possibly, t). Vector representation (2) is used for entering a system of ordinary differential equations in Mathcad.

To define a Cauchy problem for a system of ordinary differential equations, it is necessary to specify N initial conditions specifying the value of each of the $y_i(t0)$ functions in the initial point of the integration interval of the system $t0$. In vector form, these initial conditions can be written as follows:

$$Y(t0) = B \tag{3}$$

Here B is the initial condition vector of dimension $N \times 1$, composed from the values $y_i(t0)$.

As can be noticed, this task is formulated for systems of ordinary differential equations of the first order. However, if the system includes equations of a higher order, it can be reduced to a system of a larger number of the first order equations in a way similar to the one demonstrated in the example of the oscillator equation (see Section 11.2).

Notice that it is necessary to use the vector form of representation for both the equation itself and its initial condition(s). If we are dealing with one ordinary differential equation, respective vectors comprise only one element; while if we are dealing with the system of $N > 1$ equations these vectors comprise N elements.

11.3.1. Built-in Functions for Solving Systems of Ordinary Differential Equations

In Mathcad 2001i there are three built-in functions that allow you to solve the Cauchy problem formulated as (2-2) equations using various numerical methods.

- `rkfixed(y0,t0,t1,M,D)` — The Runge-Kutta method with a fixed step
- `Rkadapt(y0,t0,t1,M,D)` — The Runge-Kutta method with a variable step
- `Bulstoer(y0,t0,t1,M,D)` — The Bulirsch-Stoer method

Here:

- `y0` — The vector of initial conditions at `t0` with the dimension of $N \times 1$.
- `t0` — The starting point of the calculation.
- `t1` — The ending point of the calculation.
- `M` — The number of steps at which the numerical method approximates the solution.
- `D` — The $N \times 1$ vector function of two arguments: scalar `t` and vector `y`. Here `y` is $N \times 1$ vector function of the `t` argument, which needs to be found.

The names of the above considered functions are sensitive to the case of the first character of the name, which influences the selection of the calculation algorithm, in contrast to other built-in Mathcad functions — for example `Find` $\equiv$ `find` (see Section 11.3.2).

Each of the functions provided above returns the solution in the form of the $(M + 1) \times (N + 1)$ matrix. The left column of this matrix contains the values of the `t` argument dividing the integration interval into equal steps, while other N columns contain the values of unknown functions $y_0(t), y_1(t), \ldots, y_{N-1}(t)$, calculated for these values of the `t` argument. Since there are M points (without considering the initial point), the solution matrix will have $M + 1$ rows.

For most cases, it is sufficient to use the first function, `rkfixed`, as shown in Listing 11.4, illustrating the solution of a system of ordinary differential equations of the damped oscillator model *(see Section 11.2)*.

LISTING 11.4. Solution of a System of Two Ordinary Differential Equations

$$
D(t,y) := \begin{pmatrix} y_1 \\ -y_0 - 0.1 \cdot y_1 \end{pmatrix}
$$

$$
y0 := \begin{pmatrix} 0.1 \\ 0 \end{pmatrix}
$$

$$
M := 100
$$

$$
u := rkfixed(y0, 0, 50, M, D)
$$

The first line of this listing is the most important one, since it defines a system of ordinary differential equations that needs to be solved. Compare the system under consideration *(see Section 11.2)* written in the standard form with its formal representation in Mathcad. This will help you avoid errors. First, the D function, included as a parameter of the built-in differential equation solvers, must necessarily be the function of two arguments. Furthermore, its second argument must be a vector of the same size as the function D itself. Finally, the initial value vector $y0$ (it is defined by the second line of the listing) must have the same dimension.

Don't forget that the vector function $D(t,y)$ must be defined via the components of vector y using the **Subscript** button of the **Calculator** toolbar or by pressing the <[> key. The third line of the listing specifies the number of steps at which the solution will be approximated, while the last line of this listing assigns the result returned by the `rkfixed` function to the matrix variable u. The solution of a system of ordinary differential equations will be obtained for the $(0,50)$ interval.

Figure 11.3 shows how the whole solution might look. The dimension of the resulting matrix is equal to $(M+1) \times (N+1)$, i.e., 101×3. To view all components of the u matrix, use the vertical scroll bar. As shown in this illustration, the value calculated for the first component of y_0 at the 12-M step of calculation $(u_{12,1} = 0.07)$ is highlighted. In terms of mathematics, this corresponds to the calculate value $y_0(6.0) = 0.07$. To display the solution elements at the last point of the interval, use expressions such as $u_{M,1} = 7.523 \times 10^{-3}$.

	0	1	2
0	0	0.1	0
1	0.5	0.088	-0.047
2	1	0.056	-0.08
3	1.5	0.011	-0.093
4	2	-0.033	-0.082
5	2.5	-0.068	-0.053
6	3	-0.084	-0.013
7	3.5	-0.08	0.029
8	4	-0.057	0.062
9	4.5	-0.021	0.078
10	5	0.018	0.075
11	5.5	0.051	0.054
12	6	0.07	0.021
13	6.5	0.071	0.015
14	7	0.056	-0.046
15	7.5	0.028	-0.064

u =

FIGURE 11.3. Matrix of solutions of the system of ordinary differential equations (Listing 11.4)

Notice the difference in the index designation for the initial condition vector and solution matrix. The first column of the solution matrix contains the values of the 0-th component of the vector being calculated, the second column contains the values of the first component, and so on.

To create a graph of the solution, it is necessary to plot the appropriate components of the solution matrix along the coordinate axes: the values of the argument $u^{<0>}$ along the x axis, and the values of $u^{<1>}$ and $u^{<2>}$ along the y-axis (Figure 11.4). As a matter of fact, there is an alternative method of representing solutions of ordinary differential equations, which is often more convenient than the one described above. Using this method, solutions are plotted in phase space, and the values of each function are plotted along each of its axes. This plot formation includes the argument as a parameter. In the case under consideration (a system of two ordinary differential equations), the *phase portrait* of the system represents a curve on the phase plane. This is shown in Figure 11.5 (left part), and one can clearly see that to plot such a graph, one only needs to change the axes labels to $u^{<1>}$ and $u^{<2>}$ respectively.

A phase portrait of the type shown in Figure 11.5, has one stationary point — the attractor — around which the solution "twists." In the dynamic systems theory, an attractor of this type is known as the focus attractor.

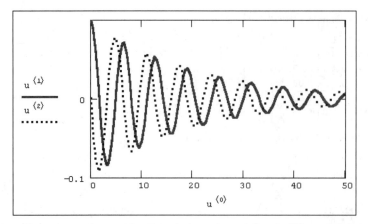

FIGURE 11.4. The graph of a solution of an ordinary differential equations system (Listing 11.4)

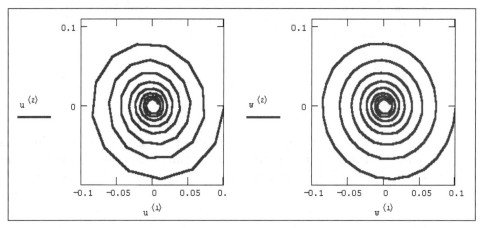

FIGURE 11.5. The phase portrait of the solution of a system of ordinary differential equations at M = 100 (left) and M = 200 (right) (Listing 11.4)

In general, if the system comprises N ordinary differential equations, the phase space is N-dimensional. If N > 3, the phase portrait loses its clearness, and one has to plot various projections in order to visualize the phase portrait.

Figure 11.5 (right) shows the results of calculating the phase portrait with a larger number of steps. A comparison of these illustrations shows that in this case, higher precision can be achieved, and, as a result, the solution is smoother. Of course, this is achieved at the expense of calculation time.

With systems of ordinary differential equations, most problems can be solved by simply increasing the number of steps used by the numeric method. In particular, this approach might prove to be helpful when you encounter an error message such as "Found a number with a magnitude greater than 10^307." This error message might mean that the numeric method uses an insufficient number of steps rather than the fact that the solution is actually divergent.

To conclude this topic, it is necessary to mention several specific features of different numeric methods. All these methods are based on approximating the differential equations with their difference analogs. Depending on the specific form of approximation, various algorithms of different precision and speed are obtained. Mathcad uses the most popular, the Runge-Kutta algorithm of the fourth order, which is described in detail in most books on numerical methods. This method provides a small calculation error for a wide range of systems of ordinary differential equations, with the exception of stiff systems. Thus, in most cases, it is recommended that you use the rkfixed function. If for any reason the calculation time becomes critical or precision ceases to be satisfactory, it would make sense to try other built-in functions instead of rkfixed. This won't present any difficulties, since these functions use the same set of parameters. To attempt to use a different function, simply change the function name in your program.

The Rkadapt function might prove to be useful when you know that the solution within the interval under consideration changes slowly, or if there are intervals where the solution changes at a different rate. The Runge-Kutta method divides the interval optimally with an adjustable step rather than by equal steps. Within the intervals where the solution changes slowly, the steps are larger, and within the regions where it changes rapidly, the steps are smaller. As a result, to achieve the same level of precision, this method uses fewer steps than the rkfixed function does. The Bulirsh-Stoer method, implemented by the Bulstoer function, often turns out to be more efficient when searching through smooth solutions.

11.3.2. Solving Systems of Ordinary Differential Equations in a Given Point

When solving differential equations, you often don't need to calculate the values of an unknown function within the whole integration interval (t0, t1). It is common to encounter a situation where you need to calculate the values of unknown functions in the last point of the interval only. For example, many are aware of the problems present in searching for attractors of dynamic systems. It is a well-known fact that for a wide range of ordinary differential equations, the same system given different (or even arbitrary, as the example of damped oscillator considered above) initial conditions comes to the same point (attractor) at $t \to \infty$. It is this point that needs to be found in most cases.

Such a problem requires fewer PC resources than solving a system of ordinary differential equations within the whole interval does. Therefore, Mathcad 2001i provides modifications of the built-in `Rkadapt` and `Bulstoer` functions. These modifications have a slightly different set of parameters and work faster than their respective analogs.

- `rkadapt(y0,t0,t1,acc,D,k,s)` — Runge-Kutta method with an adaptable step
- `bulstoer(y0,t0,t1,acc,D,k,s)` — Bulirsh-Stoer method

Here:

- `y0` — The vector of initial values in the point `t0`
- `t0, t1` — Starting and ending points of the calculation
- `acc` — Calculation accuracy (the smaller this value, the better the precision of the final result; recommended value is `0.001`)
- `D` — The vector function specifying the system of ordinary differential equations
- `k` — The maximum number of steps at which the selected method will calculate the solution
- `s` — Minimal value of the step allowed

Clearly, instead of the number of steps within the integration interval, these functions require you to specify the precision of the calculated function values at the last point. In this sense, the `acc` parameter is similar to the `TOL` constant that influences most built-in Mathcad numerical algorithms. In this case, numerical algorithms automatically determine the number of steps and their positions to provide this precision. The last two parameters are needed to enable the user to adjust the way in which the interval is divided into steps. The `k` parameter is necessary to avoid situations where the number of steps becomes too large. Notice that you can't set `k > 1000`. The `s` parameter prevents you from making steps too small in order to eliminate significant errors of difference in approximation of the equations within the algorithm. These parameters must be set explicitly, based on the properties of the specific system of ordinary differential equations. As a rule, after conducting several test calculations, you can adjust their optimal values for each specific case.

Listing 11.5 provides an example of the `bulstoer` function usage for solving the same example. As usual, the first two lines of this listing define a system of equations and initial conditions; the next line assigns the solution returned by the bulstoer function to the u matrix. The structure of this matrix is exactly the same as it is when solving the same system of ordinary differential equations using the built-in functions considered previously *(see Section 11.3.1)*. However, in this case, we are only interested in the last point of the interval. Since the number of steps used by a numeric algorithm is not preset, which means the dimension of the u matrix is not predefined, it is necessary to perform its preliminary evaluation. This is accomplished in the fourth line of this listing, which assigns this number to the M variable and displays it on the screen. A vector is displayed at the end of the interval (i.e., at the point $t = 50$) on the next to last line of the listing. The last line of this listing displays the calculated value of the first function from a system of ordinary differential equations (compare this value to the same position within the vector from the previous line).

LISTING 11.5. Solution of a System of Two Ordinary Differential Equations

$$D(t,y) := \begin{pmatrix} y_1 \\ -y_0 - 0.1 \cdot y_1 \end{pmatrix}$$

$$y0 := \begin{pmatrix} 0.1 \\ 0 \end{pmatrix}$$

$$u := bulstoer\left(y0, 0, 50, 10^{-5}, D, 300, 0.0001\right)$$

$$M := length\left(u^{\langle 1 \rangle}\right) - 1 \qquad\qquad M = 21$$

$$\left(u^T\right)^{\langle M \rangle} = \begin{pmatrix} 50 \\ 7.638 \times 10^{-3} \\ 2.648 \times 10^{-3} \end{pmatrix}$$

$$u_{M,1} = 7.638 \times 10^{-3}$$

To try an alternative numerical method, replace the bulstoer name in this listing with rkadapt.

NOTE

The bulstoer and rkadapt functions (the ones with the names starting with a lowercase letter) are not intended for finding the solution in the intermediate points of the interval, although they display these values in the resulting matrix. Figure 11.6 shows phase portraits of the system of ordinary differential equations under consideration obtained using the bulstoer function (Listing 11.5) and the *rkadapt function (obtained if we replace the third line of Listing 11.5). Notice that despite a high level of accuracy (10^{-5}) and the correct result obtained at the end of the interval, the left graph is very far from being a correct phase portrait (see, for example, Figure 11.5 or the right graph in Figure 11.6). It starts to be acceptable only at the minimum allowed value of accuracy acc = 10^{-16}.*

Finally, let us consider the influence of the acc parameter on the calculations. For this purpose, let us use a simple program provided in Listing 11.6. In this program, we have only looked at the value of one of the functions on the right boundary of the interval for the Cauchy problem. However, this result is formulated as a custom function y(ε), where the acc parameter of the bulstoer function is selected as an argument.

LISTING 11.6. Using the Solution of a System of Ordinary Differential Equations for Defining the Custom Function

$$D(t,y) := \begin{pmatrix} y_1 \\ -y_0 - 0.1 \cdot y_1 \end{pmatrix} \qquad y0 := \begin{pmatrix} 0.1 \\ 0 \end{pmatrix}$$

$$y(\varepsilon) := \begin{vmatrix} u \leftarrow \text{bulstoer} \left(y0, 0, 50, \varepsilon, D, 30, 0.0001 \right) \\ M \leftarrow \text{length} \left(u^{\langle 1 \rangle} \right) - 1 \\ u_{M,1} \end{vmatrix}$$

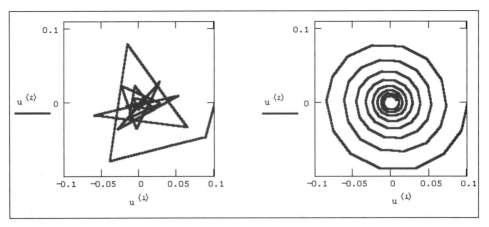

FIGURE 11.6. Phase portrait obtained using bulstoer (left) and rkadapt (right) (Listing 11.5)

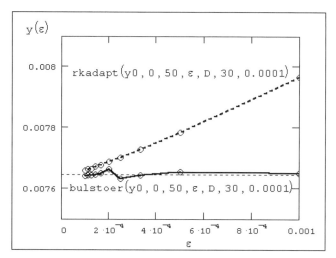

FIGURE 11.7. The dependence of the calculated value of one of the equations of a system of ordinary differential equations on the acc parameter (Listing 11.6)

The calculated solution of y(ε) is shown in the graph in Figure 11.7, along with a similar result obtained for the rkadapt function. As you can see, in this example, numeric methods work somewhat differently. The closer the result the Runge-Kutta method produces is to the actual solution, the smaller the ε = acc value. The Bulirsch-Stoer method demonstrates a more natural dependence y(ε): even with a relatively large value of ε, the results' precision remains acceptable (actually, at a much higher level than with the Runge-Kutta method). Thus, to speed up the calculations (notice that this is only true for the current specific task), the bulstoer function allows you to set larger values of acc.

To provide a specified level of precision, algorithms implemented by built-in functions can change both the number of steps dividing the (t0,t1) interval and their positions within that interval. To determine the number of steps to which the interval is subdivided when calculating y(ε) dependence (shown in Figure 11.7 for each ε), it is necessary to calculate the dimension of the resulting matrix. For this purpose, it is possible to define similar functions:

$$M(\varepsilon) := \text{length}\left(\text{rkadapt}\left(y0\,,\,0\,,\,50\,,\,\varepsilon\,,\,D\,,\,100\,,\,0.0001\right)^{\langle 1 \rangle}\right).$$

By comparing the two results of applying rkadapt for k = 30 and for k = 100, notice (Figure 11.8) the maximum number of steps k influences the M(ε) dependence. Also, notice that the same changes apparent in the k parameter have a very insignificant influence on the calculation of M(ε) using the bulstoer function.

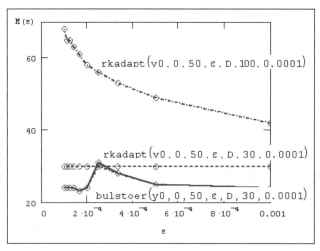

FIGURE 11.8. The dependence of the step number on the `acc` parameter for numerical methods

Thus, by performing test calculations for different tasks and selecting an optimum set of parameters, you can succeed in making the most of PC resources. Obviously, such an analysis is needed for those cases where calculation time plays a critically important role.

11.3.3. Several Examples

In the previous section of this chapter, we used some examples of linear equations — equations containing only the first power of unknown functions and their variables. However, most nonlinear equations have rather unusual and surprising properties and solutions because most of them can only be obtained numerically. In this section, we are going to consider several classical examples of systems of ordinary differential equations, bearing in mind that they might end up being useful both for educational and practical purposes. These include the models of population dynamics (Vito Volterra and Alfred J. Lotka), the generator of self-excited oscillations (Van der Pol), turbulent convection (Lorenz), and chemical reaction with diffusion (Ilya Prigogine). All these models (like the ones that were already considered in this chapter) contain derivatives in time t and describe the dynamics of various physical parameters. Cauchy problems for such models are known of as dynamic systems, and to study such systems, it is necessary to analyze their phase portraits (i. e., solutions obtained for various selections of initial conditions).

In most of the following examples, we calculate several solutions for different initial conditions.

Limiting ourselves to just a few comments, let us consider some listings and graphs of the solutions without diving into details.

The "Predator – Victim" Model

The model of interaction known as "predator – victim" was suggested independently by Alfred J. Lotka and Vito Volterra between 1925 and 1927. Two differential equations (Listing 11.7) model the time dynamics of the sizes of two biological populations of the victim y_0 and predator y_1. It is supposed that victims reproduce at a constant rate C, and the size of their population decreases as a consequence of predators. Predators, on the other hand, reproduce at a rate proportional to the amount of food consumed (with the coefficient r), and die as the result of natural causes (the death rate is determined by the D constant). The listing calculates three solutions in D, G, and P for different initial conditions.

LISTING 11.7. The "Predator – Victim" Model

$$C := 0.1 \qquad D := 1 \qquad r := 0.1$$

$$F(t, y) := \begin{pmatrix} C \cdot y_0 - r \cdot y_0 \cdot y_1 \\ -D \cdot y_1 + r \cdot y_0 \cdot y_1 \end{pmatrix}$$

$$M := 500$$

$$t0 := 0 \qquad t1 := 100$$

$$D := \text{rkfixed}\left[\begin{pmatrix} 10 \\ 8 \end{pmatrix}, t0, t1, M, F\right]$$

$$G := \text{rkfixed}\left[\begin{pmatrix} 10 \\ 4 \end{pmatrix}, t0, t1, M, F\right]$$

$$P := \text{rkfixed}\left[\begin{pmatrix} 10 \\ 1.5 \end{pmatrix}, t0, t1, M, F\right]$$

The model is interesting in that in such a system, both the predator and victim populations grow and decrease cyclically (Figure 11.9). This phenomenon is often exhibited in nature. The phase portrait of this system represents concentric closed

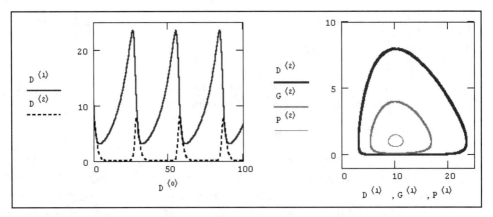

FIGURE 11.9. A graph of the solution (left) and a phase portrait (right) of the "predator – victim" model (Listing 11.7)

curves surrounding one stationary point known as the center. As you can see, the cyclic oscillations of the size of both populations significantly depend on the initial conditions — after each cycle, the system returns to the same point. Dynamic systems with such behavior are known as non-rough.

Self-Excited Oscillations

Let us consider the Van der Pol equation's solution, which describes electric oscillations in the closed circuit composed of a sequentially connected capacitor, an inductor, a non-linear resistor, and an external energy supply (Listing 11.8). The unknown time function $y(t)$ represents an electrical current, while the μ parameter reflects relationships between components of an electric circuit, including non-linear resistance.

LISTING 11.8. Van der Pol Model ($\mu = 1$)

```
µ := 1

Given
```

$$\frac{d^2}{dt^2}y(t) - \mu \cdot \left(1 - y(t)^2\right) \cdot \frac{d}{dt}y(t) + y(t) = 0$$

```
y(0) = 0.01

y'(0) = 0

y := Odesolve (t, 30)
```

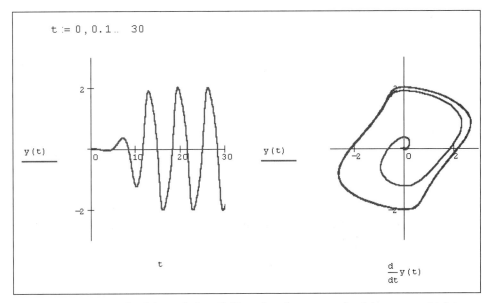

FIGURE 11.10. A graph of the solution (left) and a phase portrait of the system (right) of the Van der Pol equation (Listing 11.8)

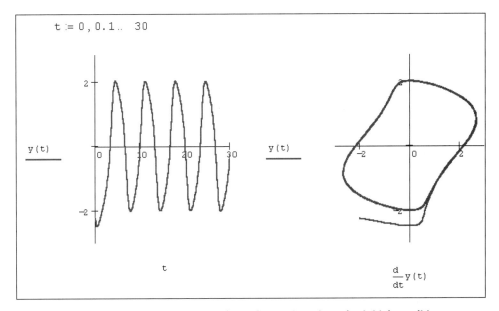

FIGURE 11.11. The solution of the Van der Pol equation given the initial conditions $y = -2$, $y' = -3$

The solution of the Van der Pol equation represents oscillations, whose graph (for $\mu = 1$) is shown in Figure 11.10. Such oscillations are known as self-excited oscillations and are fundamentally different from all the types of oscillations that we have already considered (for example, pendulum oscillations, considered in Section 11.3.2). The difference is that self-excited oscillations' characteristics (amplitude, frequency, spectrum) don't depend on the initial conditions but are defined exclusively by the properties of the dynamic system itself. After some time elapses from the initial point, the solution goes to the same repeated cycle of oscillations known as the limiting cycle. An attractor of the limiting cycle type is a closed curve on the phase plane. All curves, starting from different initial points located both inside (Figure 11.10) and outside (Figure 11.11) the limiting cycle, tend to it asymptotically.

NOTE

If your computer isn't particularly high-powered, then the phase portraits' calculations, shown in Figures 11.10 and 11.11 using Mathcad, might last for quite a long time. This is because of the numeric calculation of the solution $y(t)$ and its derivative. The calculation time can be significantly reduced if you use one of the built-in functions to return the solution matrix (rkfixed, for example) instead of the Given/Odesolve solve block.

Lorenz Attractor

One of the most famous dynamic systems was introduced in 1963 by Lorenz to represent a simplified model of the convective turbulent flow of liquid in a heated vessel of a thoroidal form. The system comprises three ordinary differential equations and has three model parameters (Listing 11.9). Since there are three unknown functions, the phase portrait of this system must be defined in 3D space.

LISTING 11.9. Lorenz Model

$$\sigma := 10 \qquad r := 27 \qquad b := \frac{8}{3}$$

$$y0 := \begin{pmatrix} 10 \\ 10 \\ 10 \end{pmatrix}$$

$$F(t, y) := \begin{pmatrix} \sigma \cdot y_1 - \sigma \cdot y_0 \\ -y_0 \cdot y_2 + r \cdot y_0 - y_1 \\ y_1 \cdot y_0 - b \cdot y_2 \end{pmatrix}$$

```
t0 := 0              t1 := 30

N := 1000

D := rkfixed (y0 , t0 , t1 , N , F)
```

With a specific combination of parameters (Figure 11.12), the solution of this system is the so-called strange attractor (or Lorenz attractor) — an attracting set of trajectories in the phase space — which is seemingly identical to the random process. In some sense, the Lorenz attractor represents stochastic self-excited oscillations sustained in a dynamic system due to an external energy supply.

A solution in the form of a strange attractor appears only with specific combinations of parameters. For example, Figure 11.13 shows the result for $r = 10$ (other parameters have the same values). As a matter of fact, the attractor in this case is the focus. Usually, the phase portrait is recombined in the area of intermediate values of r. The critical combination of parameters, at which the phase portrait of the system undergoes a radical change, is known in the theory of dynamic systems as the bifurcation point. The physical principle of the bifurcation in the Lorenz model, according to present-day ideas, describes the transition of a laminar flow to a turbulent flow.

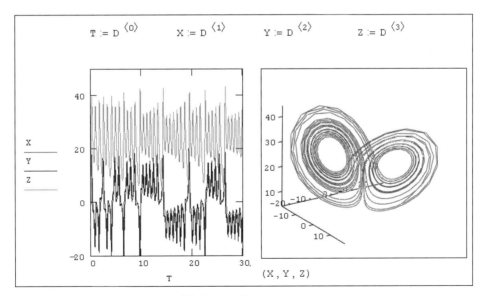

FIGURE 11.12. Lorenz attractor (Listing 11.9)

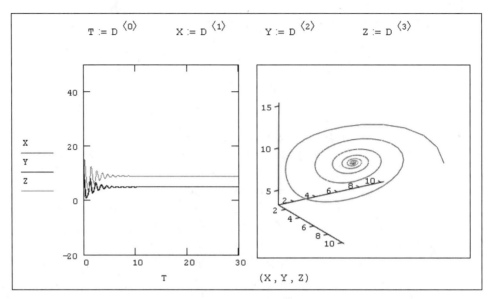

FIGURE 11.13. The solution of the Lorenz system with the changed parameter $r = 10$

It should be mentioned that the solution of this and similar non-linear dynamic systems can be obtained only numerically. Therefore, investigations of these problems began to develop with the growth of modern computers' capabilities.

11.4. THE PHASE PORTRAIT OF A DYNAMIC SYSTEM

Up to now, we provided the graphs of 1–3 trajectories in the phase plane as examples of dynamic system calculations. However, to perform a fundamental investigation of the phase portrait, it is necessary to solve a system of ordinary differential equations many times with different initial conditions (and possibly, different sets of model parameters), to determine those attractors with which different trajectories converge. In Mathcad, you can implement this task in the form of an algorithm. For example, Listing 11.10 shows an algorithm for solving a system of equations describing a self-catalytic chemical reaction with diffusion. This model, also known as the brusselator model, was suggested in 1968 by Lefever and Ilya Prigogine. Unknown functions reflect the dynamics of concentration of intermediate products of some chemical reaction. The parameter of the model is equal to the initial concentration of the catalyst.

LISTING 11.10. Constructing the Phase Portrait for the Brusselator Model

$$v := \begin{pmatrix} 0 & 0 & 2.5 & 1.5 & 0.5 & 1 & 1 & 1.5 & 0.1 & 0.5 \\ 0.5 & 1.5 & 0 & 0 & 1 & 0 & 1 & 2 & 0.1 & 0.2 \end{pmatrix}$$

$B := 0.5$

$$D(t,y) := \begin{bmatrix} -(B+1) \cdot y_0 + (y_0)^2 \cdot y_1 + 1 \\ B \cdot y_0 - (y_0)^2 \cdot y_1 \end{bmatrix}$$

$t0 := 0 \qquad t1 := 10$

$M := 100$

$$U := \begin{vmatrix} y \leftarrow v^{\langle 0 \rangle} \\ Z \leftarrow \text{rkfixed}(y, t0, t1, M, D) \\ Z1^{\langle 0 \rangle} \leftarrow Z^{\langle 0 \rangle} \\ Z1^{\langle 1 \rangle} \leftarrow Z^{\langle 1 \rangle} \\ Z1^{\langle 2 \rangle} \leftarrow Z^{\langle 2 \rangle} \\ \text{for } k \in 1.. \ \text{last}\left\lfloor \left(v^T \right)^{\langle 1 \rangle} \right\rfloor \\ \quad \begin{vmatrix} y \leftarrow v^{\langle k \rangle} \\ Z \leftarrow \text{rkfixed}(y, t0, t1, M, D) \\ Z2^{\langle 0 \rangle} \leftarrow Z^{\langle 0 \rangle} \\ Z2^{\langle 1 \rangle} \leftarrow Z^{\langle 1 \rangle} \\ Z2^{\langle 2 \rangle} \leftarrow Z^{\langle 2 \rangle} \\ Z1 \leftarrow \text{stack}(Z1, Z2) \end{vmatrix} \\ Z1 \end{vmatrix}$$

The algorithm proposed here combines specific solution matrices of a system of ordinary differential equations with different initial conditions and the joined matrix U. The pairs of initial conditions are specified in the first line of this listing as a matrix v of dimension 2×10. This means that 10 trajectories will be generated. To change the number of trajectories, change the dimension of this matrix as is appropriate. Then the elements of the matrix U are plotted on the graph as separate points (Figure 11.14). The drawback of this algorithm is its inability to connect these points with lines. Notice, however, that this drawback is insignificant, since Mathcad has allowed you to easily represent a large number of trajectories on the phase plane.

As Figure 11.14 shows, all trajectories originating in different points tend asymptotically to the same attractor (1,0.5). Within the theory of dynamic systems, this kind of attractor is known as the node (we have already encountered nodes in examples provided in Section 11.1). When analyzing the phase portrait, it is undoubtedly desirable to investigate as many trajectories as possible by specifying a wider range of initial conditions. It is possible that in other regions of the phase plane, trajectories will converge with other attractor(s).

You can investigate the evolution of the brusselator phase portrait by performing calculations with different values of the B parameter. With the increase of this parameter, the node will gradually move to the point with (1,B) coordinates until it reaches the bifurcation value B = 2. At this point, the portrait is reorganized qualitatively, and the limiting cycle appears. A further increase of the B parameter results only in quantitative changes of the parameters of this cycle. The solution obtained at B = 2.5 is shown in Figure 11.15.

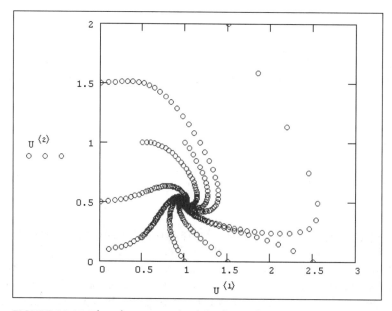

FIGURE 11.14. The phase portrait of the brusselator at B = 0.5 (Listing 11.10)

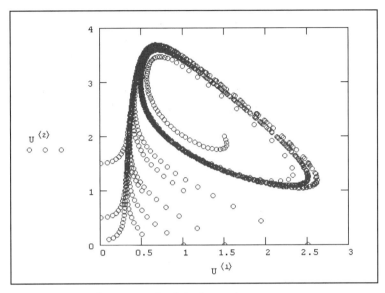

FIGURE 11.15. Phase portrait of the brusselator at $B = 2.5$

To find attractors of the dynamical system, it is necessary to solve the system of algebraic equations that will be obtained if the left parts of the ordinary differential equations are replaced by zeroes. The most convenient way to solve such problems is by using Mathcad tools (see Chapter 8). In particular, it is possible to investigate the dependence of the phase portrait on the system parameters and to search for bifurcations using the method of continuation by parameter (see Section 8.8, "Method of Continuation by Parameter," in Chapter 8).

Those readers dealing with the calculation of dynamical systems will certainly appreciate the capabilities provided by Mathcad for constructing a phase portrait and investigating bifurcations. It is possible that they will find better programmatic solutions than the algorithm suggested by the author in this section.

11.5. STIFF SYSTEMS OF ORDINARY DIFFERENTIAL EQUATIONS

Before now, we were only dealing with "good" equations that are reliably solved using Runge-Kutta numerical methods. However, there is a range of systems of ordinary differential equations (the so-called "stiff" systems), for which standard numerical methods are practically inapplicable, since their solution requires one to use very small step values. Some special algorithms developed explicitly for such systems are also implemented in Mathcad.

11.5.1. What Are Stiff Ordinary Differential Equations?

There is no strict mathematical definition of the stiff ordinary differential equation. For the purposes of this book, let us consider stiff systems to be equations for which it is much simpler to obtain a solution using determined implicit methods rather than explicit methods similar to the ones considered in the previous sections. A similar definition of stiff systems was put forth in 1950s by Curtiss and Hirshfelder, pioneers in this field. Let us start our discussion of stiff ordinary differential equations with the example of a non-stiff equation (Listing 11.11), whose solution is shown in Figure 11.16. The same graph shows a solution of a similar ordinary differential equation with another coefficient on the right, which is equal to -30 rather than to -10. Solving both equations didn't present any problems, and the numeric Runge-Kutta method produced the correct result.

LISTING 11.11. Solution of a Non-Stiff Ordinary Differential Equation

```
Given

d
──y(t) = -10·(y(t) - cos(t))
dt

y(0) = 1

y := Odesolve(t, 1)
```

Figure 11.17 shows the solution of the same ordinary differential equation using the coefficient -50. The result produced by Mathcad seems rather strange. The characteristic divergence of this solution proves that the algorithm has become unstable. The first thing that you can do to fix this is increase the number of steps in the Runge-Kutta method. To do so, add the step parameter to the Odesolve(t,1,step) function. After several experiments, you'll be able to adjust the step value to provide solution stability. When you do these experiments yourself, you will need to make sure that the divergence disappears at step > 20 and that the solution becomes similar to the graphs shown in Figure 11.16.

Thus, it is clear that, first, the same equations with different parameters can be stiff or non-stiff; and second, the "stiffer" the equation is, the more steps are required in normal numeric methods to get the stable solution. For example, we succeeded in obtaining a stable solution for the classic example of the differential equation from Listing 11.11 because this equation was not too stiff and the problem could be solved by increasing the number of steps. However, to obtain solutions of a stiffer equation, we may require millions of steps (or even more).

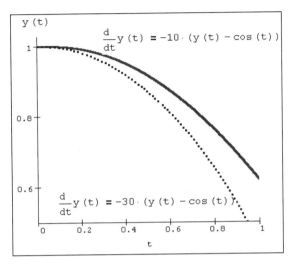

FIGURE 11.16. Solution of a non-stiff ordinary differential equations using the Runge-Kutta method (Listing 11.11)

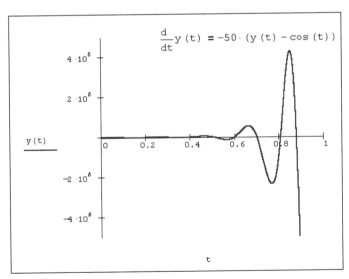

FIGURE 11.17. Incorrect solution of stiff a differential equation obtained using Runge-Kutta method

Some scientists have noticed that during the last several years, the Runge-Kutta methods are losing their leading positions among algorithms for solving ordinary differential equations. Currently, methods allowing one to solve stiff tasks are becoming more and more popular.

Historically, the interest in stiff systems arose in the mid-20th century in the course of investigations in the field of chemical kinetics. These studies focused on the problem of systems that simultaneously exhibited very slow and very fast chemical reactions. At that time, Runge-Kutta methods, which up until that point had been considered very reliable, began to fail when solving such tasks. For example, let us consider the classical model of interaction between three substances (Robertson, 1966) that best illustrates the idea of ordinary differential equation stiffness.

Let us imagine that the substance 0 slowly turns into 1 :$0 \rightarrow 1$ (at the speed of 0.1), the substance 1 serves as a catalyst for itself and turns into the substance 2 very quickly: $1 + 1 \rightarrow 2 + 1$ (10^3). Finally, the substances 2 and 1 interact with each other in a similar way, but at a medium rate: $1 + 2 \rightarrow 0 + 2$ (10^2). Listing 11.12 shows the system of ordinary differential equations describing the concentration dynamics of the reacting mixture, and attempts to solve it using Runge-Kutta method.

LISTING 11.12. A Stiff System of Ordinary Differential Equations Describing the Chemical Kinetics of the Reacting Mixture of Three Substances

$$F(t,y) := \begin{pmatrix} -0.1 \cdot y_0 + 10^2 \cdot y_1 \cdot y_2 \\ 0.1 \cdot y_0 - 10^2 \cdot y_1 \cdot y_2 - 10^3 \cdot y_1 \\ 10^3 \cdot y_1 \end{pmatrix} \qquad y0 := \begin{pmatrix} 1 \\ 0 \\ 0 \end{pmatrix}$$

$$D := \text{rkfixed}(y0, 0, 50, 20000, F)$$

A significantly different order of these equations' coefficients immediately catches the eye. It is the degree of this difference that primarily influences system stiffness. As the stiffness characteristic, one selects the Jacobi matrix of the vector function $F(t,y)$, i.e., the functional matrix composed of the derivatives of the $F(t,y)$ function *(see Section 7.2.3, "Partial Derivatives," in Chapter 7)*. The more degenerate the Jacobi matrix is, the stiffer the system of ordinary differential equations. In the example provided above, the determinant of the Jacobian is equal to zero at any of the values of y_0, y_1, and y_2 (Listing 11.13, the second line). The first line of Listing 11.13 serves as a reminder of how to calculate the Jacobi matrix by using Mathcad. The example used is that of defining the elements of its first row.

LISTING 11.13. Jacobian of the Previously Described System of Chemical Kinetics Differential Equations

$$\frac{\partial}{\partial x}F\left[t,\begin{pmatrix}x\\y_1\\y_2\end{pmatrix}\right]_0 \rightarrow -.1 \qquad \frac{\partial}{\partial x}F\left[t,\begin{pmatrix}y_0\\x\\y_2\end{pmatrix}\right]_0 \rightarrow 100\cdot y_2$$

$$\frac{\partial}{\partial x}F\left[t,\begin{pmatrix}y_0\\y_1\\x\end{pmatrix}\right]_0 \rightarrow 100\cdot y_1$$

$$\left|\begin{pmatrix}-0.1 & 10^2\cdot y_2 & 10^2\cdot y_1\\0.1 & -10^2\cdot y_2 - 10^3 & -10^2\cdot y_1\\0 & 10^3 & 0\end{pmatrix}\right| \rightarrow 0$$

For the example provided in Listing 11.12, the standard Runge-Kutta method still succeeds in finding the solution (it is shown in Figure 11.18). However, because this solution requires several steps (M=20000), the calculations are very slow. If there are fewer steps, the numeric algorithm will fail to find the solution — rather, it will diverge, and Mathcad will finally display an error message instead of a result.

There is another fact that deserves to be mentioned: the difference in the order of the resulting value. As is evident from Figure 11.18, the concentration of the first reacting substance y_1 is significantly larger (actually, several thousand times larger) than the concentration of other reacting components. This property is also characteristic of stiff systems.

Principally, one can also decrease the stiffness of a system manually by using the re-scaling approach. For this purpose, one can manipulate the values so that they decrease the values of the unknown function y_1, by means of dividing all system components containing y_1 by 1000. After re-scaling, the Runge-Kutta method will require only 20 steps to successfully find the solution.

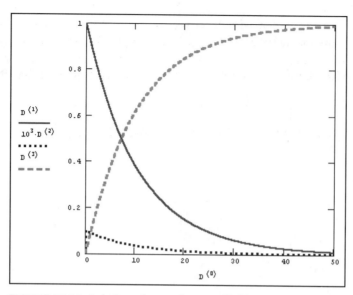

FIGURE 11.18. Solution of an ordinary, stiff differential equation system of chemical kinetics obtained using the Runge-Kutta method (Listing 11.12)

11.5.2. Functions for Solving Ordinary Stiff Differential Equations

Stiff systems of ordinary differential equations can be solved using built-in functions similar to the above considered family of functions for ordinary differential equations. The only difference is that the additional matrix function specifying the Jacobian of the ordinary differential equation system has been added to the standard set of parameters. The solution is returned in the form of a matrix similar to the solution functions of non-stiff Cauchy problems.

- ■ Stiffb(y0,t0,t1,M,F,J) — The Bulirsh-Stoer method for stiff ordinary differential equation systems
- ■ Stiffr(y0,t0,t1,M,F,J) — The Rosenbrock method for stiff ordinary differential equation systems

Here:

- ● y0 — The vector of initial values t0
- ● t0,t1 — The starting and ending points of the calculation
- ● M — The number of steps used by the numerical method
- ● F — The vector function F(t,y) of dimension 1×N, specifying the ordinary differential equation system
- ● J — The matrix function J(t,y) of the size (N+1)×N, composed of the vector of derivatives of the F(t,y) function and its Jacobian (N left columns)

Let us demonstrate the effect of these algorithms on the example of the same stiff system of ordinary differential equations (Listing 11.14). Notice how, in this case, it is necessary to represent the Jacobian by comparing the specification of the matrix function in the next to last line of Listing 11.14 and the specification of the Jacobian in Listing 11.13.

LISTING 11.14. Solution of a Stiff System of Ordinary Differential Equations

$$y0 := \begin{pmatrix} 1 \\ 0 \\ 0 \end{pmatrix}$$

$$F(t, y) := \begin{pmatrix} -0.1 \cdot y_0 + 10^2 \cdot y_1 \cdot y_2 \\ 0.1 \cdot y_0 - 10^2 \cdot y_1 \cdot y_2 - 10^3 \cdot y_1 \\ 10^3 \cdot y_1 \end{pmatrix}$$

$$J(t, y) := \begin{pmatrix} 0 & -0.1 & 10^2 \cdot y_2 & 10^2 \cdot y_1 \\ 0 & 0.1 & -10^2 \cdot y_2 - 10^3 & -10^2 \cdot y_1 \\ 0 & 0 & 10^3 & 0 \end{pmatrix}$$

```
D := Stiffb (y0 , 0 , 50 , 20 , F , J)
```

The calculations show that to obtain the same result (see Figure 11.18), it is now sufficient to work through a thousand fewer steps of the numeric algorithm in comparison to the standard Runge-Kutta method. Consequently, the calculations will be significantly faster. Thus, if you are dealing with stiff systems, it is necessary to use special algorithms.

It is important to notice that until now, we were dealing with a system that was not particularly stiff. For example, try to replace the reaction rates mentioned above *(see Section 11.5.1)* — 0.1, 10^3, and 10^2 — with other numbers, such as 0.05, 10^4, and 10^7, respectively. Notice that such a relationship between reaction rates is rather common for applied chemical kinetics tasks, and it defines a rather stiff system of ordinary differential equations. This system can never be solved using standard methods, since the number of steps required by the numeric method would be astronomical. On the other hand, algorithms specially developed for stiff systems of differential equations solve this task easily (Figure 11.19), and require approximately the same number of steps taken in Listing 11.14.

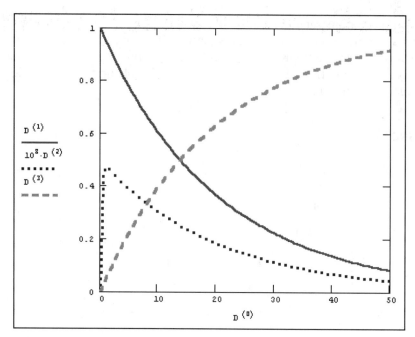

FIGURE 11.19. Solution of a stiffer system of ordinary differential equations using the Rosenbrock method

Take note of the fact that the orders of solution values for the concentration of various components shown in Figure 11.19 differ more significantly than in the previous (less stiff) example.

This example demonstrates once again, that the same system with different coefficients can be stiffer or less stiff. For example, the above described example of the Van der Pol generator with parameter $\mu = 5000$ is rather stiff.

Finally, let us describe the built-in functions used for solving stiff systems of ordinary differential equations at the specified point `t1` rather than on the whole interval.

■ `stiffb(y0,t0,t1,acc,F,J,k,s)` — The Bulirsch-Stoer method
■ `stiffr(y0,t0,t1,acc,F,J,k,s)` — The Rosenbrock method

The names of these functions start with a lowercase letter. Their working principle and set of parameters are the same as the ones for the functions considered above for finding solutions of non-stiff systems in a specified point *(see Section 11.3.2)*. The difference is due to the specific features of the algorithm usage and the need to specify the matrix function of the Jacobian `J(t,y)`.

12 Boundary Value Problems

I n this chapter we will consider boundary value problems for ordinary differential equations and for partial differential equations. Mathcad's tools enable you to solve boundary value problems for systems of ordinary differential equations, where part of the boundary conditions are set at the starting point of the interval, and the remaining conditions are set at its ending point *(see Section 12.1)*. Furthermore, it is also possible to solve equations with boundary conditions specified at some internal point of the interval. Mathcad provides appropriate built-in functions for solving boundary value problems. These functions implement the shooting method algorithm *(see Section 12.1.2)*.

Boundary value problems for most practical applications often depend on some numeric parameter. In this case, the solution exists only for some countable number of parameter values rather than for all its values. Such problems are also known as eigenvalue problems *(see Section 12.2)*.

Despite the fact that while in contrast to Cauchy problems for ordinary differential equations, Mathcad doesn't provide built-in functions for solving stiff boundary value problems, these problems can still be solved using difference scheme programming *(see Section 12.3)*. Possible variations of difference scheme implementations are also provided for some boundary value problems with partial differential equations.

Furthermore, some particular cases of partial differential equations, such as Poisson's equations, can even be solved using built-in functions provided by Mathcad *(see Section 12.4).*

12.1. BOUNDARY VALUE PROBLEMS FOR ORDINARY DIFFERENTIAL EQUATIONS

Formulation of the boundary value problems for ordinary differential equations are different from the Cauchy problems considered in Chapter 11, in that the boundary conditions for them are set on both boundaries of the integration interval rather than at a single initial point. For a system of N ordinary differential equations of the first order, part of the boundary conditions can be set at the first boundary of the interval, and the remaining conditions can be specified at the opposite boundary of the interval.

Differential equations of higher orders can be reduced to an equivalent system of ordinary differential equations of the first order (see Chapter 11).

12.1.1. About Formulations of the Boundary Value Problems

To get a better understanding of boundary value problems, let us consider their formulation on a specific physical example of the model of interaction between two oncoming beams of light rays. Let us suppose that we need to determine the intensity distribution of light radiation in the space between the source (laser) and the mirror (Figure 12.1). Naturally, this space is filled by some medium. Let us suppose that the mirror reflects the R-th part of the incoming radiation (i.e., its reflection coefficient is equal to R), and the medium both absorbs the radiation with the absorption factor $a(x)$ and dissipates it. Under these conditions the backward dissipation coefficient is equal to $r(x)$. In this case, the law according to which the intensity $y_0(x)$ of the radiation propagating to the right and intensity $y_1(x)$ of the radiation propagating to the left is determined by a system of two ordinary differential equations of the first order:

$$\frac{dy_0(x)}{dx} = -a(x) \cdot y_0(x) + r(x) \cdot y_1(x)$$

$$\frac{dy_1(x)}{dx} = a(x) \cdot y_1(x) - r(x) \cdot y_0(x)$$

(1)

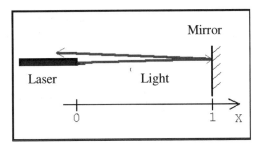

FIGURE 12.1. Model for formulation of the boundary value problem

For correct formulation of the task, you need to specify the same number of boundary conditions. One of them will express the given intensity I0 or the radiation incoming from the left boundary x = 0, while the second boundary condition will represent the reflection law at the right boundary x = 1:

$$
\begin{aligned}
y_0(0) &= \text{I0} \\
y_1(1) &= R \cdot y_0(1)
\end{aligned}
\tag{2}
$$

The problem that we have just formulated is known as the *boundary value problem*, since the constraints are specified for both boundaries of the interval (0,1) rather than for a single boundary. As a result, such problems can't be solved using the methods described in the previous chapter and those that are intended for initial value problems. From now on, to demonstrate Mathcad's capabilities, we'll use the previously-described example with R = 1 and specific dependencies a(x) = const = 1 and r(x) = const = 0.1, which is characteristic of the case of isotropic (i.e., independent from the x coordinate) dissipation.

The model shown in Figure 12.1 has led to the boundary value problem for the system of linear differential equations. This model has an analytical solution in the form of a combination of exponential functions. Non-linear problems are more complicated, and can only be solved numerically. Obviously, the above-described model that we used to formulate the boundary value problem will also become non-linear if we suppose that the absorption coefficient and dissipation coefficient depend on radiation intensity. In terms of physics, this will reflect the dependence of the medium optical properties on the intensity of powerful radiation.

The model of the oncoming beams of light has resulted in a system of equations (1), which includes the derivatives of only the x variable (which means that dissipation takes place only in a forward and backward direction). If we were considering more complex effects of dissipation in other directions of the 3D space, we'd get a system of partial differential equations including partial derivatives of other variables including y and z. In this case, we'd get a boundary value problem for partial differential equations, which is much more complex than the boundary value problem for a system of an ordinary differential equation.

12.1.2. Shooting Method

Mathcad implements the most popular algorithm for solving boundary value problems, known as the *shooting method*. Actually, this method reduces the solution of the boundary value problem to that of a series of Cauchy problems with different initial conditions. Let us consider its main principle with the example of the model described in the previous section (Figure 12.1). In the next section, we will discuss the built-in Mathcad functions that implement this algorithm.

   The core idea of the shooting method lies in specifying the test values of lacking boundary conditions at the left boundary of the integration interval and solving the resulting Cauchy problem using well-known methods *(see Chapter 11)*. In our example, the initial value for $y_1(0)$ is lacking, therefore, let us specify some value for it, for example, $y_1(0) = 10$. This selection of the initial value is definitely not absolutely arbitrary, since it is clear that first, radiation intensity is always positive and, second, the intensity of reflected radiation must be significantly less than that of the incoming radiation. Listing 12.1 shows the solution of the Cauchy problem using the rkfixed function.

LISTING 12.1. Solving the Test Cauchy Problem for the Model (*See (1) in Section 12.1*)

```
r (x) := .1     a (x) := 1

R := 1
```

$$D (x, y) := \begin{pmatrix} -a (x) \cdot y_0 + r (x) \cdot y_1 \\ a (x) \cdot y_1 - r (x) \cdot y_0 \end{pmatrix}$$

$$I0 := 100 \qquad y := \begin{pmatrix} I0 \\ 10 \end{pmatrix}$$

```
M := 10

S := rkfixed (y, 0, 1, M, D)
```

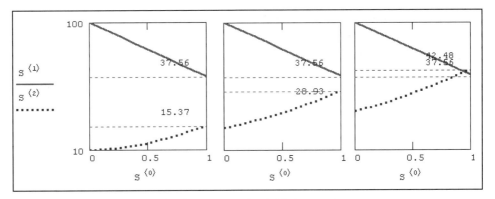

FIGURE 12.2. Illustration of the shooting method (Listing 12.1)

The graph of the obtained solutions is shown in Figure 12.2 (left). When looking at this graph, note that it is obvious that the second initial condition taken as a guess didn't provide a solution that would satisfy the boundary condition at x = 1. It is clear that in order to satisfy this boundary condition better, we need to take a larger value of $y_1(0)$. For example, let us assume that $y_1(0) = 15$, and solve the Cauchy problem once again. The result of this solution is shown in Figure 12.2 (center). The boundary condition is satisfied with better accuracy; however, this is still insufficient. For the next value, $y_1(0) = 20$, we'll get the solution shown in Figure 12.2 (right). Comparing the center and right graphs, one can conclude that the missing initial condition is greater than 15 but less than 20. Proceeding with this "shooting," it is possible to find the missing initial condition and to obtain the correct solution of the boundary value task.

This is the main principle of the shooting algorithm. By selecting initial test conditions and solving an appropriate series of Cauchy problems, one can find a solution of a system of ordinary differential equations that would satisfy the boundary condition(s) at the other boundary of the calculation interval.

Of course, you can easily implement the method described above yourself — by programming the solution of the system of equations expressing the boundary conditions at the second boundary in relation to the unknown initial conditions. However, you don't need to do this, since this method is implemented in Mathcad in the form of built-in functions.

12.1.3. Solving Two-Point Boundary Value Tasks

For solving boundary value tasks for systems of ordinary differential equations using the shooting method, Mathcad implements two built-in functions. The first of these functions is intended for two-point boundary value problems with boundary values specified at the boundaries of the interval.

The sbval(z,x0,x1,D,load,score) function searches for the vector of missing L initial conditions for a two-point boundary value problem formulated for the system of N ordinary differential equations, where:

- z — A vector with dimension $L \times 1$, that assigns initial values to the missing initial conditions at the left boundary of the interval.
- x0 — The left boundary of the integration interval.
- x1 — The right boundary of the integration interval.
- load(x0,z) — A vector function with dimension $N \times 1$ that expresses the left boundary conditions, where missing initial conditions are named by appropriate components of the z vector argument.
- score(x1,y) — A vector function with dimension $L \times 1$, which expresses L right boundary conditions for the vector function y at the point x1.
- D(x,y) — A vector function that describes the system of N ordinary differential equations with the dimension $N \times 1$. This function accepts two arguments — scalar x and vector y. Here y is an unknown vector function of the x argument of the same size $N \times 1$.

NOTE

Solution of the boundary value problems in Mathcad, in contrast to most other operations, is implemented in a way that can't be classified as "obvious," "clear," or "self-evident." For example, you must remember that the number of elements of the vectors D and load is equal to the number of equations N. The number of elements of the vectors z, score, and the result returned by the function sbval is equal to the number of the right boundary conditions L. Consequently, there must be (N - L) left boundary conditions in this problem.

As you can see, the sbval function is meant to find the missing initial conditions on the first point of the interval (i.e., $y_i(x0)$) rather than the solution itself (unknown functions $y_i(x)$). To calculate $y_i(x)$ for the whole interval, it is necessary to additionally solve the Cauchy problem.

In Listing 12.2, we discuss some specific features of the sbval function usage, which was first described in Section 12.1.1. The boundary value task comprises a system of two equations ($N = 2$): one left boundary condition ($L = 1$), and one right boundary condition ($N - L = 2 - 1 = 1$).

LISTING 12.2. Solving the Boundary Value Problem

$$r(x) := .1 \qquad a(x) := 1$$

$$R := 1 \qquad I0 := 100$$

$$D(x, y) := \begin{pmatrix} -a(x) \cdot y_0 + r(x) \cdot y_1 \\ a(x) \cdot y_1 - r(x) \cdot y_0 \end{pmatrix}$$

$$z_0 := 10$$

$$\text{load}(x0, z) := \begin{pmatrix} 100 \\ z_0 \end{pmatrix}$$

$$\text{score}(x1, y) := R \cdot y_0 - y_1$$

$$I1 := \text{sbval}(z, 0, 1, D, \text{load}, \text{score})$$

$$I1 = (\ 18.555\ )$$

$$S := \text{rkfixed}\left[\begin{pmatrix} I0 \\ I1_0 \end{pmatrix}, 0, 1, 10, D\right]$$

The first three lines of this listing specify the required parameters of the problem and the system of ordinary differential equations itself. The fourth line specifies vector z. Since we have only one right boundary condition, there is only one missing initial condition, and, correspondingly, vector z has only one element — z_0. To start the shooting algorithm *(see Section 12.1.2)*, we need to assign it an initial guess value (here we have assumed $z_0 = 10$, as in Listing 12.1).

The initial guess is actually the parameter of the numeric method, and, therefore, can significantly influence the solution of the boundary value problem.

NOTE

In the next line of the listing, the vector function load(x, z) is assigned the left boundary conditions. This function is similar to the vector variable defining initial values for the built-in functions that solve Cauchy problems. The difference is in the designation of missing conditions. Instead of specific numbers, appropriate positions are filled with the names of the vector z elements. In our example, instead of the second initial condition, the z_0 argument of the load function is specified. The first argument of the function load is the point in which the left boundary condition is specified. Its specific value is determined directly in the list of arguments of the sbval function.

The next line of the listing defines the right boundary condition, for which the score(x,y) function is used. This condition is written the same way as the system of equations in function D. The x argument of the score function is similar to the load function and is required for cases where the boundary condition explicitly depends on the x coordinate. The score vector must contain the same number of elements as z.

The shooting algorithm implemented in sbval function searches the missing initial conditions in such a way as to make the solution of the resulting Cauchy problem make the score(x,y) function as close to zero as possible. As you can see from the listing, the result of applying sbval for the (0,1) interval is assigned to the vector variable I1. This vector is similar to the z vector. However, it contains the sought initial conditions instead of the guessed initial conditions specified in z. In our example, vector I1, similarly to z, contains only one element — $I1_0$. Using it, one can define the boundary value problem solution y(x) (the last line of the listing). Thus, the sbval function reduces the boundary value problems to Cauchy problems. Figure 12.3 shows a graph of the solution of the boundary value problem.

Figure 12.4 shows the solution of the same boundary value problem with another right boundary condition corresponding to R = 0 (i.e., without the mirror at the right boundary). In this case, the weak returning beam is formed exclusively due to the inverse dissipation of laser radiation. Most readers will, no doubt, point out that a real physical medium can't produce such a significant dissipation in a backward direction. In other words, the values r(x) << a(x) are more realistic.

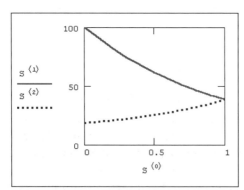

FIGURE 12.3. Solution to the boundary value problem for R = 1 (Listing 12.2)

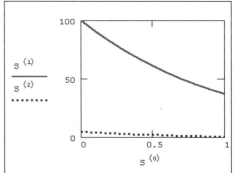

FIGURE 12.4. Solution to the boundary value problem for R = 0

However, when the coefficients in a system of ordinary differential equations at different y_i differ significantly (by several orders of magnitude), the system of ordinary differential equations becomes stiff, and the `sbval` function is unable to find a solution. If this happens, an error message is displayed, most often informing you that a solution couldn't be found ("**Could not find a solution**").

 The shooting method is not suitable for solving stiff boundary value problems. Therefore, users must create custom algorithms for solving stiff boundary value problems in Mathcad (see Section 12.3).

12.1.4. Solving Boundary Value Problems with Additional Conditions at an Intermediate Point

Sometimes differential equations are defined with boundary conditions set not only at the ends of the interval, but also with an additional condition at some internal point of the calculation interval. Most often, such additional conditions contain data on solutions that aren't smooth within some internal point of the interval. For solving such tasks, Mathcad provides the `bvalfit` function, which implements the shooting algorithm.

The `bvalfit(z1,z2,x0,x1,xf,D,load1,load2,score)` function searches for the vector of missing boundary conditions for the boundary value problem with the additional condition at the intermediate point for a system of N ordinary differential equations where:

- `z1` — The vector that assigns initial values to the missing initial conditions at the left boundary of the interval.
- `z2` — The vector of the same size, which assigns initial values to the missing boundary conditions at the right boundary of the interval.
- `x0` — The left boundary of the calculation interval.
- `x1` — The right boundary of the calculation interval.
- `xf` — The internal point located within the calculation interval.
- `D(x,y)` — The vector function that describes the system of N ordinary differential equations of the size N × 1. Accepts two arguments: scalar `x` and vector `y`. Here `y` is an unknown vector function of the `x` argument of the same size N × 1.
- `load1(x0,z)` — N×1 vector function of left boundary conditions. Missing boundary conditions are named by appropriate components of the `z` vector argument.

- `load2 (x1,z)` — $N \times 1$ vector function of right boundary conditions. Where missing boundary conditions are named by appropriate components of the z vector argument.
- `score (xf,y)` — $N \times 1$ vector function that expresses internal condition for the vector function y at the point xf.

Normally, the `bvalfit` function is used for problems where the derivative of the solution has a discontinuity in the internal point xf. Some of these tasks can't be solved using a normal shooting method; therefore, one has to perform shooting from both boundary points simultaneously. In this case, the additional internal condition in the xf point represents the matching conditions for the left and right solutions. Consequently, for such a formulation of the task, the internal condition is written in the following form: `score (xf,y) := y`.

Let us consider the working principle of the `bvalfit` function on the previously considered example of interaction between two oncoming beams of light (see Figure 12.1). Let us assume that the interval between $xf = 0.5$ and $x1 = 1.0$ is filled with another, more optically dense medium characterized by another coefficient of light attenuation $a(x) = 3$. The boundary value problem describing this case is solved in Listing 12.3. Notice that the attenuation coefficient at the discontinuity point is defined on the second line of this listing.

LISTING 12.3. A Boundary Value Problem with the Boundary Condition on the Internal Point

$$I0 := 100 \qquad r(x) := .1 \qquad R := 1$$

$$a(x) := \begin{cases} 1 & \text{if } x < 0.5 \\ 3 & \text{otherwise} \end{cases}$$

$$D(x,y) := \begin{pmatrix} -a(x) \cdot y_0 + r(x) \cdot y_1 \\ a(x) \cdot y_1 - r(x) \cdot y_0 \end{pmatrix}$$

$$z1_0 := 20$$

$$load1(x0,z1) := \begin{pmatrix} 100 \\ z1_0 \end{pmatrix}$$

$$z2_0 := 30$$

$$load2(x1,z2) := \begin{pmatrix} z2_0 \\ R \cdot z2_0 \end{pmatrix}$$

```
score (xf , y) := y
```

$$I1 := bvalfit (z1 , z2 , 0 , 1 , 0.5 , D , load1 , load2 , score)^T$$

$$I1 = \begin{pmatrix} 5.618 \\ 13.801 \end{pmatrix}$$

The system of equations and the left boundary condition are entered in the same way as in the previous listing illustrating the usage of the sbval function. Notice that the right boundary condition is written in the same way. To introduce the condition determining the reflection at the right boundary, we had to determine one more unknown shooting parameter — $z2_0$. The line defining the score function specifies the shooting condition that matches of the two functions on the point xf. The last line of the listing returns the answer that the values of both shooting parameters determined numerically and joined into the vector I1 (in the next to last line of the listing we used the transpose operation to represent the result as a vector rather than as a row matrix).

To construct a correct graph of the solution, it is preferable to construct it from two parts: the solution of the Cauchy problem at the (x0,xf) interval, and another Cauchy problem for the interval (xf,x1). Implementation of this method is provided in Listing 12.4, which actually continues Listing 12.3. In the last line of Listing 12.4, the value of the second sought-for function at the right boundary of the interval is displayed. It is always useful to check to see if it matches the appropriate shooting parameter (displayed on the last line of Listing 12.3).

LISTING 12.4. Solution of the Boundary Value Problem *(Listing 12.3 Continued)*

$$M := 10$$

$$S0 := rkfixed \left[\begin{pmatrix} I0 \\ I1_0 \end{pmatrix} , 0 , 0.5 , M , D \right]$$

$$S1 := rkfixed \left[\begin{pmatrix} S0_{M,1} \\ S0_{M,2} \end{pmatrix} , 0.5 , 1 , M , D \right]$$

$$S1_{M,2} = 13.801$$

The solution of the boundary value problem is shown in Figure 12.5. From the physicist's point of view, it is natural for the light intensity in a denser medium (in the right part of the calculation interval) to decrease more rapidly. In the intermediate point xf = 0.5, as expected, the derivatives of both solutions have discontinuity.

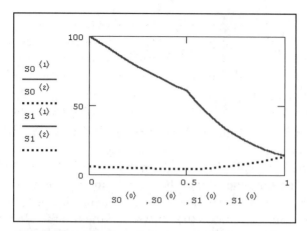

FIGURE 12.5. Solution of the boundary value problem with discontinuity at $xf = 0.5$ (Listings 12.3 and 12.4)

The Mathcad on-line Help system provides just another example of how to solve boundary value problems with discontinuities.

NOTE

   It should be noticed that the boundary value problem considered above can also be solved using the sbval function. To do so, it is sufficient to replace the a(x) dependence in Listing 12.2 with the third line from Listing 12.3. In this case, Listing 12.2 will give exactly the same answer as shown in Figure 12.5. However, sometimes (especially when you need to make calculations faster), it is preferable and more convenient to use the bvalfit function (i.e., perform the shooting from both boundaries of the integration interval).

If you are dealing with similar equations, first try to solve them as normal boundary value problems using the simpler and more reliable sbval function.

TIP

12.2. EIGENVALUE PROBLEMS FOR ORDINARY DIFFERENTIAL EQUATIONS

Eigenvalue problems are boundary value problems for systems of ordinary differential equations where the right parts of equations depend on one or more parameters λ. Parameter values are unknown, and the boundary value solution exists for only specific λ_k, which are known as eigenvalues) of the problem. Solutions corresponding to these eigenvalues λ_k, are known as eigenfunctions of the problem. Correct formulation of such problems requires you to specify boundary conditions, the total number of which would be equal to the sum of the number of equations and the number of eigenvalues. Physical examples of eigenvalue problems are, for example, the string oscillation equation, the Shroedinger equation in quantum mechanics, wave equations in resonators, and many more.

In respect to calculation, eigenvalue problems are very similar to the above-described boundary value problems. In particular, the shooting method *(see Section 12.1.2)* is applicable to most of them. The difference is in the fact that when solving eigenvalue problems, it is necessary to use shooting to match not only missing left boundary values, but also the sought eigenvalues. For solving eigenvalue problems, Mathcad provides the same sbval and bvalfit functions. The initial guess for an eigenvalue must be included in the first argument of these functions — the vector assigning initial values to the missing initial conditions.

Let us consider this method through an example of determining the eigenvalues of string oscillations. The profile of the string oscillations y(x) is described by a linear differential equation of the second order:

$$\frac{d}{dx}\left[p(x) \cdot \frac{dy(x)}{dx} \right] + \lambda \cdot q(x) \cdot y(x) = 0 \tag{1}$$

Here, p(x) and q(x) are rigidity and density that, generally speaking, can vary along the string. If the string is fastened at both ends, then boundary values are specified in the form of the following equation: y(0) = y(1) = 0. The task formulated in this form represents a particular case of the Sturm-Liouville problem. Since here we are solving the system of two ordinary differential equations containing one eigenvalue λ, the problem, principally, requires three (2 + 1) conditions to be formulated. However, as you can easily see, the string oscillations equation is linear and uniform, therefore, in any case, the solution y(x) will be determined with precision up to the multiplier. This means that the solution derivative can be specified arbitrarily, for example, y'(0) = 1. This equality will be used as the third condition. After that, the boundary value problem can be solved like the Cauchy problem, and it can perform shooting to match only one parameter — namely, the eigenvalue.

Listing 12.5 presents the procedure involved in finding the first eigenvalue.

LISTING 12.5. Solving the String Oscillations Eigenvalues Problem

$$p(x) := 1 \qquad q(x) := 1 \qquad p'(x) := 0$$

$$a := 0 \qquad b := 1$$

$$\lambda_0 := 0.5$$

$$D(x, y) := \begin{bmatrix} y_1 \\ -\dfrac{1}{p(x)} \cdot \left(p'(x) \cdot y_1 + y_2 \cdot q(x) \cdot y_0\right) \\ 0 \end{bmatrix}$$

$$\text{load}(a, \lambda) := \begin{pmatrix} 0 \\ 1 \\ \lambda_0 \end{pmatrix}$$

$$\text{score}(b, y) := y_0$$

$$\Lambda := \text{sbval}\left(\lambda, a, b, D, \text{load}, \text{score}\right)$$

$$\Lambda = (\ 9.87\ ) \qquad 1^2 \cdot \pi^2 = 9.87$$

The first two lines of this listing determine the functions that make up the task formulation, including $p'(x) := 0$, and the boundaries of the integration interval $(0, 1)$. The third line provides the guess value for the eigenvalue λ_0, while the fourth line defines the system of ordinary differential equations. Pay special attention to the fact that this system comprises three equations rather than two, as might be expected. The first two equations determine the system of ordinary differential equations of the first order equivalent to the equation (1), while the third equation is required for specifying the eigenvalue in the form of the additional component y_2 of the sought vector y. Since, according to the definition, the eigenvalue is constant for all x, its derivative must be equal to zero (this is the fact specified by the last equation). It is important to notice that the second equation includes the eigenvalue as y_2, since it is one of the unknowns.

The next two lines of the listing specify the left boundary condition including the missing initial condition for the eigenvalue for the third equation and the right boundary condition $y_0 = 0$. In the next to last line of the listing, the sbval function is applied in a normal way, and the last line returns its result along with the analytically defined eigenvalue $n^2 \cdot \pi^2$. As you can easily see, we have found the first eigenvalue for $n = 1$. To find other eigenvalues, it is necessary to specify other guess values (in the third line of the Listing 12.5). For example, the selection of $\lambda_0 = 50$ will result in returning the second eigenvalue $2^2 \cdot \pi^2$, and if $\lambda_0 = 100$, the function will return the third eigenvalue $3^2 \cdot \pi^2$.

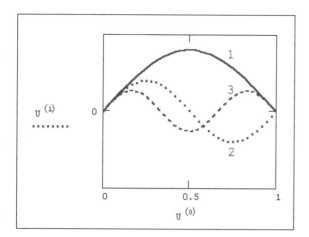

FIGURE 12.6. The first three eigenfunctions of the string oscillations problem

To construct a graph of the appropriate eigenfunction, you need to add a new line to the listing. This additional line must program the solution of the Cauchy problem, for example, as follows: `U := rkfixed(load(a,Λ),a,b,100,D)`. The resulting curves are shown in Figure 12.6 as three graphs are calculated for three eigenvalues.

The Mathcad Resource Center provides several examples that illustrate the procedures for solving eigenvalue problems.

12.3. DIFFERENCE SCHEMES FOR ORDINARY DIFFERENTIAL EQUATIONS

Quite a large number of boundary value problems can't be solved using the shooting method. However, Mathcad 2001i doesn't provide other built-in algorithms. Still, this doesn't mean that boundary value problems can't be solved using other methods, since the user can implement other numeric algorithms programmatically. For example, let us consider the possible implementation of the finite difference method that allows one to solve boundary value problems both for ordinary and partial differential equations.

12.3.1. About the Difference Method when Solving Ordinary Differential Equations

Let us consider difference method implementation for using a boundary value problem on the example of the model of interaction between light beams considered earlier (see Figure 12.1). To do this, let us consider the system (refer to (1) in Section 12.1) and reassign some values by designating the intensity in the right-ward direction as Y, and by identifying the intensity of radiation in the leftward di-rection with y (this is done for the sake of simplicity, to avoid writing subscripts). When using the difference method, we cover the calculation interval with a mesh of N points. Thus, we are defining (N - 1) steps (Figure 12.7). After that, it is nec-essary to replace differential equations of the original boundary value problem by approximating equations in finite differences. For this purpose, let us write ap-propriate difference equations for each i-th step. In our case, it is sufficient to re-place the first derivatives from (1) in Section 12.1 with their difference analogs (this method is also known as *Euler's method*):

$$\frac{Y_{i+1} - Y_i}{\Delta} = -a(x_i) \cdot Y_i + r(x_i) \cdot y_i \tag{1}$$

$$\frac{y_{i+1} - y_i}{\Delta} = a(x_i) \cdot y_i - r(x_i) \cdot Y_i$$

There are lots of methods that approximate differential equations with difference equations. Selection of a specific method influences the simplicity of implementa-tion, convenience, and speed of calculations, as well whether or not one will ob-tain the correct solution.

In the example under consideration, we have a system of $2 \times (N - 1)$ difference algebraic equations in $2 \cdot N$ unknowns Y_i and y_i. For this system to have a unique solution, one needs to complement the number of equations up to $2 \cdot N$. This can be done by writing both boundary conditions in the form of difference equations:

$$Y_0 = I0, \quad y_N = R \cdot Y_N \tag{2}$$

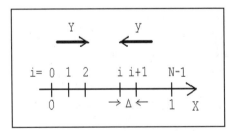

FIGURE 12.7. The mesh covering the calculation interval

Formulating a system of algebraic equations in such a way is known as differ-ence scheme approximating the initial boundary value problem. Take note of the fact that the right parts of difference equations of system (1) at each step are writ-ten for the left boundary of the step. Such difference schemes are known as ex-plicit, since all values Y_{i+1} and y_{i+1} are in the left parts of the equations. The re-sulting explicit difference scheme can be easily written in matrix form:

$$A \cdot z = B \tag{3}$$

where z is an unknown vector obtained by joining vectors Y and y. By solving the system (3), we'll get the solution of a boundary value problem.

In practice, the situation is more complicated, since, generally speaking, it is nec-essary to first prove that the resulting difference scheme actually approximates the differential equations and, second, that the difference solution actually converges with the differential solution at $N \to \infty$.

The process of solving the system of difference equations is also known as the implementation of the difference scheme. The program that solves the boundary value problem described earlier using the difference method is provided in List-ing 12.6.

LISTING 12.6. Implementation of the Explicit Difference Scheme

$$a(x) := 1 \qquad r(x) := 0.01 \qquad I0 := 100 \qquad R := 1$$

$$N := 5 \qquad\qquad \Delta := \frac{1}{N-1}$$

$$i := 1 .. \quad N-1$$

$$A_{i,i} := a(i \cdot \Delta - \Delta) \cdot \Delta - 1 \qquad A_{i,i+1} := 1 \qquad A_{i,i+N} := -\Delta \cdot r(i \cdot \Delta - \Delta)$$

$$i := N+1 .. \quad 2 \cdot N-1$$

$$A_{i,i} := -a(i \cdot \Delta - \Delta) \cdot \Delta - 1 \qquad A_{i,i+1} := 1 \qquad A_{i,i-N} := \Delta \cdot r(i \cdot \Delta - \Delta)$$

$$A_{0,1} := 1$$

$$A_{N,N \cdot 2} := 1 \qquad A_{N,N} := -R$$

$$i := 1 .. \quad 2 \cdot N-1$$

$$B_i := 0$$

```
B₀ := I0

A := submatrix (A, 0, 2·N − 1, 1, 2·N)

U := lsolve (A, B)

Uɴ = 13.453          Uɴ₋₁ = 31.78
```

Let us provide just a few comments, hoping that the interested reader will perform all the required math conversions and calculations to get a sound understanding of indices and matrix elements. Perhaps that such readers will even provide better implementation of the program, than the one presented here.

The first line of Listing 12.6 defines functions and constants included in the model. The second line specifies the number of grid points $N = 5$ and its uniform step. The next two lines define matrix coefficients approximating equations for Y_i, while the fifth and the sixth lines perform the same operations for y_i. The seventh and eighth lines of the listing specify the left and right boundary conditions, respectively, and the 9-th — 11-th lines specify the right parts of the system (3). The next line is involved in constructing the matrix A by extracting its leftmost 0-th column. The next to last line of the listing applies the built-in lsolve function in order to solve the system (3), while the last line returns the calculated boundary values. The graphs of the solution are provided in Figure 12.8. Here the first N elements of the resulting vector represent the calculated radiation in a forward direction, while the last N elements represent the radiation in a backward direction.

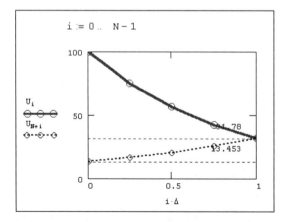

FIGURE 12.8. The solution of the boundary value problem obtained using the difference method (Listing 12.6)

As we have seen, implementation of difference schemes in Mathcad is possible. Furthermore, it doesn't present serious difficulties. For example, the program discussed earlier comprises only a couple dozen math expressions. Indeed, to implement such a program, it is necessary to spend both time and effort to perform the calculations, but, after all, this is the job of mathematicians. By the way, if there are not many steps, the calculations with difference schemes won't be time-consuming (for example, the program provided in Listing 12.6 works faster than the shooting method implemented in the sbval function). Furthermore, there are methods of speeding up the calculations by means of using more appropriate methods for solving systems linear equations with a sparse matrix.

12.3.2. Stiff Boundary Value Problems

Solving stiff boundary value problems is one of the cases where the usage of difference schemes might prove to be very useful (more detailed information on the stiff ordinary differential equations is provided in *Chapter 11*). In particular, the problem of oncoming beams of light becomes stiff when the attenuation coefficient a(x) increases significantly (several times tenfold or more). For example, an attempt to solve this problem for a(x) := 100 using Listing 12.2 will result in the error message "Can't converge to a solution. **Encountered too many integration steps.**" This is not surprising; the requirement to use extremely small step values in standard algorithms is characteristic of stiff systems.

Explicit difference schemes discussed in the previous section are also not applicable to solutions of stiff systems. For example, the result of calculations using the program provided in Listing 12.6 for a(x) := 20 (Figure 12.9), shows a divergence characteristic for unstable difference schemes — it produces oscillations with increasing amplitude that have nothing in common with the actual solution.

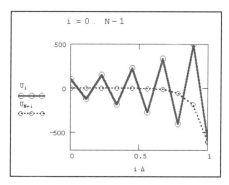

FIGURE 12.9. Incorrect solution of the stiff boundary value problem using an unstable explicit difference scheme

To overcome this problem, it is necessary to use implicit difference schemes. For example, to solve our problem, it is sufficient to replace the right parts of the equations (1) with the values at the right boundary of each step:

$$\frac{Y_{i+1} - Y_i}{\Delta} = -a(x_{i+1}) \cdot Y_{i+1} + r(x_{i+1}) \cdot y_{i+1}$$

$$\frac{y_{i+1} - y_i}{\Delta} = a(x_{i+1}) \cdot y_{i+1} - r(x_{i+1}) \cdot Y_{i+1}$$

(4)

Boundary conditions can be left as is (2). Since we are dealing with linear differential equations, the scheme (4) can be easily written as matrix equality (3) by re-grouping the expression (4) as appropriate, and collecting the terms. Certainly, the resulting matrix A will differ from the matrix A obtained for the explicit scheme (1). As a result, the solution (implementation of the implicit scheme) might be different from the result of calculations by the explicit scheme shown in Figure 12.9. The program created for solving the system (4) is provided in Listing 12.7.

LISTING 12.7. Implementation of the Implicit Difference Scheme for a Stiff Boundary Value Problem

$$a(x) := 20 \qquad r(x) := 0.01 \qquad I0 := 100 \qquad R := 1$$

$$N := 20 \qquad \Delta := \frac{1}{N-1}$$

$$i := 1 .. \quad N-1$$

$$A_{i,i} := 0 - 1 \qquad A_{i,i+N+1} := -\Delta \cdot r(i \cdot \Delta - \Delta) \qquad A_{i,i+1} := a(i \cdot \Delta - \Delta) \cdot \Delta + 1$$

$$i := N+1 .. \quad 2 \cdot N - 1$$

$$A_{i,i} := 0 - 1 \qquad A_{i,i+1} := 1 - a(i \cdot \Delta - \Delta) \cdot \Delta \qquad A_{i,i-N+1} := \Delta \cdot r(i \cdot \Delta - \Delta)$$

$$A_{0,1} := 1$$

$$A_{N,N\cdot2} := 1 \qquad A_{N,N} := -R$$

$$i := 1 .. \quad 2 \cdot N - 1$$

$$B_i := 0$$

$$B_0 := I0$$

$$A := \text{submatrix}(A, 0, 2 \cdot N - 1, 1, 2 \cdot N)$$

$$U := \text{lsolve}(A, B)$$

$$U_N = 0.017 \qquad U_{N-1} = 1.523 \times 10^{-6}$$

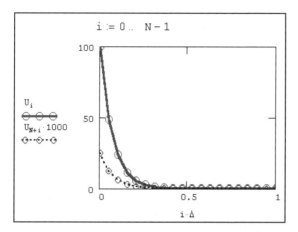

FIGURE 12.10. Solution of the boundary value problem using the implicit difference scheme (Listing 12.7)

We won't specially concentrate on a discussion of Listing 12.7, since it is very similar to the previous one. The only difference is the method used for constructing the matrix A, which, in this case, corresponds to the implicit scheme. The solution shown in Figure 12.10 demonstrates that the miraculous has happened: the divergence has disappeared, and the intensity distribution has become physically predictable. Notice that as a result of the large attenuation coefficient, the reflected beam has a very low intensity (to make it visible on the graph, it is necessary to re-scale it by 1000-fold magnification).

All examples considered in this section are related to boundary value problems for linear ordinary differential equations. Because the initial equations are linear, the resulting system of difference equations is also linear. However, if the differential equations are non-linear, the solution of a system using difference schemes becomes more complicated. A detailed description of Mathcad's programs that are appropriate for such cases goes far beyond the scope of this book.

12.4. PARTIAL DIFFERENTIAL EQUATIONS

In contrast to ordinary differential equations, which require you to find an unknown function of one variable, partial differential equations require that you find functions of several variables (two or more). These equations include partial derivatives of different variables. Partial differential equations describe a wide range of various physical phenomena and can be successfully used for modeling rather complex processes, such as diffusion, fluid and gas dynamics, quantum mechanics, ecology, and so on. Mathcad provides limited capabilities in relation to partial differential equations. Its built-in functions allow you to solve only some problems describing particular cases of specific phenomena.

12.4.1. Solving the Poisson's Equation

The two-dimensional Poisson's equation represents an example of a partial differential equation of the elliptic type, which includes second derivatives of the $T(x,y)$ function of two variables:

$$\frac{\partial^2 T(x, y)}{\partial x^2} + \frac{\partial^2 T(x, y)}{\partial y^2} = -f(x, y) \qquad (1)$$

For example, the Poisson's equation can be used to describe the distribution of the electrostatic field $T(x,y)$ in a 2D area with the charge density $f(x,y)$, or a stationary distribution of temperature $T(x,y)$ on the plane containing sources (or absorbers) with $f(x,y)$ intensity. We'll consider Poisson's equation in the latter physical interpretation (that's why we have designated the sought function by the symbol generally used to designate temperature).

The correct formulation of the boundary value problem for Poisson's equation requires that you specify boundary conditions. In Mathcad, the solution is sought at the plane of the quadrangular area covered with the grid including $(M + 1) \times (M + 1)$ points. Therefore, the boundary conditions must be defined for all sides of the above-mentioned square. The simplest method is to specify zero boundary conditions, i.e., the constant temperature of the whole perimeter of the calculation area. In this case, it is possible to use the built-in `multigrid` function.

The `multigrid(F,ncycle)` function specifies the solution matrix of Poisson's equation with dimension $(M + 1) \times (M + 1)$ at the square area with zero boundary conditions, where:

- ■ `F` — $(M + 1) \times (M + 1)$ matrix, specifying the right part of Poisson's equation
- ■ `ncycle` — Parameter of the numeric algorithm (number of cycles within each iteration)

The side of the square calculation area must comprise $M = 2^n$ points, where n is an integer.

TIP

In most cases, it is sufficient to set the value of the `ncycle` numeric method parameter to 2. Listing 12.8 contains an example of the `multigrid` built-in function for calculating the boundary value problem at the area comprising 33×33 points and containing the point heat source positioned at the point with coordinates (15,20) within that area.

LISTING 12.8. Solution of Poisson's Equation with Zero Boundary Conditions

```
M := 32

F_{M,M} := 0

F_{15,20} := 10^4

G := multigrid (−F , 2)
```

The first line of this listing specifies the value M=32; the next two lines create the matrix of the right part of Poisson's equation. All elements of this matrix are equal to zero, except for the one specifying the position of the point heat source. The last line of the listing assigns the result returned by the `multigrid` function to the G matrix. Notice that the first argument of this function is preceded by the minus sign, which corresponds to the right part of Poisson's equation (1). The graphs of this solution are shown in Figures 12.11 and 12.12 in the form of a 3D surface and level curves, respectively.

In more complex cases, for example, when solving a boundary value problem with nonzero boundary conditions, it is necessary to use another built-in function provided by Mathcad, namely, the `relax` function.

The `relax(a,b,c,d,e,F,v,rjac)` function specifies the solution matrix of a partial differential equation of the square area obtained by the relaxation algorithm for the grid method, where:

- `a,b,c,d,e` — Square coefficient matrices of the difference scheme approximating the differential equation
- `F` — Square matrix specifying the right part of the differential equation
- `v` — Square matrix of boundary conditions and initial guess values
- `rjac` — Parameter of the numeric algorithm (spectral radius of the Jacobi iterations)

The parameter of the numeric algorithm characterizes the convergence rate of the iterations. It must be a number from the range from 0 to 1. As for the boundary value matrix v, here you must specify only boundary elements, based on the boundary

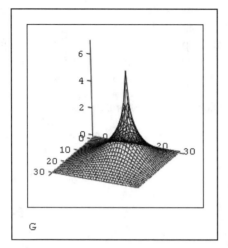

FIGURE 12.11. 3D surface graph representing the solution to Poisson's equation (Listing 12.8)

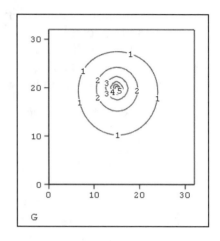

FIGURE 12.12. Graph of level curves representing the solution to Poisson's equation (Listing 12.8)

conditions at the perimeter of the calculation area. Other (internal) elements of this matrix specify initial guess values for the solution. The idea behind the relaxation algorithm is as follows: in the course of iterations, the equations are checked and the values of the unknown function are corrected appropriately at each point. If the initial guess was selected satisfactorily, we can hope that the algorithm will converge ("relax") with the correct solution.

NOTE

All matrices specifying the difference scheme coefficients $a, b, c, d, e,$ *boundary values* $v,$ *and the solution* $F,$ *must have the same dimension* $(M + 1) \times (M + 1),$ *corresponding to the dimension of the calculation area. The integer number* M *must be a power of* 2 : $M = 2^n.$

The solution to Poisson's equation with three sources of different intensity using the `relax` function is provided in Listing 12.9.

LISTING 12.9. Solution of Poisson's Equation Using the `relax` Function

$M := 32$

$F_{M, M} := 0$

$F_{15, 20} := 10 \qquad F_{25, 10} := 5 \qquad F_{10, 10} := -5$

```
i := 0 .. M        k := 0 .. M

a_{i,k} := 1

b := a

c := a

d := a

e := -4 · a

v_{i,k} := 0

G := relax (a, b, c, d, e, -F, v, .95)
```

The first three lines have the same meaning as the first lines of the previous listing. In this case, however, we have more than one heat source: one powerful source, another less powerful source, and one heat absorber. The next six lines specify the coefficients of the difference scheme. These coefficients will be discussed in the last section of this chapter. Here we'll limit ourselves to stating that in order to solve Poisson's equation, one must choose the coefficients shown in Listing 12.9. The next to last line of the listing specifies the matrix of zero boundary conditions and zero initial guess values. The last line assigns the result returned by the `relax` function to the G matrix. Figure 12.13 presents a graph of the obtained solution represented as level curves.

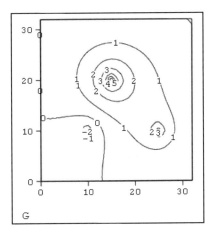

Figure 12.13. Solution to Poisson's equation using the `relax` function (Listing 12.9)

12.4.2. Difference Schemes

Mathcad's built-in functions allow you to solve only the simplest partial differential equations. Generally, users who need to work through these equations have a good understanding of appropriate numeric algorithms, most often related to calculations of difference schemes. Therefore, in this section we'll limit ourselves to a demonstration of the program that implements the grid method for solving the heat conduction (or heat diffusion) equation:

$$\frac{\partial T(x, y)}{\partial t} = D \frac{\partial^2 T(x, y)}{\partial x^2} + \phi(x, T(x, y), t) \tag{2}$$

This is an equation of the *parabolic type*, containing the first derivative of time t and the second derivative of the space coordinate x. It describes the temperature dynamics $T(x,t)$ in the presence of heat sources $\phi(x,T,t)$, for example, when heating the metal bar. The difference scheme for the equation (2) appears as presented below:

$$K \cdot \left(T_{i,k+1} - T_{i,k}\right) = T_{i-1,k} - 2 \cdot T_{i,k} + T_{i+1,k} + \phi_{i,k} \tag{3}$$

Here the coefficient K characterizes the ratio of the difference scheme steps by space and by time $K = \dfrac{\Delta_x^2}{\Delta_t \cdot D}$. To construct the difference scheme (3), we have covered the calculation area (x,t) with the grid and used node configuration (template), shown in Figure 12.14, to approximate the equation (2) (The principles of construction difference schemes are covered in Section 12.3.1.) In this illustration, the values of the coefficients of the difference scheme obtained for respective grid nodes are shown near each point. These coefficients are obtained after collecting like terms in scheme (3).

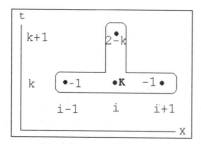

FIGURE 12.14. The template used for approximating the heat conduction equation

Implementation of the difference scheme (3) for the model without heat sources $\phi(x, T, t) = 0$ is provided in Listing 12.10. The first three lines of this listing specify the steps (τ and Δ) by time and space variables and the diffusion coefficient D, which is equal to 1. The next two lines specify initial (heated center of the area) and boundary (constant temperature at the edges) conditions, respectively. These lines are followed by one of the possible programmatic solutions of the difference scheme, where the custom function V(t) specifies the distribution vector of the sought temperature for each time moment specified by the integer value t.

LISTING 12.10. Solution of the Linear Heat Condition Equation

$$\tau := 0.0005 \qquad M := 20$$

$$\Delta := \frac{1}{M}$$

$$D(u) := 1 \qquad \phi(x, u) := 0$$

$$\text{Init}(x) := \Phi(x - 0.45) - \Phi(x - 0.55)$$

$$\text{Border}(t) := 0$$

$$m := 0 \ .. \ M$$

$$u_m := \text{Init}(m \cdot \Delta)$$

$$F(v) := \begin{vmatrix} v1_0 \leftarrow \text{Border}(\tau \cdot T) \\ v1_M \leftarrow \text{Border}(\tau \cdot T) \\ \text{for } m \in 1 \ .. \ M-1 \\ \qquad v1_m \leftarrow \phi(m \cdot \Delta, v_m) \cdot \tau + v_{m-1} \cdot \dfrac{D(v_{m-1}) \cdot \tau}{\Delta^2} + v_m \cdot \left(1 - \dfrac{2 \cdot D(v_m) \cdot \tau}{\Delta^2}\right) + v_{m+1} \cdot \dfrac{D(v_{m-1}) \cdot \tau}{\Delta^2} \\ v1 \end{vmatrix}$$

$$V(t) := \begin{vmatrix} u & \text{if } t = 0 \\ F(V(t-1)) & \text{otherwise} \end{vmatrix}$$

Initial distribution of the temperature along the calculation area and solutions for two time moments are shown in Figure 12.15 by solid, dashed, and dotted lines, respectively. In respect to physics, such behavior is quite natural, since with time, heat is transferred from a hotter area to a colder area, and the high-temperature zone cools and smears.

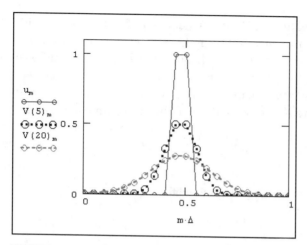

FIGURE 12.15. Solution of the linear heat conductivity equation (Listing 12.10)

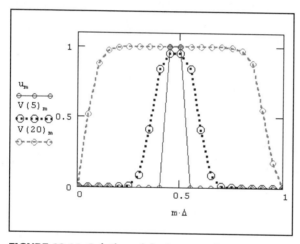

FIGURE 12.16. Solution of the heat conductivity equation with a non-linear heat source (heat front)

Far more interesting results can be obtained for the heat conduction equation, for example, with a non-linear heat source $\phi(x,t) = 10^3 \cdot (T - T^3)$. If we specify the presence of such a source on the third line of Listing 12.10, we'll obtain the solution in the form of heat fronts propagating in both directions from the zone of initial heating (Figure 12.16). If the diffusion coefficient is also non-linear, the solution becomes even more unusual. For example, if we assume that $D(x,T) = T^2$, and $\phi(x,t) = 10^3 \cdot T^{3.5}$, we'll be able to model the effect of burning localized in the area of initial heating. The interested reader will certainly experiment with this and other non-linear variations of the heat conductivity equation. What's more important, such results can only be obtained numerically (in Mathcad obtaining such results requires programming).

12.4.3. Solving Other Partial Differential Equations Using the *relax* Function

Despite the fact that the Mathcad on-line Help system lacks information on how to solve partial linear differential equations other than Poisson's equation, you are able to obtain these solutions using the `relax` function *(see Section 12.4.1)*. For this purpose, you need to correctly specify the coefficients of the difference scheme.

Let us start with an explanation of the selection of these coefficients (see Listing 12.9) for a solution to Poisson's equation. According to the ideas discussed in the previous section, Poisson's equation (1) can be written in the difference form using the "cross" template (Figure 12.17). In this case, after collecting the terms in difference equations, the coefficients of the difference scheme will look like those shown in the illustrations next to the appropriate template nodes (similar coefficients for the heat conductivity equations are shown in Figure 12.14). Now, if you compare the obtained results to the constants assigned to the elements of the argument matrices in the `relax` function (see Listing 12.9), you'll notice that they describe the coefficients of the "cross" difference scheme we have just described.

Thus, you can easily guess that while using the built-in `relax` function, one can also solve other partial differential equations that can be approximated by the "cross" scheme or another scheme representing a part of the "cross" scheme. For example, it is possible to implement the explicit difference scheme for the uniform heat conductivity equation *(see Section 12.4.2)*. For this purpose, it is necessary to specify the coefficients shown in the template (see Figure 12.14) as the arguments of the `relax` function (Listing 12.11).

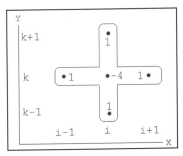

FIGURE 12.17. The "cross" difference scheme template

LISTING 12.11. Solution of the Heat Conductivity Equation Using the `relax` Function

```
M := 16

K := 2.2

i := 0 .. M        k := 0 .. M

a_{i,k} := -1      b_{i,k} := -1

c_{i,k} := 0       d_{i,k} := 2 - K

e_{i,k} := K

F_{i,k} := 0

v_{i,k} := 0       v_{0,5} := 1

G := relax (a, b, c, d, e, F, v, .9)
```

In all other respects, this program works just like the one represented in Listing 12.9. Its result is represented in Figure 12.18 in the form of a 3D surface. If we compare Figure 12.18 to Figure 12.15, whose solution was obtained by calculations with the programmed difference scheme, you'll easily recognize the graphs shown in Figure 12.15 as a cross section of this surface by t=const planes.

To conclude this discussion of partial differential equations, let us say a few words about visualizing the results. Solving dynamic equations depending on time t looks much more impressive if you animate it. To create an animated clip, express the calculated time via the FRAME constant and then apply the View | **Animate** command (more detailed information on this topic will be provided in *Chapter 15*).

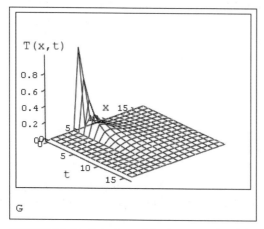

FIGURE 12.18. Solution of the heat conductivity equation using the `relax` function (Listing 12.11)

13 | Mathematical Statistics

Mathcad provides powerful capabilities that work with the problems of mathematical statistics. First, there are several built-in special functions that allow you to calculate probability densities and other basic characteristics of the distribution laws of random values *(see Section 13.1)*. In addition, Mathcad implements an impressive number of pseudo-random value generators for each distribution law *(see Section 13.1)*, which enable you to perform efficient modeling using Monte-Carlo methods. You can create bar charts and calculate statistic characteristics of random value samples and random processes such as averages, dispersions, correlations, etc. *(see Section 13.2)*. Notice that random sequences can be created using random number generators or imported from files.

The concluding sections of this chapter go over some methods of solving several typical problems related to random process analysis *(see Section 13.3)* and mathematical statistics *(see Section 13.4)*.

13.1. RANDOM VALUES

The methods known as Monte-Carlo methods are those most widely used for modeling various physical, economical, and other effects. The main idea of these methods lies in the creation of specific sequences of random values modeling specific effects, such as noise in a physical experiment, random dynamics of stokes indices, etc. For this purpose, Mathcad provides a range of pseudo-random number generators.

According to the definition, a random variable can take some value, but the specific value it takes depends on random circumstances of the experiment and can't be predicted for certain. It is only possible to speak about the *probability* $P(x_K)$ that a discrete random value takes a specific value x_K, or the probability that a continuous random value fits within a specific interval $(x, x + \Delta x)$. The probability $P(x_K)$ or $P(x) \cdot (\Delta x)$ respectively can take values from 0 (which means that this value of the random value is absolutely improbable) to 1 (the random variable will almost certainly take a number from x to x + Δx). The relation $P(x_K)$ is known as the distribution law of the random value, and the dependence $P(x)$ between possible values of continuous random value and the probabilities of their taking values in close proximity is known as its probability density.

Mathcad provides a range of built-in functions specifying distribution laws widely used in mathematical statistics. They calculate both the probability densities of various distributions by using the value of the random variable x and several other functions. Both these functions, in principle, are either built-in analytical dependencies or special functions. The most useful and interesting are generators of pseudo-random numbers that create samples of pseudo-random data with the appropriate distribution law. Let us consider the statistical capabilities of Mathcad with the examples of several of the most popular distribution laws, and then we will provide a listing of all distributions built into Mathcad.

13.1.1. Normal (Gauss) Distribution

It is a proven fact of the probability theory that the sum of different independent random components (independent from their distribution law) is a random value distributed according to the normal distribution law (the so-called *central limit theorem*). As a result, normal distribution is suitable for modeling a large variety of phenomena which are influenced by several independent random factors.

Listed below are Mathcad's built-in functions that are used for describing normal probability distribution:

- dnorm(x,μ,σ) — The probability density of normal distribution
- pnorm(x,μ,σ) — Cumulative probability distribution
- cnorm(x) — Cumulative probability distribution with μ = 0, σ = 1
- qnorm(P,μ,σ) — Inverse cumulative probability distribution

■ rnorm(M,μ,σ) — The vector of M random numbers having a normal distribution

Here:

- x — A random value
- p — The probability
- μ — The expectation or mean value
- σ — Standard deviation

The mean value and standard deviation are principally distribution parameters. Distribution densities for three pairs of parameter values are shown in Figure 13.1. Notice that the distribution density dnorm specifies the probability that the random value x takes the value belonging to the small interval from x to x + Δx. Thus, for example, for the first graph (a solid line) the probability that the random variable x will take the value in close proximity to zero is approximately three times greater than the probability that it will take the value in proximity to the point x = 2. Furthermore, values greater than 5 or less than −5 are extremely unlikely.

The distribution function F(x) (cumulative probability) is the probability that a random variable will take a value less than or equal to x. As follows from its math sense, it represents an integral from the probability density within the limits from -∞ to x. Distribution functions for the normal laws mentioned earlier are shown in Figure 13.2. The inverse function of F(x) (inverse cumulative probability), also known as the distribution quantile, enables you to determine the x value with the specified argument p, where a random value is less than or equal to x with the probability p.

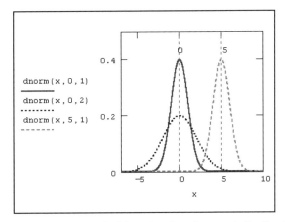

FIGURE 13.1. Probability density of normal distributions

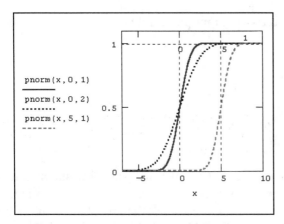

FIGURE 13.2. Normal distribution functions

From now on, all graphs of various statistical functions shown in illustrations will be constructed using Mathcad without any additional expressions in the work area.

Let us consider several calculations illustrating the mathematical idea of the considered functions just discussed in the example of the random value x distributed according to the normal distribution law with μ=0 and σ=1 (Listings 13.1–13.5).

LISTING 13.1. The Probability That x Will Be Less than 1.881

    pnorm (1.881 , 0 , 1) = 0.97

LISTING 13.2. 97% Quantile of the Normal Distribution

    qnorm (0.97 , 0 , 1) = 1.881

LISTING 13.3. The Probability That x Will Be Greater than 2

    1 − pnorm (2 , 0 , 1) = 0.02275

LISTING 13.4. The Probability That x Will Fit within the Interval (2,3)

    pnorm (3 , 0 , 1) − pnorm (2 , 0 , 1) = 0.021

$$\frac{1}{2} \cdot \left(\text{erf}\left(\frac{3}{\sqrt{2}}\right) - \text{erf}\left(\frac{2}{\sqrt{2}}\right) \right) = 0.021$$

LISTING 13.5. The Probability That $|x| < 2$

```
pnorm (2 , 0 , 1) - pnorm (-2 , 0 , 1) = 0.954
```

$$\text{erf}\left(\frac{2}{\sqrt{2}}\right) = 0.954$$

Notice that the problems solved in the two last listings are solved using different methods. The second method is related to another built-in function `erf`, also known as error function or probability integral.

- ◼ `erf(x)` — Error function
- ◼ `erfc(x)` ≡ 1 - `erf(x)`

The mathematical sense of the error function is clear from Listing 13.5. The probability integral, in contrast to the normal distribution function, has only one argument. Historically, it has just become common practice to calculate the normal distribution function via the tabulated probability integral with the formulae provided in Listing 13.6 for arbitrary values for μ and σ parameters (Listing 13.6).

LISTING 13.6. The Probability That x Will Belong to the Interval $(2,3)$

$$\mu := 5 \qquad \sigma := 2$$

$$\text{pnorm}\left(3 , \mu , \sigma\right) - \text{pnorm}\left(2 , \mu , \sigma\right) = 0.092$$

$$\frac{1}{2}\cdot\left(\text{erf}\left(\frac{3-\mu}{\sigma\cdot\sqrt{2}}\right) - \text{erf}\left(\frac{2-\mu}{\sigma\cdot\sqrt{2}}\right)\right) = 0.092$$

If you are dealing with modeling using Monte-Carlo methods, use the built-in `rnorm` function as a generator of random numbers with the normal distribution law. Listing 13.7 illustrates the principle of this function using the example of creating two vectors comprising $M = 500$ elements, each with independent pseudo-random numbers $x1_i$ and $x2_i$, distributed according to the normal law. Figure 13.3 illustrates the character of the distribution of random elements of these vectors. Later on, we'll often deal with the generation of pseudo-random numbers and the calculation of their different mean characteristics.

LISTING 13.7. Generating Two Vectors Distributed According to the Normal Law

$$\sigma := 1 \qquad \mu := 0$$

$$M := 500$$

$$x1 := \text{rnorm}\left(M , \mu , \sigma\right)$$

$$x2 := \text{rnorm}\left(M , \mu , \sigma\right)$$

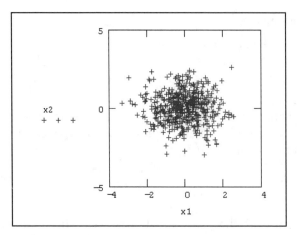

FIGURE 13.3. Pseudo-random numbers with the normal distribution law (Listing 13.7)

13.1.2. Uniform Distribution

The simplest distribution of a random value is the uniform distribution with a constant probability. The probability is evaluated as $P = const = 1/(b-a)$ for $x \in (a,b)$, and as $P = 0$ for x outside the interval (a,b). This probability density, along with other statistical characteristics, are specified by the following functions:

■ dunif(x,a,b) — The probability density of uniform distribution.
■ punif(x,a,b) — The cumulative probability of uniform distribution
■ qunif(P,a,b) — The quantile of uniform distribution
■ runif(M,a,b) — The vector of M independent random values having random distribution
■ rnd(x) — The random value uniformly distributed between (0,x)

Here:

● x — The random variable
● P — The probability
● (a,b) — The interval within which the random value is uniformly distributed

The last function from the list above, which generates a single pseudo-random number, is the one most frequently used in simple programs. The presence of such a built-in function in Mathcad is due to a tradition existing in most programming environments. Figure 13.4 provides an example illustrating the usage

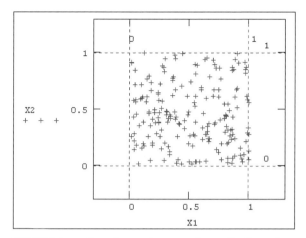

FIGURE 13.4. Pseudo-random numbers distributed according to the uniform law

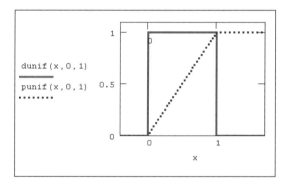

FIGURE 13.5. Probability density and uniform distribution function

of the random generator generating the vector, which comprises M random numbers. As you can see, it is obtained by replacing the normal generator in the last two lines of Listing 13.7 with the `runif(M,0,1)` function. Probability density and function of uniform distribution are shown in Figure 13.5.

13.1.3. Binomial Distribution

Let us discuss the built-in functions describing just another distribution of the random value, which, in contrast to the previous two distributions, is not continuous. This distribution can take only discrete values. Binomial distribution describes the sequence of independent tests, each of which can result in the generation of a specific event with the constant probability p.

The list of built-in Mathcad functions for describing binomial distribution is provided below:

- dbinom(k,n,p) — The probability density of binomial distribution (Figure 13.6)
- pbinom(k,n,p) — The binomial distribution function
- qbinom(P,n,p) — The binomial distribution quantile
- rbinom(M,n,p) — The vector comprising M independent random values, each of which has binomial distribution

Here:

- k — The discrete value of a specific random variable
- P — The probability value
- n — The distribution parameter (the number of independent tests)
- p — The distribution parameter (the probability of each single random event)

As an example of binomial distribution, one can consider the tossing of a coin n-times. The probability of heads or tails in each trial is equal to p = 0.5, and the total sum of trials, resulting, for example, in heads, is characterized by binomial probability density. For example, if we tossed the coin 50 times, then the most probable number of getting heads would be 25, as shown in Figure 13.6, according to the maximum of the probability density. The probability of getting heads 25 times makes up dbinom(25,50,0.5) = 0.112, and, for example, the probability of getting heads 15 times is equal to dbinom(15,50,0.5) = 0.002.

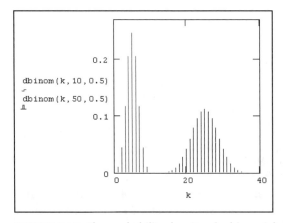

FIGURE 13.6. The probability density of a binomial distribution

13.1.4. Other Statistical Distributions

As can be easily noticed by the three distributions described previously, Mathcad has the following four categories of built-in statistical functions. They differ by the first character of their name, while the remaining part of the function name (in the list below it is designated by the asterisk for generality) identifies the specific type of distribution.

- ▨ d*(x,par) — The probability density
- ▨ p*(x,par) — The distribution function
- ▨ q*(P,par) — The distribution quantile
- ▨ r*(M,par) — The vector comprising M independent random values, each of which has an appropriate distribution type

Here:

- ● x — The random value (function argument)
- ● P — The probability
- ● par — The list of distribution parameters

To produce the functions related, for example, to uniform distribution, replace the * character with unif and enter the appropriate list of parameters par. In this case, the list of parameters will comprise two numbers — a, and b — which specify the interval of the distribution of the random variable.

Now, let us list all types of distributions implemented in Mathcad, along with their parameters. This time, we'll use the asterisk * to designate the missing first letter of the built-in functions. Some of the probability densities are shown in Figure 13.7.

- ▨ *beta(x,s1,s2) — Beta-distribution ($s1, s2 > 0$ — Parameters, $0 < x < 1$)
- ▨ *binom(k,n,p) — Binomial distribution (n — Integer parameter, $0 \le k \le n$, $0 \le p \le 1$ — Parameter equal to the probability of success of each single trial)
- ▨ *cauchy(x,l,s) — Cauchy distribution (l — Location parameter, $s > 0$ — Scale parameter)
- ▨ *chisq(x,d) — χ^2 ("Chi-squared") distribution ($d > 0$ — Number of degrees of freedom)
- ▨ *exp(x,r) — Exponential distribution ($r > 0$ — Exponential rate)
- ▨ *F(x,d1,d2) — Fisher's distribution ($d1, d2 > 0$ — Numbers of degrees of freedom)
- ▨ *gamma(x,s) — Gamma distribution ($s > 0$ — Shape parameter)

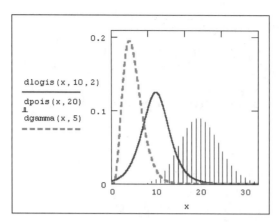

FIGURE 13.7. Probability density of some distributions

- ■ *geom(k,p) — Geometric distribution ($0 \leq p \leq 1$ — Probability of success for a specific trial)
- ■ *hypergeom(k,a,b,n) — Hypergeometric distribution (a,b,n — Integer parameters)
- ■ *lnorm(x,μ,σ) — Log normal distribution (μ — Natural log of the mean, σ>0 — Natural log of the standard deviation)
- ■ *logis(x,l,s) — Logistic distribution (l — Location parameter, s > 0 — Scale parameter)
- ■ *nbinom(k,n,p) — Negative binomial distribution ($n > 0$ — Integer parameter, $0 < p \leq 1$)
- ■ *norm(x,μ,σ) — Normal distribution (μ — Mean, σ>0 — Standard deviation)
- ■ *pois(k,λ) — Poisson distribution ($λ > 0$ — Parameter)
- ■ *t(x,d) — Student's distribution ($d > 0$ — Number of degrees of freedom)
- ■ *unif(x,a,b) — Uniform distribution ($a < b$ — Boundaries of the interval)
- ■ *weibull(x,s) — Weibull distribution ($s > 0$ — Parameter)

The most convenient way to insert the statistical functions we just considered into the programs is to use the **Insert Function** dialog. To achieve this, proceed as follows:

1. Position the cursor in the desired position within a document, where you are going to insert the function.
2. Open the **Insert Function** dialog by clicking the f(x) button on the standard toolbar. Alternately, you can select the **Insert | Function** commands from the menu or press the <Ctrl>+<E> keys.

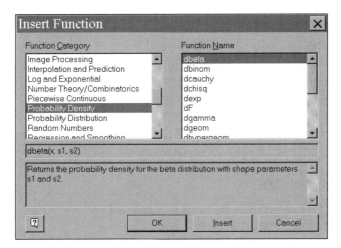

Figure 13.8. The **Insert Function** dialog

3. From the Function Category list (Figure 13.8), select one of the categories of statistic functions. The **Probability Density** category contains the built-in functions for probability densities, the **Probability Distribution** category contains the built-in functions for distribution functions and quantiles, the **Random Numbers** category provides the list of random number generator functions.

4. Select the name of the required function from the **Function Name** list, depending on the desired distribution law. When you select a list item, information on the selected function will be displayed in the text fields at the bottom of this window.

5. Click OK to insert the selected function into the document.

13.2. STATISTICAL CHARACTERISTICS

In most statistical calculations, you are dealing with either random data obtained experimentally (data input is performed from the file or the user enters data directly into the document), or with the results of the generation of a set of random values using the built-in functions considered in the preceding section. These functions enable you to model specific phenomenon using the Monte-Carlo method. Let us consider Mathcad capabilities of evaluating distribution functions and calculating numeric characteristics of the random data.

13.2.1. Constructing Histograms

A *histogram* is the graph that approximates the density of the distribution of random data. When constructing the histogram, the range of values of the random variable (a,b) is divided into the specified number of segments (bin), then the percentage of values that fit within each segment is calculated. Mathcad 2001i provides several built-in functions for constructing histograms. Let us consider these functions, starting with the one that is the most difficult to use, to get a better understanding of the capabilities of each function.

A Histogram with Arbitrary Segments

The hist(intvls,x) function is the vector representing the frequencies with which the data falls into intervals of the histogram, where:

- intvls — The vector whose elements specify the intervals used to construct a histogram, where $a \le \text{intvls}_i < b$
- x — The vector of random data

If the intvls vector has bin elements, the result of applying the hist function will have the same number of elements. Listing 13.8 and Figure 13.9 illustrate the procedure of constructing the histogram.

LISTING 13.8. Constructing a Histogram

```
N := 1000

bin := 30

x := rnorm (N , 0 , 1)

lower := floor (min (x) )

upper := ceil (max (x) )
```

$$h := \frac{\text{upper} - \text{lower}}{\text{bin}}$$

```
j := 0 .. bin

int_j := lower + h · j
```

$$f := \frac{1}{N \cdot h} \cdot \text{hist (int , x)}$$

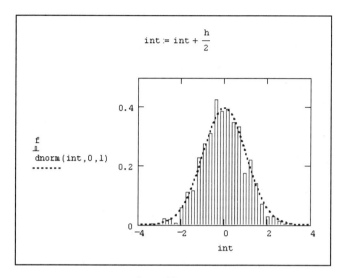

FIGURE 13.9. Constructing a histogram (Listing 13.8)

For analysis, we have taken N = 1000 elements of random data with the normal distribution law. This random data was generated using the random number generator (the third line of the listing). The next lines determine the boundaries of the interval (upper, lower) that contains all the random values within it. Then this interval is divided into the specified number of the same segments (bin), initial points of which are written into the vector int (the next to last line of the listing).

When constructing the int vector, it is possible to specify different boundaries of the segments into which the interval is divided. Thus, it is possible to use segments of different widths.

Notice that the last line of the listing normalizes the histogram values so that they correctly approximate the probability density, also shown on the graph. Redefinition of the int vector at the top of Figure 13.9 is required for transition from the left boundary of each elementary segment to its center.

Histogram with Equal Segment

If there is no need to specify histogram segments of different widths, then it is much more convenient to use the simplified variation of the hist function.

■ hist(bin,x) — The vector specifying the frequency of the data falling within histogram intervals:
■ bin — The number of segments of the histogram
■ x — The vector of random data

To use this variation of the hist function instead of the previous one, it is sufficient to replace the first of its arguments in Listing 13.8 as follows:

$$f := \frac{1}{N \cdot h} \cdot hist(int, x)$$

The drawback of the simplified form of the hist function is that it still requires that one define the segment vector needed to construct the histogram. The new histogram function first introduced with Mathcad 2001, however, does not require this.

The histogram(bin,x) function is the matrix of the histogram with the size bin × 2, comprising the column composed of the segments into which the interval is divided and the column containing frequencies with which the data falls within the respective segment, where:

■ bin — Number of segments
■ x — Vector of random data

Listing 13.9 and Figure 13.10 give examples illustrating the usage of the histogram function. In comparison to the previous listing, this method of constructing the histogram is very simple and easy (notice that in Listing 13.9, in contrast to the previous one, we don't need to normalize the histogram).

LISTING 13.9. A Simplified Method of Constructing a Histogram

```
N := 1000

bin := 10

x := rnorm(N, 0, 1)

f := histogram(bin, x)
```

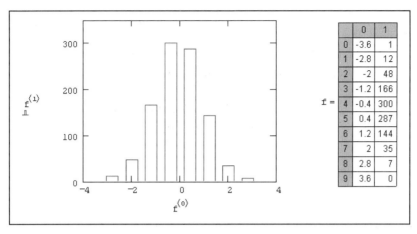

FIGURE 13.10. Graph and matrix of the histogram (Listing 13.9)

Constructing Bar Charts

To represent a graph in the form of a histogram, proceed as follows:

1. Create a 2D graph and specify axis variables and interval boundaries along the x-axis (in the example provided in Listing 13.9 these are lower and upper numbers).

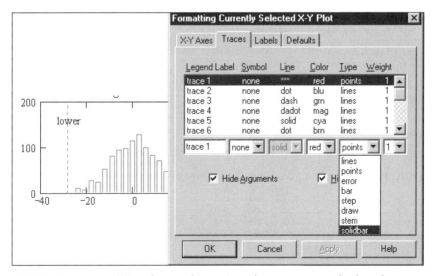

FIGURE 13.11. Specifying the graph type in order to construct the bar chart

2. Open the **Formatting Currently Selected Graph** dialog (for example, by double-clicking the graph) and go to the **Traces** tab.
3. Go to the **Type** field and select the **bar** (columns) or **solidbar** (histogram) types for the data series, as shown in Figure 13.11.
4. Click **OK**.

For the histograms shown in Figures 13.9 and 13.10, the **bar** setting was selected. Mathcad 2001i introduced the new feature of constructing the histogram in a more customary form — as a bar chart (solidbar). Figure 13.11 illustrates such a graph.

13.2.2. Mean Value and Dispersion

Mathcad 2001i provides a range of built-in functions for calculating statistical characteristics of random data sequences:

- `mean(x)` — The arithmetic means of the data sample
- `median(x)` — The median of the data sample, which represents the argument value that divides the probability density histogram into two equal parts
- `var(x)` — The variance of the data sample
- `stdev(x)` — The root mean square or «standard» deviation of the set of data
- `max(x)`, `min(x)` — The maximum and minimum values of the data sample
- `mode(x)` — The value within the data sample that occurs most often
- `Var(x)`, `Stdev(x)` — The sample variance and standard deviation using another normalization scheme

Here:

- x — The vector (or matrix) containing a sample of random data

Listing 13.10 provides some examples illustrating the usage of the first four functions.

LISTING 13.10. Calculating Numeric Characteristics of a Random Vector

```
x := rweibull (1000 , 1.5)

N := length (x)       N = 1 × 10³

mean (x) = 0.917

median (x) = 0.781
```

$\text{var}(x) = 0.402$

$\text{stdev}(x) = 0.634$ $\sqrt{\text{var}(x)} = 0.634$

$\text{hi} := \text{mean}(x) + \text{stdev}(x)$ $\text{lo} := \text{mean}(x) - \text{stdev}(x)$

Figure 13.12 shows the histogram of the sample of random numbers distributed according to the Weibull law. The values corresponding to dashed vertical lines shown in this graph are calculated in the last line of the listing and designate the standard deviation from the mean value. This histogram was obtained using Listing 13.8 considered in the previous section. Notice that since in contrast to, say, a Gauss distribution, a Weibull distribution is asymmetric, the median value doesn't coincide with the mean value.

The definition of statistic characteristics of random values is provided in Listing 13.11 with the example of processing of a small sample (five data elements). The same listing illustrates the usage of two other functions that return dispersion and standard deviation with a somewhat different normalization method. By comparing different expressions, you'll easily notice the difference between the built-in functions.

Be very careful when selecting the first character (upper- or lowercase) in the names of these functions, especially when processing small samples (Listing 13.11).

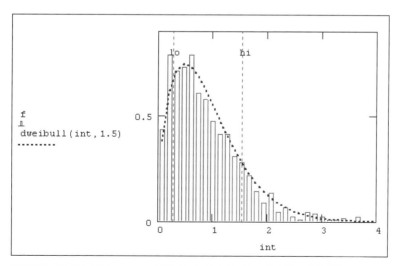

FIGURE 13.12. Histogram of Weibull distribution (Listing 13.10)

LISTING 13.11. Illustration of the Usage of Statistical Characteristics

$$x := (\begin{array}{ccccc} 5 & 2 & 14 & 3 & 2 \end{array})^T$$

$$N := \text{length}(x) \qquad N = 5$$

$$\frac{1}{N} \cdot \sum_{i=0}^{N-1} x_i = 5.2 \qquad m := \text{mean}(x) \qquad m = 5.2$$

$$\text{median}(x) = 3 \qquad \text{mode}(x) = 2 \qquad \text{max}(x) = 14 \qquad \text{min}(x) = 2$$

$$\frac{1}{N} \cdot \sum_{i=0}^{N-1} (x_i - m)^2 = 20.56$$

$$\text{var}(x) = 20.56 \qquad \text{Var}(x) \cdot \frac{N-1}{N} = 20.56$$

$$\sqrt{\text{var}(x)} = 4.534 \qquad \text{stdev}(x) = 4.534$$

$$\text{Stdev}(x) \cdot \sqrt{\frac{N-1}{N}} = 4.534$$

$$\frac{1}{N-1} \cdot \sum_{i=0}^{N-1} (x_i - m)^2 = 25.7$$

$$\text{Var}(x) = 25.7 \qquad \text{var}(x) \cdot \frac{N}{N-1} = 25.7$$

$$\sqrt{\text{Var}(x)} = 5.07 \qquad \text{Stdev}(x) = 5.07$$

$$\text{stdev}(x) \cdot \sqrt{\frac{N}{N-1}} = 5.07$$

13.2.3. Generating Correlated Random Numbers

Up to now, we have considered the simplest method of using independent random number generators. In the Monte-Carlo method, it is often necessary to generate random numbers with a specific correlation. Let us provide an example program creating two vectors, x1 and x2, of the same dimension and with the same distribution, random numbers of which are mutually correlated with the correlation factor R (Listing 13.12).

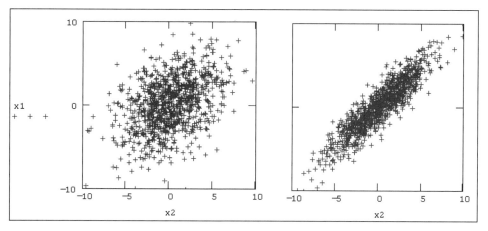

FIGURE 13.13. Pseudo-random numbers with correlation $R = 0.4$ (Listing 13.12) and $R = 0.9$

LISTING 13.12. Generating Mutually Correlated Random Numbers

$$\sigma := 3 \qquad\qquad N := 1000$$

$$R := 0.4$$

$$x1 := \text{rnorm}(N, 0, \sigma)$$

$$x2 := R \cdot x1 + \sqrt{1 - R^2} \cdot \text{rnorm}(N, 0, \sigma)$$

$$\text{corr}(x1, x2) = 0.415$$

The result of this program for $R = 0.4$ is shown in Figure 13.13 (left). Compare this sample to the right graph obtained for the high correlation ($R = 0.9$) and to Figure 13.3 *(see Section 13.1.1)* for independent data, i.e., $R = 0$.

13.2.4. Covariance and Correlation

Functions that establish the relationship between pairs of values of two random vectors are known as covariance and correlation (correlation function is also known as correlation factor, since it returns the Pearson's r correlation coefficient). As follows from their definitions, they differ by their normalization (Listing 13.13).

■   corr(x) — Coefficient of correlation of the two samples

■ cvar (x) — Covariance of the two samples

Here:

- x1, x2 — Vectors (or matrices) of the same size with samples of random data

LISTING 13.13. Calculation of Covariance and Correlation (*Listing 13.12, Continued*)

$$m1 := mean\ (x1) \qquad m2 := mean\ (x2)$$

$$\sigma1 := stdev\ (x1) \qquad \sigma2 := stdev\ (x2)$$

$$\frac{1}{N} \cdot \sum_{i\ =\ 0}^{N-1} \left(x1_i - m1\right) \cdot \left(x2_i - m2\right) = 3.823$$

$$cvar\ (x1,\ x2) = 3.823$$

$$\frac{cvar\ (x1,\ x2)}{\sigma1 \cdot \sigma2} = 0.415$$

$$corr\ (x1,\ x2) = 0.415$$

13.2.5. Skewness and Kurtosis Coefficients

The *Skewness coefficient* specifies the degree of asymmetry of the probability density in relation to the axis containing its center of gravity. The skewness coefficient is defined by the third central momentum of the distribution. In any symmetric distribution with a mean of zero (in normal distribution, for example), all odd momentums, including the third, are equal to zero, therefore, the skewness coefficient is also equal to zero.

There is yet another value that determines the level of smoothness of the probability density in proximity to the main maximum — kurtosis. It shows how pointed the peak of the probability density of the distribution is in comparison to normal distribution. If kurtosis is greater than zero, this means that the distribution's peak is sharper than that of the Gauss distribution. On the other hand, if kurtosis is negative, the distribution's peak is smoother than that of the Gauss distribution.

For calculating the skewness coefficient and kurtosis, Mathcad provides two built-in functions:

- ◼ kurt(x) — Kurtosis of the random data sample x
- ◼ skew(x) — Skewness of the random data sample x

Listing 13.14 gives some examples illustrating the calculation of skewness and kurtosis for a Weibull distribution (see Figure 13.10).

LISTING 13.14. Calculation of Skewness and Kurtosis

```
x := rweibull (1000 , 1.5)

skew (x) = 1.216

kurt (x) = 1.89
```

13.2.6. Other Statistical Characteristics

In previous sections we have considered built-in functions intended to calculate the most frequently used statistical characteristics of random data samples. Sometimes, however, statistics requires using other functions. For example, besides the arithmetic mean value, there are other mean values:

- ◼ gmean(x) — The geometric mean of the random data sample
- ◼ hmean(x) — The harmonic mean of the random data sample

The mathematic definitions of these functions along with an example of their usage in Mathcad are provided in Listing 13.15.

LISTING 13.15. Calculating Geometric and Harmonic Mean Values

```
N := 10

x := runif (N , 0 , 1)
```

$$\frac{1}{N} \cdot \sum_{i\,=\,0}^{N-1} x_i = 0.338 \qquad\qquad \text{mean} (x) = 0.338$$

$$\left(\frac{1}{N} \cdot \sum_{i=0}^{N-1} \frac{1}{x_i} \right)^{-1} = 0.012$$

$$\text{hmean}(x) = 0.012$$

$$\sqrt[N]{\prod_{i=0}^{N-1} x_i} = 0.171$$

$$\text{gmean}(x) = 0.171$$

13.2.7. Statistic Operations on Matrices

All examples of statistic functions considered above were related to vectors whose elements were random numbers. However, all of these functions are also applicable to random data grouped in matrices. In this situation, statistical characteristics are calculated for the set of all matrix elements, without grouping them by rows and columns. For example, if the matrix dimension is M×N, the size of the sample will be equal to $M \cdot N$.

Listing 13.16 shows an appropriate example of calculating the mean value. The first line of this listing defines the matrix of data x with the dimension 4×2. The built-in mean function (the next to last line of the listing) explicitly calculates the sum of all the elements of the x matrix (the last line). All the other built-in functions operate on matrices in the same way as they do on vectors (Listing 13.17).

LISTING 13.16. Calculating the Mean Value of the Matrix Elements

$$x := \begin{pmatrix} 1.0 & 4 \\ 1.5 & 7 \\ 0.9 & 1.2 \\ 1.2 & 12 \end{pmatrix}$$

$M := \text{rows}(x) \qquad M = 4$

$N := \text{cols}(x) \qquad N = 2$

$$\frac{1}{M \cdot N} \cdot \sum_{i=0}^{M-1} \sum_{j=0}^{N-1} x_{i,j} = 3.6$$

$\text{mean}(x) = 3.6$

LISTING 13.17. Statistical Operations on Matrices

```
median (x) = 1.35

mode (x) = 1.2

var (x) = 14.033

Var (x) = 16.037

stdev (x) = 3.746

Stdev (x) = 4.005
```

Some statistical functions (for example, calculation of the covariance) accept two arguments. These arguments can also represent matrices. However, according this function's inherent nature, both matrices must have the same dimension.

Most statistic functions can accept any number of matrices, vectors, and scalars as their arguments. Numeric characteristics will be calculated for the whole set of the function argument values. An appropriate example is provided in Listing 13.18.

LISTING 13.18. Statistical Functions of Several Arguments

```
x := ( 1   −1   4 )

y := ⎛ 3 ⎞
     ⎜   ⎟
     ⎝ 8 ⎠

z := ⎛ 3    7 ⎞
     ⎜       ⎟
     ⎝ 11   4 ⎠

mean (x, y, z) = 4.444

stdev (x, 5, 77) = 29.976

median (2) = 2

mode (y, z) = 3
```

13.3. RANDOM PROCESSES

Built-in Mathcad random number generators create a sample of random data A_i. Quite often, it is necessary to create a continuous or discrete random function $A(t)$ of one or more variables (random process or random field), the values of which would be ordered in relation to the function variables. Listing 13.19 presents an example of creating a pseudo-random process.

LISTING 13.19. Generation of a Pseudo-Random Process

```
N := 20

τ := 0.5

Tmax := (N − 1) · τ

j := 0 ..  N − 1

Tⱼ := j · τ

x := rnorm (N , 0 , 1)

KS1 := cspline (T , x)

A (t) := interp (KS1 , T , x , t)
```

The first line of Listing 13.19 defines the number N of independent random numbers that will be generated later. The second line determines the time correlation radius τ, while the next three lines define time moments T_j, to which random values $A(t_j)$ will correspond. The creation of the normal random process is reduced to the generation of a vector comprising a specified number of independent random numbers x and creating interpolating dependence between them. In Listing 13.19, we have used spline interpolation for this purpose *(see Chapter 14).*

As a result, we obtained the random process $A(t)$, whose correlation radius is determined by the distance τ between the points for which interpolation is performed. Figure 13.14 provides a graph showing the $A(t)$ random process along with the original random numbers. A random field can be created using a somewhat more complex method of multidimensional interpolation.

Random processes generated in such a way, as well as experimental data, can be processed using any statistical processing methods such as correlation or spectral analysis. Consider, for example, Listing 13.20, which illustrates the calculation of the correlation function of a random process.

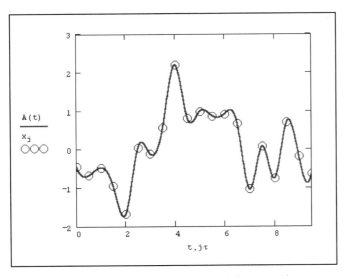

Figure 13.14. Pseudo-random process (Listing 13.19)

LISTING 13.20. Calculating a Correlation Function of a Random Process
(Listing 13.19, Continued)

$$\Delta := 0.02$$

$$M := 20$$

$$n := \text{floor}\left(\frac{T_{max}}{\Delta}\right) \qquad\qquad n = 475$$

$$j := 0 .. \ n$$

$$Y_j := A(\Delta \cdot j)$$

$$m := \text{mean}(Y)$$

$$D := \text{var}(Y) \qquad\qquad D = 0.785$$

$$R(j) := \frac{1}{(n - 2 \cdot M)} \cdot \sum_{i \ = \ M}^{n-M} \frac{(Y_{i+j} - m) \cdot (Y_i - m)}{D}$$

$$R(0) = 1.025$$

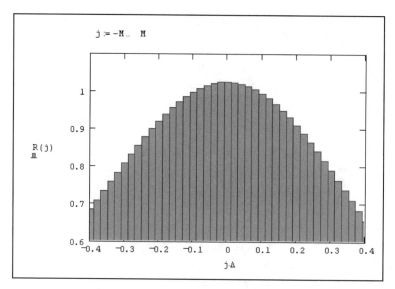

FIGURE 13.15. The correlation function (Listings 13.19 and 13.20)

Time sampling of the interval (0,Tmax) for the random process A(t) is performed using different elementary intervals Δ (the first line of the listing). Depending on the value of Δ, we get a different size n of the sample of random numbers Yᵢ, which are values of the A(t) random function in sampling points. In the last four lines of the listing, we define various characteristics of the random value Y, which, actually, are characteristics of the random process A(t). The graph calculated in 2·M + 1 points of the correlation function R(j) is shown in Figure 13.15.

Answer the following question: Why is the value of the correlation function R(0) calculated in such a way that is not equal to 1, as would be expected according to its definition?

13.4. SEVERAL EXAMPLES

Let us provide two characteristic statistical examples and illustrate their solutions using Mathcad 2001i.

13.4.1. Interval Evaluation of Variance

It is necessary to determine the interval (L,U) within which the variance of a normal random value will fall with the probability of 1 - α = 75%, based on the size of the data sample comprising N numbers. In statistics, this problem is solved using χ^2 distribution (Listing 13.21).

LISTING 13.21. The Interval Evaluation of Variance

```
N := 50

x := rnorm (N, 0, 1)

α := 0.25                                    1 - α = 0.75
```

$$\chi2_0 := qchisq\left(\frac{\alpha}{2}, N - 1\right) \qquad \chi2_0 = 37.901$$

$$\chi2_1 := qchisq\left(1 - \frac{\alpha}{2}, N - 1\right) \qquad \chi2_1 = 60.53$$

$$L := \frac{(N - 1) \cdot Stdev(x)^2}{\chi2_1} \qquad U := \frac{(N - 1) \cdot Stdev(x)^2}{\chi2_0}$$

```
L = 0.662                                    U = 1.057
```

The specified interval is known as the $(1 - \alpha)$% trust interval. Notice the usage of the `Stdev` function when solving this task for calculating the standard deviation (also notice, that the function name starts with an uppercase letter). In statistics, the expressions that are more conveniently written in this normalization method are quite common, therefore, such functions were introduced in Mathcad.

13.4.2. Testing Statistical Hypotheses

In statistics, there are quite a few problems related to testing specific hypotheses. Let us consider an example of a simple hypothesis. Let us suppose that we have a sample of N numbers with the normal distribution law and unknown variance and mean. It is necessary to accept or reject the hypothesis, assuming that the mean of the distribution law is equal to some value $\mu0 = 0.2$.

The problems of testing hypotheses require you to specify the level criterion α, which describes the probability of erroneous deviation of a true hypothesis. If the α value is very small, the hypothesis, even if it is false, will almost always be accepted. On the contrary, if α is close to 1, the criterion will be very stringent and the hypothesis, even if it is true, will most likely be rejected. In our case, the hypothesis states that $\mu0 = 0.2$, and its alternative states that $\mu0 \neq 0.2$. The evaluation of the mean, as follows from classical statistics, is solved using the Student's distribution with the parameter N − 1 (this parameter is known as the degree of freedom of the distribution).

To check the hypothesis (Listing 13.22), we calculate $(\alpha/2)$ — the quantile of the Student's distribution T, which serves as a critical value for accepting or rejecting the hypothesis. If the appropriate sample value t is by module (absolute value) less than T, the hypothesis is accepted, otherwise the hypothesis is false and must be rejected.

LISTING 13.22. Checking the Hypothesis on the Mean at an Unknown Variance

$N := 50$

$x := \text{rnorm}(N, 0, 1)$

$\alpha := 0.1$ $1 - \alpha = 0.9$

$\mu0 := 0.2$

$$t := \frac{\text{mean}(x) - \mu0}{\left(\dfrac{\text{Stdev}(x)}{\sqrt{N}}\right)}$$ $t = -1.883$

$$T := \text{qt}\left(1 - \frac{\alpha}{2}, N - 1\right)$$ $T = 1.677$

$|t| < T = 0$

In the last line of the listing, we calculated the criterion that determines if the condition representing the problem solution is true or false. Since the condition proved to be false (equal to 0 rather than to 1), this hypothesis must be rejected.

Figure 13.16 shows the Student's distribution with N − 1 degrees of freedom, along with critical values determining the trust interval. If t (this value is also shown in the graph) falls within this interval, the hypothesis is accepted. If, as in this case, the value of t doesn't fall within the trust interval, the hypothesis is rejected. If we increase α by making the criterion more stringent, the trust interval will be smaller than the one shown in this illustration.

Listing 13.23 demonstrates an alternative method of checking the same hypothesis based on the calculation of the Student's distribution itself rather than its quantile.

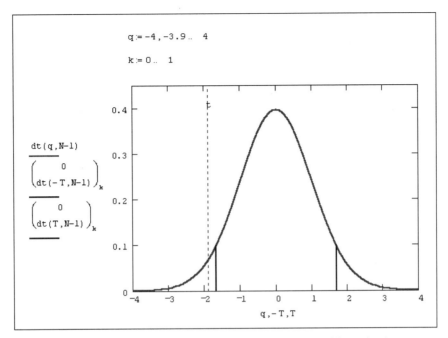

FIGURE 13.16. Illustration of the problem of testing statistical hypothesis (Listing 13.22)

LISTING 13.23. An Alternative Method of Testing a Statistical Hypothesis
(Listing 13.22, Continued)

$$\text{pt}(t, N-1) = 0.033$$

$$\frac{\alpha}{2} < \text{pt}(t, N-1) < 1 - \frac{\alpha}{2} = 0$$

We have considered only two characteristic examples of statistical calculations. However, using Mathcad, it is possible to easily solve many problems of mathematical statistics.

Lots of examples illustrating the solutions of mathematical statistics problems are provided in the Resource Center, under the Statistics topic.

14 Data Analysis

W hen you are dealing with a sample of experimental data, most often this data is represented in the form of an array composed of pairs of numbers (x_i, y_i). Subsequently, the problem of approximating discrete dependence $y(x_i)$ by using a continuous function $f(x)$ arises. The function $f(x)$, depending on the specific characteristics of the problem, might need to satisfy various requirements:

- $f(x)$ must contain all the points (x_i, y_i), i.e., $f(x_i) = y_i, i = 1 \ldots n$. In this case *(see Section 14.1)*, we are dealing with data interpolation of the function $f(x)$ in the internal points between x_i, or with data extrapolation (outside the interval containing all x_i).
- $f(x)$ must approximate $y(x_i)$ in a specified manner (for example, as a predefined analytical dependence), not necessarily containing all the points with coordinates (x_i, y_i). In this case, we are dealing with a regression problem *(see Section 14.2)*, which, in most cases, can be considered to be data smoothing.
- $f(x)$ must approximate the experimental dependence $y(x_i)$, taking into account some facts, for example, considering that the (x_i, y_i) data was measured with some error representing the noise component of the measurements.

In this case, the function f(x) uses a specific algorithm to minimize the measurement error present in the data (x$_i$,y$_i$). Problems of this type are known as filtering problems *(see Section 14.3)*. Smoothing is a particular case of filtering.

The various ways to generate the approximating dependence f(x) are shown in Figure 14.1. In this illustration, circles designate the source data, the curve is interpolated by sections of straight lines, and by dashed lines; linear regressions is indicated by a slanted straight line; and the filtering is indicated by a thick smooth curve. These dependencies are presented as an example and reflect only a small part of Mathcad's powerful data processing capabilities. Generally speaking, Mathcad provides a wide range of built-in functions that enable the user to perform various types of regression, interpolation, extrapolation and smoothing.

Various integral transforms are widely used for noise suppression and for solving other problems of data processing and adjustment. These transforms map the whole data set y(x) to some specific function F(ω) of another coordinate or coordinates. The Fourier transform *(see Section 14.4.1)* and wavelet transform *(see Section 14.4.2)* are examples of integral transforms. Notice that some transforms, such as the Fourier or Laplace transforms, can be performed in symbolic mode *(see Chapter 5)*. Each integral transform is efficient for a specific range of data analysis problems.

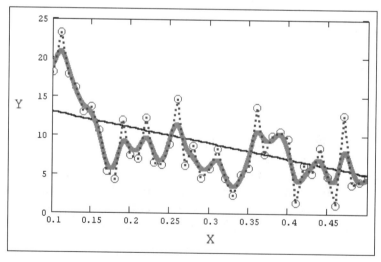

FIGURE 14.1. Various types of data approximation

14.1. INTERPOLATION

Mathcad provides several built-in functions for performing interpolation and extrapolation. These functions allow the user to "connect" the data sample points (x_i, y_i) by sections of the curves with various degrees of smoothness. By definition, *interpolation* is the construction of the $A(x)$ function that approximates the dependence $y(x)$ in the intermediate points belonging to the intervals between x_i.points. As a result, interpolation is also known as approximation. At the x_i points, the values of the interpolating function must coincide with the initial data, i.e., $A(x_i) = y(x_i)$.

Anywhere within this section, when discussing various types of interpolation instead of the $A(x)$ designation, we'll use another name for the interpolation function's argument – $A(t)$, to avoid confusing the data vector x and scalar variable t.

NOTE

14.1.1. Linear Interpolation

Linear interpolation is the simplest type of interpolation procedure. It represents the sought dependence $A(x)$ as a broken line composed of sections of straight lines that connect the initial data points (Figure 14.2).

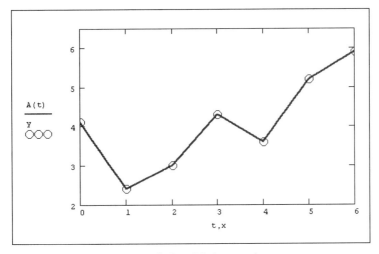

FIGURE 14.2. Linear interpolation (Listing 14.1)

To construct linear interpolation, use the `linterp` built-in function (Listing 14.1).

`linterp(x,y,t)` is the function approximating the data of vectors x and y by partially-linear dependence, where:

- x — A vector of real data values (argument)
- y — A vector of real data values (of the same size as the x vector)
- t — Argument values, at which interpolating function is calculated

Elements of the x vector must be defined in ascending order, i.e.,
$x_1 < x_2 < x_3 < \ldots < x_N.$

LISTING 14.1. Linear Interpolation

$$x := (0 \quad 1 \quad 2 \quad 3 \quad 4 \quad 5 \quad 6)^T$$

$$y := (4.1 \quad 2.4 \quad 3 \quad 4.3 \quad 3.6 \quad 5.2 \quad 5.9)^T$$

$$A(t) := linterp(x, y, t)$$

As you can see from this listing, to accomplish linear interpolation it is necessary to perform the following steps:

1. Enter the data vectors x and y (the first two lines of the listing).
2. Define the `linterp(x,y,t)` function.
3. Calculate the values of this function at the required points, for example `linterp(x,y,2.4) = 3.52` or `linterp(x,y,6) = 5.9`, or plot the graph of this function as shown in Figure 14.2.

Notice that the A(t) function shown in this graph accepts the t argument rather than x. This means that Mathcad automatically calculates the A(t) function within the whole interval (0,6) rather than only at the argument values (i.e., seven points). In this case, the difference is not noticeable, since when constructing the A(x) graph (Figure 14.3), Mathcad simply connects the initial data points by sections of a straight line (which means, that it implicitly performs linear interpolation).

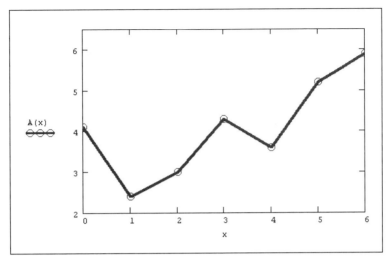

FIGURE 14.3. Normal plotting of the graph of the function of vector variable x (Listing 14.1)

14.1.2. Cubic Spline Interpolation

For most practical applications, it is desirable to connect experimental points with a smooth curve rather than with a broken line. Cubic spline interpolation, i.e., interpolation of the curve by sections of cubic polynomials (Figure 14.4), is known as one of the most suitable and convenient methods for this purpose.

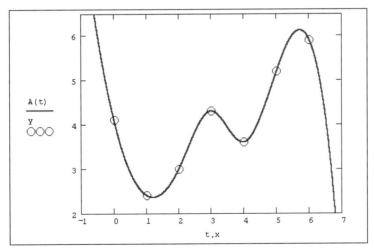

FIGURE 14.4. Spline interpolation (Listing 14.2)

`interp(s,x,y,t)` is the function approximating the data of x and y vectors by cubic splines, where:

- s — A vector of second derivatives, generated by one of the following functions: `cspline`, `pspline`, or `lspline`
- x — A vector of real data values in ascending order (the argument)
- y — A vector of real data values, of the same size as x
- t — An argument value, for which the interpolating function is calculated

Spline interpolation implementation in Mathcad is somewhat more complex than that of linear interpolation. Before applying the `interp` function, it is necessary to define the first of its arguments — namely, the vector variable s. This can be accomplished by using one of the following three built-in functions accepting the same arguments (x,y).

- `lspline(x,y)` — The vector the linear spline coefficients' values
- `pspline(x,y)` — The vector of the square spline coefficients' values
- `cspline(x,y)` — The vector of cubic spline coefficients' values

Here:

- x,y — The data vectors

Selection of the specific function of spline coefficients influences the interpolation in the proximity of the endpoints of the interval. Listing 14.2 represents the example of spline interpolation.

LISTING 14.2. Cubic Spline Interpolation

$$x := (\, 0 \quad 1 \quad 2 \quad 3 \quad 4 \quad 5 \quad 6 \,)^T$$

$$y := (\, 4.1 \quad 2.4 \quad 3 \quad 4.3 \quad 3.6 \quad 5.2 \quad 5.9 \,)^T$$

$$s := cspline\,(\,x\,,\,y\,)$$

$$A(\,t\,) := interp\,(\,s\,,\,x\,,\,y\,,\,t\,)$$

The idea of spline interpolation lies in the fact that in the intervals between the data points, the dependence is approximated by the following polynomial: $A(t) = a \cdot t^3 + b \cdot t^2 + c \cdot t + d$. The coefficients $a, b, c,$ and d are calculated

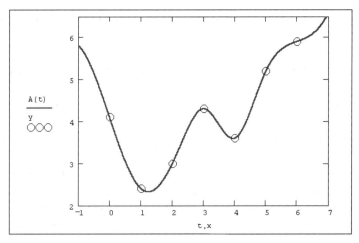

FIGURE 14.5. Spline interpolation with linear spline coefficients `lspline`

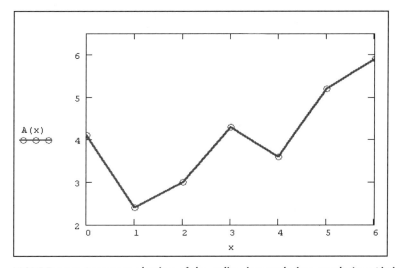

FIGURE 14.6. Incorrect plotting of the spline interpolation graph (see Listing 14.2)

independ ently for each interval, based on the values of y_i at the nearest points. This process is hidden from the user, since the key to the interpolation problem is returning the value of the A(t) function at any point t (Figure 14.4).

To emphasize the differences that correspond to the various helper functions cspline, pspline, and lspline, let us demonstrate the result that will be returned by Listing 14.2 after the replacement of the cspline function in the last line by the linear function `lspline` (Figure 14.5). As you can see, the set of helper functions significantly influences the behavior of `A(t)` in the proximity of the endpoints of the interval under consideration — $(0,6)$ in our example. This difference is especially striking at the endpoints of the intervals and outside it.

To conclude our discussion, let us consider a common error encountered when plotting graphs of the interpolating function (see Figure 14.3). If we specify construction of the `A(x)` function instead of `A(t)` on the graph, illustrating Listing 14.2, the initial points will simply be connected by a broken line (Figure 14.6). This happens because the interpolating function is not calculated in the intervals between the initial data points.

14.1.3. Polynomial Spline Interpolation

Let us consider a more complex type of interpolation — so-called B-spline interpolation. In contrast to normal spline interpolation *(see Section 14.1.2)*, elementary B-splines are not joined at the x_i points, but, rather, in other points u_i, whose coordinates must be entered by the user. Splines may represent polynomials of the 1, 2, or 3 order (linear, square, or cubic). B-spline interpolation is used just like normal spline interpolation; the only difference lies in the definition of the helper function calculating spline coefficients.

■ `interp(s,x,y,t)` — A function approximating the data of vectors x and y using B-splines

■ `bspline(x,y,u,n)` — A vector of the B-spline coefficient values

Here:

● s — A vector of second derivatives generated by the `bspline` function
● x — A vector of real data values (the argument), where the values follow in ascending order
● y — A vector of real data values; must have the same dimension as x
● t — An argument value at which the interpolating function is calculated
● u — A vector of the argument values specifying the points at which B-splines are joined
● n — The order of the polynomials used for spline interpolation (1, 2, or 3)

The dimension of the u vector must be less than that of the x and y vectors by 1, 2, or 3. The first element of the u vector must be less than or equal to the first element of the x vector, while the last element of u must be greater than or equal to the last element of x.

NOTE

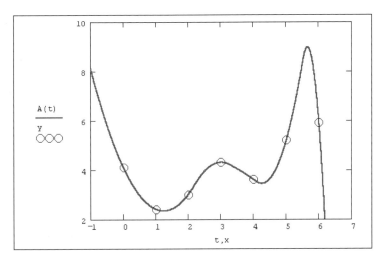

FIGURE 14.7. B-spline interpolation (Listing 14.3)

B-spline interpolation is illustrated by Listing 14.3 and Figure 14.7.

LISTING 14.3. B-Spline Interpolation

```
x := ( 0   1   2   3   4   5   6 ) T

y := ( 4.1   2.4   3   4.3   3.6   5.2   5.9 ) T

u := ( -0.5   2.2   3.3   4.1   5.5   7 ) T

s := bspline ( x , y , u , 2 )

A ( t ) := interp ( s , x , y , t )
```

14.1.4. Extrapolation with the Prediction Function

All functions considered above *(see Sections 14.1.1–14.1.3)* extrapolated the data outside the interpolation interval using an appropriate dependence based on the analysis of the location of several initial data points at the interval boundaries. Mathcad provides a more advanced extrapolation tool, which takes into account data distribution along the whole interval. The `predict` function implements a linear algorithm predicting the function behavior based on the oscillation analysis.

predict(y,m,n) is the prediction function for the vector extrapolating the data sample, where:

- ■ y — A vector of real values taken at equal intervals
- ■ m — A number of sequential elements of the y vector according to which extrapolation is performed
- ■ n — A number of predicted values

Listing 14.4 gives an example illustrating the usage of the prediction function in an example of extrapolating the y_j data oscillating with variable amplitude. The extrapolated graph, along with the function itself, is presented in Figure 14.8. Arguments and the working principle of the predict function are different from the previously considered built-in interpolation and extrapolation function. Argument values for the data are not required, since, by definition, this function influences the data that follow each other at constant intervals. Notice that the result returned by the predict function is added to the "tail" of the initial data.

LISTING 14.4. Extrapolation Using the Prediction Function

$$k := 100 \qquad\qquad j := 0 .. \ k$$

$$y_j := e^{\frac{-j}{100}} \cdot \sin\left(\frac{j}{10}\right)$$

$$m := 50 \qquad\qquad n := 150$$

$$A := predict \ (y, m, n)$$

As shown in Figure 14.9, the prediction function might prove to be useful when extrapolating data at small distances. As the distance from the initial data grows, the result often becomes unsatisfactory. Furthermore, the predict function works well when dealing with the problems of analyzing detailed data with a clear behavior pattern (like the one shown in Figure 14.8), mainly of the oscillating type.

If the amount of data is insufficient, the prediction might prove to be useless. For example, in Listing 14.5, a small data sample (taken from the examples considered in previous sections) is extrapolated. The result of this operation is shown in Figure 14.9 for different endpoints of the array of initial data for which extrapolation is constructed.

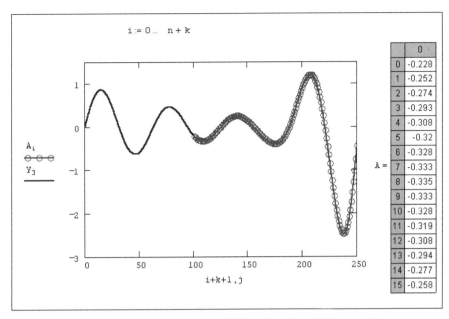

FIGURE 14.8. Extrapolation using the prediction function (Listing 14.4)

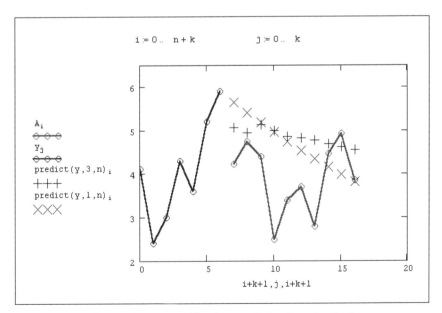

FIGURE 14.9. The results returned by the prediction function in the case of an insufficient amount of data (Listing 14.5)

LISTING 14.5. Extrapolation Using the Prediction Function

$$y := \begin{pmatrix} 4.1 & 2.4 & 3 & 4.3 & 3.6 & 5.2 & 5.9 \end{pmatrix}^T$$

$$k := \text{rows}(y) - 1$$

$$m := 5 \qquad\qquad n := 10$$

$$A := \text{predict}(y, m, n)$$

14.1.5. Multidimensional Interpolation

Two-dimensional spline interpolation results in the construction of the surface $z(x,y)$ containing an array of points describing the grid at the coordinate plane (x,y). The surface is created by the fragments of two-dimensional cubic splines representing the function of (x,y) and having continuous first and second derivatives of both coordinates.

Multidimensional interpolation is performed using the same built-in functions as one-dimensional interpolation *(see Section 14.1.2)*, but accepts matrices rather than vectors as its arguments. There is an important limitation of multidimensional interpolation related to the fact that only square $N{\times}N$ data arrays can be interpolated.

$\text{interp}(S, X, Z, V)$ is a scalar function approximating the data from the sample of a two-dimensional field with x and y coordinates using cubic splines, where:

- S — A vector of second derivatives generated by one of the built-in helper functions: `cspline`, `pspline`, or `lspline`
- X — A $N \times 2$ matrix defining the diagonal of the argument value grid (elements of both columns correspond to x and y coordinates, and are sorted in ascending order)
- Z — A $N \times N$ matrix composed of real values
- V — A vector whose two elements contain the values of x and y arguments for which it is necessary to calculate the interpolation

Auxiliary functions for calculating the values of the second derivatives have the same matrix arguments as `interp`: `lspline(X, Y)`, `pspline(X, Y)`, *and* `cspline(X, Y)`.

NOTE

An example of initial data is provided in Figure 14.10 in the form of level curves. Listing 14.6 presents the programmatic implementation of two-dimensional spline interpolation, while the results of this operation are shown in Figure 14.11.

LISTING 14.6. Two-Dimensional Interpolation

$$X := \begin{pmatrix} 0 & 0 \\ 1 & 10 \\ 2 & 20 \\ 3 & 30 \\ 4 & 40 \end{pmatrix} \qquad Y := \begin{pmatrix} 1 & 1 & 0 & 1.1 & 1.2 \\ 1 & 2 & 3 & 2.1 & 1.5 \\ 1.3 & 3.3 & 5 & 1.7 & 2 \\ 1.3 & 3 & 3.7 & 2.1 & 2.9 \\ 1.5 & 2 & 2.5 & 2.8 & 4 \end{pmatrix}$$

$$S := cspline(X, Y)$$

$$V := \begin{pmatrix} 3.7 \\ 2.2 \end{pmatrix} \qquad\qquad interp(S, X, Y, V) = 1.636$$

$$k := 30$$

$$i := 0 .. \ k \qquad\qquad j := 0 .. \ k$$

$$A_{i,j} := interp \left[S, X, Y, \begin{pmatrix} \dfrac{i}{k} \cdot 4 \\ \dfrac{j}{k} \cdot 40 \end{pmatrix} \right]$$

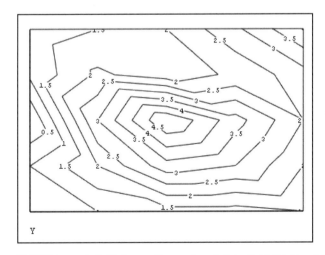

FIGURE 14.10. Initial two-dimensional data field (Listing 14.6)

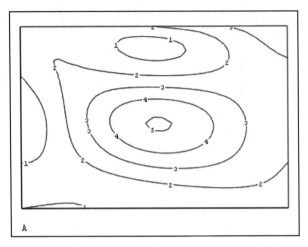

FIGURE 14.11. The result of two-dimensional interpolation (Listing 14.6)

14.2. REGRESSION

The problems of mathematical regression are aimed at approximating the data sample (x_i, y_i) with some function $f(x)$, which minimizes the total error $|f(x_i) - y_i|$ in a specific way. Regression is reduced to the selection of unknown coefficients that determine analytical dependence $f(x)$. Because of specific features of the operation itself, most regression problems represent particular cases of more general problems of data smoothing.

As a rule, regression is quite efficient when the data distribution law is predefined or can at least be reliably predicted.

14.2.1. Linear Regression

Linear regression is the simplest and most frequently used type of regression. As its implies, the data (x_i, y_i) is approximated by the linear function $y(x) = b + a \cdot x$. The linear function can be represented as a straight line on the (x, y) plane (Figure 14.12). Linear regression is also known as the method of least squares, since the coefficients a and b are calculated based on the condition of minimizing the sum of error squares $|b + a \cdot x_i - y_i|$.

Quite often, the same condition is set in other regression problems, where the data array (x_i, y_i) is approximated by other dependencies $y(x)$. An exception, however, is illustrated in Listing 14.9

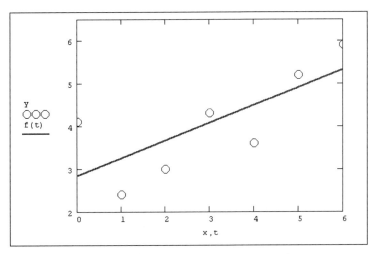

FIGURE 14.12. Linear regression (Listing 14.7 or 14.8)

Mathcad provides two mutually duplicating methods for calculating linear regression. Listings 14.7 and 14.8 illustrate the rules of their usage. Notice that both methods produce the same result (Figure 14.12).

- `line(x,y)` — A vector whose two elements `(b,a)` represent coefficients of linear regression $b + a \cdot x$
- `intercept(x,y)` — The coefficient b of linear regression
- `slope(x,y)` — The coefficient a (slope) of linear regression

Here:

- x — A vector of real data of the argument
- y — A vector of real data (must be the same size as x)

LISTING 14.7. Linear Regression

$$x := (\, 0 \quad 1 \quad 2 \quad 3 \quad 4 \quad 5 \quad 6 \,)^T$$

$$y := (\, 4.1 \quad 2.4 \quad 3 \quad 4.3 \quad 3.6 \quad 5.2 \quad 5.9 \,)^T$$

$$\text{line}(x, y) = \begin{pmatrix} 2.829 \\ 0.414 \end{pmatrix}$$

$$f(t) := \text{line}(x, y)_0 + \text{line}(x, y)_1 \cdot t$$

LISTING 14.8. Another Form of Linear Regression

$$x := (0 \quad 1 \quad 2 \quad 3 \quad 4 \quad 5 \quad 6)^T$$

$$y := (4.1 \quad 2.4 \quad 3 \quad 4.3 \quad 3.6 \quad 5.2 \quad 5.9)^T$$

$$\text{intercept} (x, y) = 2.829$$

$$\text{slope} (x, y) = 0.414$$

$$f(t) := \text{intercept} (x, y) + \text{slope} (x, y) \cdot t$$

Mathcad has an alternative algorithm implementing median-median linear regression for calculating a and b coefficients (Listing 14.9) rather than the least squares method.

medfit (x, y) is a vector of two elements (b, a) of the coefficients of the linear median-median regression b + a · x, where:

■ x, y — Vectors of real values must have the same dimension.

LISTING 14.9. An Alternative Method of Linear Regression

$$\text{medfit} (x, y) = \begin{pmatrix} 2.517 \\ 0.55 \end{pmatrix}$$

$$g(t) := \text{medfit} (x, y)_0 + \text{medfit} (x, y)_1 \cdot t$$

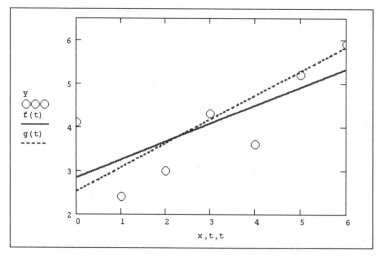

FIGURE 14.13. Linear regression using the least squares method and median-median method (Listings 14.7 and 14.9)

The difference between linear regressions obtained using the least squares method and median-median regression is illustrated by Figure 14.13.

14.2.2. Polynomial Regression

Mathcad implements several methods of polynomial regression, including regression of one polynomial, sections of several polynomials, and two-dimensional regression of the data array.

Polynomial Regression

Polynomial regression approximates a data array (x_i, y_i) with the k-th order polynomial $A(x) = a + b \cdot x + c \cdot x^2 + d \cdot x^3 + \ldots + h \cdot x^k$ (Figure 14.14). When $k = 1$ the polynomial represents a straight line, at $k = 2$ it is a square parabola, at $k = 3$ it represents a cubic parabola, and so on. As a general rule, $k < 5$ are used in practice.

To construct a regression using a k-th order polynomial, it is necessary to have at least $(k + 1)$ data points.

NOTE

Polynomial regression in Mathcad is performed with the combination of the built-in `regress` function and a polynomial interpolation *(see Section 14.1.2).*

- `regress(x,y,k)` — A vector of coefficients for constructing a polynomial data regression
- `interp(s,x,y,t)` — The result of the polynomial regression

Here:

- `s = regress(x,y,k)`
- `x` — A vector of real argument data whose elements follow in ascending order
- `y` — A vector of real data values (must have the same dimension as `x`)
- `k` — The order of the regression polynomial (positive integer value)
- `t` — The argument value for the regression polynomial

To construct a polynomial regression, you must use the `interp` function after the `regress` function.

NOTE

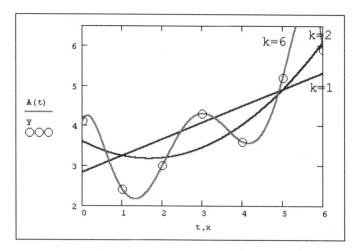

FIGURE 14.14. Polynomial regression using polynomials of different orders (the results of Listing 14.10 for different values of k)

Listing 14.10 provides an example illustrating polynomial regression using square parabola.

LISTING 14.10. Polynomial Regression

```
x := ( 0   1   2   3   4   5   6 )^T

y := ( 4.1   2.4   3   4.3   3.6   5.2   5.9 )^T

k := 2

s := regress ( x , y , k )

A ( t ) := interp ( s , x , y , t )
```

Regression by Polynomial Fragments

Besides approximation of the data array with a single polynomial, you can perform the regression by joining fragments of several polynomials. For this purpose, Mathcad provides the `loess` built-in function, which is used in a similar way to the `regress` function (Listing 14.11 and Figure 14.15).

■ `loess(x,y,span)` — The vector of coefficients for construction data regression by fragments of polynomials

■   `interp(s,x,y,t)` — The result of polynomial regression

Here:

- `s = loess(x,y,span)`
- `x` — A vector of real argument values whose elements are in ascending order
- `y` — A vector of real data (must have the same dimension as `x`)
- `span` — The parameter defining the size of a polynomial section (a positive number; the best results are obtained when `span = 0.75`)

The `span` parameter specifies the degree of data smoothing. With large values of `span`, the regression has no practical difference from the regression by a single polynomial (for example, the value `span = 2` provides nearly the same result as that of the approximation of the points of a parabola).

LISTING 14.11. Regression of Polynomial Fragments

```
x := ( 0   1   2   3   4   5   6 )ᵀ

y := ( 4.1   2.4   3   4.3   3.6   5.2   5.9 )ᵀ

s := loess ( x , y , 0.9 )

A ( t ) := interp ( s , x , y , t )
```

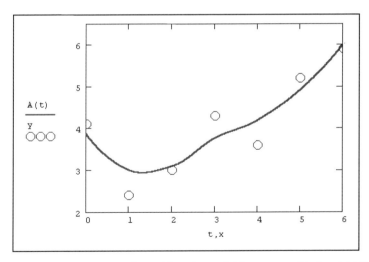

FIGURE 14.15. Regression with polynomial fragments (Listing 14.11)

Regression of a single polynomial is efficient when a set of data points looks like a polynomial, otherwise regression of polynomial fragments is more efficient.

TIP

Two Dimensional Polynomial Regression

Besides one-dimensional polynomial regression and two-dimensional interpolation *(see Section 14.1.5)*, Mathcad provides the capability to approximate the set of data points $z_{i,j}(x_i, y_j)$ with a surface determined by two-dimensional polynomial dependence. In this case, built-in functions for constructing polynomial regression must accept appropriate matrices rather than vectors as its arguments.

- ■ `regress(X,Z,k)` — The vector of coefficients for construction of polynomial data regression
- ■ `loess(X,Z,span)` — The vector of coefficients for constructing data regression by polynomial fragments
- ■ `interp(S,X,Z,V)` — A scalar function approximating the data of the two-dimensional field sample by x and y coordinates using cubic splines

Here:

- ● S — A vector of second derivatives generated by one of the auxiliary built-in functions: `loess` or `regress`
- ● X — A matrix of the dimension $N \times 2$, specifying pairs of argument values (the columns correspond to x and y)
- ● Z — A vector of real values with the dimension N
- ● `span` — A parameter defining the size of the polynomial fragment
- ● k — The order of the regression polynomial (positive integer number)
- ● V — A vector of two elements, containing the values of x and y, for which interpolation is calculated

You don't need to perform any previous sorting or ordering of data to construct a regression (in contrast to two-dimensional interpolation, for example, which requires the data to be represented as $N \times N$ matrix). Due to this fact, the data can be represented in vector form.

NOTE

Listing 14.12 and Figure 14.16 provide examples that illustrating two-dimensional polynomial regression. Compare the style of data representation for two-dimensional regression to a representation of the same data for two-dimensional spline interpolation (see Listing 14.6) and its result to the initial data (see Figure 14.10) and respective spline interpolation (see Figure 14.11).

LISTING 14.12. Two-Dimensional Polynomial Regression

$$X := \begin{pmatrix} 0 & 0 & 0 & 0 & 0 & 1 & 1 & 1 & 1 & 1 & 2 & 2 & 2 & 2 & 2 & 3 & 3 & 3 & 3 & 3 & 4 & 4 & 4 & 4 & 4 \\ 0 & 10 & 20 & 30 & 40 & 0 & 10 & 20 & 30 & 40 & 0 & 10 & 20 & 30 & 40 & 0 & 10 & 20 & 30 & 40 & 0 & 10 & 20 & 30 & 40 \end{pmatrix}^T$$

$$Z := (1\ 1\ 0\ 1.11.21\ 2\ 3\ 2.11.51.33.35\ 1.72\ 1.33\ 3.72.12.91.52\ 2.52.84)^T$$

$$S := regress(X,Z,3)$$

$$k := 30$$

$$i := 0..k \qquad\qquad j := 0..k$$

$$A_{i,j} := interp \left[S, X, Z, \begin{pmatrix} \dfrac{i}{k} \cdot 4 \\[2mm] \dfrac{j}{k} \cdot 40 \end{pmatrix} \right]$$

Notice the transpose signs in the listing. This sign is used to correctly represent the arguments (such as z) in the form of a vector rather than in string form.

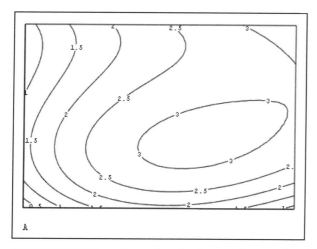

FIGURE 14.16. Two-dimensional polynomial regression (Listing 14.12)

14.2.3. Special Types of Regression

Besides regression types covered in the previous sections, Mathcad provides several other types of three-parameter regression. Their implementation is different from the regression variations considered above in that for this case, besides the data array, it is necessary to specify some initial values for the coefficients a, b, and c. Use the appropriate type of regression if you can guess what type of dependency is the best to describe your data array. If the regression type poorly reflects the data sequence, the result most often proves to be unsatisfactory or even significantly different from the actual dependence (this depends on the selection of initial value). Each of these functions returns the vector of corrected parameters a, b, c.

■ expfit(x,y,g) — Regression by exponential function $f(x) = a \cdot e^{bx} + c$
■ lgsfit(x,y,g) — Regression by logistic function $f(x) = a / (1 + b \cdot e^{-cx})$
■ sinfit(x,y,g) — Regression by sinusoid $f(x) = a \cdot \sin(x + b) + c$
■ pwrfit(x,y,g) — Regression by power curve $f(x) = a \cdot x^b + c$
■ logfit(x,y,g) — Regression by logarithmic curve $f(x) = a \cdot \ln(x + b) + c$
■ lnfit(x,y) — Regression by logarithmic curve with two parameters $f(x) = a \cdot \ln(x) + b$

Here:

- x — A vector of real argument data
- y — A vector of real values (must be the same dimension as x)
- g — A vector whose three elements specify initial values for a, b, c

Accuracy of the initial value selection can be evaluated by the result of regression — if the function returned by Mathcad approximates the y(x) dependence satisfactorily, this means that initial values were selected well.

Listing 14.13 and Figure 14.17 provide examples of the calculation using one of the three-parametric regression types (by exponential function). In the next to last line of the listing, the calculated coefficients a, b, c are returned in vector form. In the last line, these coefficients are used to define the sought function f(x).

LISTING 14.13. Exponential Regression

$$x := (\begin{matrix} 0 & 1 & 2 & 3 & 4 & 5 & 6 \end{matrix})^T$$

$$y := (\begin{matrix} 4.1 & 2.4 & 3 & 4.3 & 3.6 & 5.2 & 5.9 \end{matrix})^T$$

$$g := \begin{pmatrix} 1 \\ 1 \\ 1 \end{pmatrix}$$

$$C := expfit(x, y, g)$$

$$C = \begin{pmatrix} 0.111 \\ 0.544 \\ 3.099 \end{pmatrix}$$

$$f(t) := C_0 \cdot exp(C_1 \cdot t) + C_2$$

Many problems of data regression using various two-parametric dependencies $y(x)$ can be reduced to more reliable (in terms of calculation) linear regression. This can be accomplished by means of variable substitution.

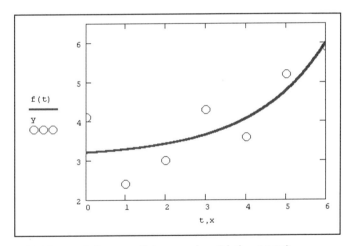

FIGURE 14.17. Exponential regression (Listing 14.13)

14.2.4. General Regression

In Mathcad, you can perform regression in the form of the linear combination $C_1 f_1(x) + C_2 f_2(x) + \ldots$, where $f_i(x)$ are any custom functions and C_i are coefficients that must be defined. Furthermore, there is the method of performing regression of a more general type, when the user himself defines the combination of functions and coefficients.

Let us discuss built-in functions for performing general regression and provide examples of their usage (Listings 14.14 and 14.15). More detailed information on these special capabilities is provided in the Mathcad on-line Help system and in the Resource Center.

- ◼ `linfit(x,y,F)` — A vector whose elements are parameters of the linear combination of custom functions intended to perform data regression
- ◼ `genfit(x,y,g,G)` — A vector of parameters implementing data regression using custom user-defined functions

Here:

- • x — A vector of real argument data, whose elements follow in ascending order
- • y — A vector of real values, must be of the same dimension as x
- • F(x) — A custom user-defined vector function of scalar argument
- • g — A vector of initial values of the regression parameter, of dimension N
- • G(x,C) — A vector function of the dimension N + 1, composed from the custom user-defined function and its N partial derivatives from each of the C parameters

LISTING 14.14. Regression by Means of Linear Combination of Custom Functions

$$x := (0 \quad 1 \quad 2 \quad 3 \quad 4 \quad 5 \quad 6)^T$$

$$y := (4.1 \quad 2.4 \quad 3 \quad 4.3 \quad 3.6 \quad 5.2 \quad 5.9)^T$$

$$F(x) := \begin{pmatrix} \dfrac{1}{x+1} \\ x \\ e^x \end{pmatrix}$$

$$C := \text{linfit}(x, y, F)$$

$$C = \begin{pmatrix} 3.957 \\ 0.854 \\ 5.605 \times 10^{-4} \end{pmatrix}$$

$$f(x) := \dfrac{C_0}{x+1} + C_1 \cdot x + C_2 \cdot e^x$$

LISTING 14.15. Regression of a General Type

$$x := (0 \quad 1 \quad 2 \quad 3 \quad 4 \quad 5 \quad 6)^T$$

$$y := (4.1 \quad 2.4 \quad 3 \quad 4.3 \quad 3.6 \quad 5.2 \quad 5.9)^T$$

$$g := \begin{pmatrix} 0.1 \\ 1 \end{pmatrix} \qquad\qquad G(x, C) := \begin{pmatrix} C_0 \cdot x^{C_1 \cdot x} \\ x^{C_1 \cdot x} \\ C_0 \cdot x^{C_1 \cdot x + 1} \end{pmatrix}$$

$$C := \text{genfit}(x, y, g, G)$$

$$C = \begin{pmatrix} 3.168 \\ 0.057 \end{pmatrix}$$

$$f(x) := C_0 \cdot x^{C_1 \cdot x}$$

14.3. SMOOTHING AND FILTERING

When analyzing data, the problem of data filtering often arises, consisting of the elimination of one of the components of $y(x_i)$ dependence. Quite often, filtering is aimed at suppression of fast variation of $y(x_i)$, which most frequently is due to noise. As a result, the frequently oscillating $y(x_i)$ dependence is transformed to another, smoothened dependence, in which the component of lower frequency is dominant.

Various types of regression *(see Section 14.2)* can be considered to be the simplest and most efficient methods of smoothing. However, regression often eliminates the information component of the data, leaving only dependency predefined by the user.

Quite often, one has to consider a problem that is the inverse of filtering — elimination of slowly changing variations in order to investigate high-frequency components. In this case we are dealing with the problem of trend elimination. Sometimes it might be necessary to solve combined problems, where only a medium-range frequency component needs to be separated by means of suppressing both low- and high-frequency components. One of the possibilities of solving such problems relates to the usage of band pass filters.

Several examples of programmatic implementation of various types of filtering will be considered in this section.

14.3.1. Built-in Functions for Smoothing

Mathcad provides several built-in functions implementing various algorithms of data smoothing.

- ▓ medsmooth(y,b) — Smoothing while using the "running medians" algorithm
- ▓ ksmooth(x,y,b) — Smoothing based on the Gauss function
- ▓ supsmooth(x,y) — Local smoothing by the adaptive algorithm based on the analysis of the closest neighbors of each pair of data points

Here:

- • x — A vector of real argument data (for supsmooth its elements must be in ascending order)
- • y — A vector of real values, must have the same dimension as x
- • b — The width of the smoothing window

All functions accept arguments that represent vectors composed from the data array and return a result that represents the vector of smoothed data of the same size. The medsmooth function supposes that the data is distributed evenly.

Detailed information on the algorithms implemented in the smoothing functions is provided in the Mathcad on-line Help system, under the Smoothing topic in the Statistics section. The Resource Center also provides some very useful information on various types of filtering.

Quite often it is useful to combine smoothing with interpolation or regression. An appropriate example is provided in Listing 14.16 for the supsmooth function. The results of this listing are shown in Figure 14.18 (initial data points are designated by circles, smoothed data by crosses, and the results of spline interpolation by a dashed line). The results of smoothing the same data using the "running medians" and the Gauss function with different values of the smoothing window width are shown in Figures 14.19 and 14.20, respectively.

LISTING 14.16. Smoothing with Subsequent Spline Interpolation

$$x := (\, 0 \quad 1 \quad 2 \quad 3 \quad 4 \quad 5 \quad 6 \quad 7 \quad 8 \quad 9 \quad 10 \quad 11 \quad 12 \quad 13 \quad 14 \quad 15 \quad 16 \,)^T$$

$$y := (\, 4.1 \quad 2.4 \quad 3 \quad 4.3 \quad 3.6 \quad 5.2 \quad 5.9 \quad 5 \quad 4.7 \quad 4 \quad 3.5 \quad 3.9 \quad 3 \quad 2.7 \quad 3.7 \quad 4.8 \quad 5.4 \,)^T$$

$$z := \text{supsmooth}\,(x,y)$$

$$s := \text{cspline}\,(x,z)$$

$$A(t) := \text{interp}\,(s,x,z,t)$$

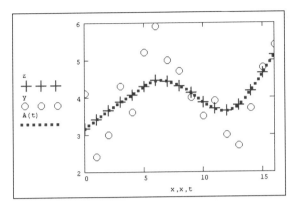

FIGURE 14.18. Adaptive smoothing (Listing 14.16)

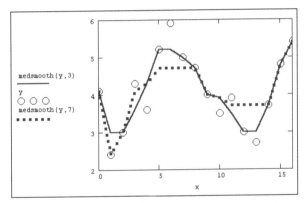

FIGURE 14.19. Smoothing using the "running medians" method

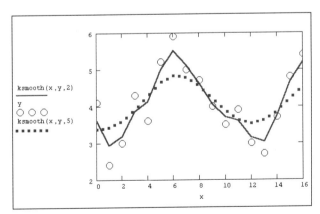

FIGURE 14.20. Smoothing using the `ksmooth` function

14.3.2. Sliding Averaging

Besides built-in smoothing methods implemented in Mathcad, there are several other popular smoothing algorithms, one of which deserves special mention. The sliding averaging is a very simple and rather efficient method. Basically, it calculates an average value for each argument value taking w neighboring data points. The w number is known as the sliding averaging window. The larger this value is, the larger the number of data points participating in calculation of the average, and the smoother the resulting curve. Figure 14.21 shows the result of the sliding averaging of the same data (shown by circles) with a different window: w = 3 (dots line), w = 5 (dashes) and w = 15 (solid line). It is easy to see that at small w, smoothed curves practically follow the changes of the initial data, and with the growth of w, they only reflect the trend of slow data variations.

To implement sliding averaging in Mathcad, it is sufficient to write a small program shown in Listing 14.17. This program uses only y values represented in the form of a vector, implicitly suggesting that they correspond to the values of the x argument following each other with constant increment. The x vector was only used to plot the graph of the result (Figure 14.21).

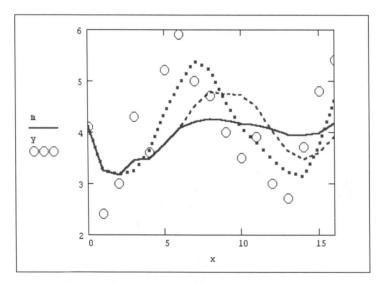

FIGURE 14.21. Sliding averaging with different values of w = 3,5,15 (Listing 14.17)

LISTING 14.17. Smoothing with Sliding Averaging

$$x := (\,0 \quad 1 \quad 2 \quad 3 \quad 4 \quad 5 \quad 6 \quad 7 \quad 8 \quad 9 \quad 10 \quad 11 \quad 12 \quad 13 \quad 14 \quad 15 \quad 16\,)^T$$

$$y := (\,4.1 \quad 2.4 \quad 3 \quad 4.3 \quad 3.6 \quad 5.2 \quad 5.9 \quad 5 \quad 4.7 \quad 4 \quad 3.5 \quad 3.9 \quad 3 \quad 2.7 \quad 3.7 \quad 4.8 \quad 5.4\,)^T$$

$$w := 15$$

$$N := \text{rows}\,(y) \qquad\qquad N = 17$$

$$i := 0\,..\ N-1$$

$$m_i := \text{if}\left(i < w,\ \frac{\displaystyle\sum_{j\,=\,0}^{i} y_j}{i+1}\,,\ \frac{\displaystyle\sum_{j\,=\,i-w+1}^{i} y_j}{w} \right)$$

NOTE

Program implementation of sliding averaging provided in Listing 14.17 is the simplest, but not the best type. You have probably noticed the fact that all curves of the sliding average shown in Figure 14.21 go slightly "ahead" of the initial data. The reason for this behavior is easily understandable: according to the algorithm implemented in the last line of the Listing 14.17, the sliding average for each point is calculated by averaging the values of the previous w points. To provide more adequate results of the sliding averaging, use the centered algorithm of calculating the average by w/2 previous and w/2 next values. This algorithm will be somewhat more complex, since now you'll need to take into account a lack of data points not only in the start of the initial data array (as this was done in the program using the condition operator if), but also at the end of this array.

14.3.3. Trend Elimination

Another typical problem arises when you need to investigate fast variations of the $y(x)$ signal rather than slow (or low frequency) variations (for which data smoothing is used). Quite often it happens that fast (or high frequency) variations are superimposed in a specific way over the slow variations, which are usually known as trends. Most often, the trend is predictable and has a specific form (linear, for example). To eliminate the trend, it is possible to use the sequence of actions implemented by Listing 14.18.

1. Calculate the regression $f(x)$; for example, linear, based on the a priori information on the trend (next to last line of the listing).
2. Subtract the trend $f(x)$ from the data $y(x)$ (the last line of the listing).

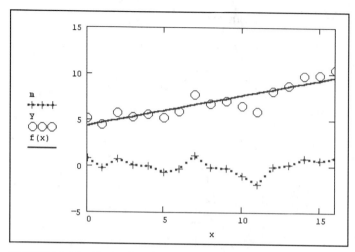

FIGURE 14.22. Eliminating the trend (Listing 14.18)

LISTING 14.18. Trend Elimination

$$x := (\;0\quad 1\quad 2\quad 3\quad 4\quad 5\quad 6\quad 7\quad 8\quad 9\quad 10\quad 11\quad 12\quad 13\quad 14\quad 15\quad 16\;)^T$$

$$y := (5.1\;4.4\;5.7\;5.3\;5.6\;5.2\;5.9\;7.7\;6.7\;7\;6.5\;5.9\;8.1\;8.7\;9.7\;9.8\;10.4)^T$$

$$N := \text{rows}\,(y) \qquad\qquad N = 17$$

$$i := 0\,..\;N-1$$

$$f(t) := \text{line}\,(x,y)_0 + \text{line}\,(x,y)_1 \cdot t$$

$$m_i := y_i - f(x_i)$$

Figure 14.22 illustrates the initial data (designated by circles), linear trend detected using regression (solid straight line), and the result of trend elimination (dashed line connecting the crosses).

14.3.4. Band Pass Filtering

In the previous sections, we have considered filtering of the fast variations of the signal (smoothing) and its slow variations (trend elimination). Sometimes, however, one needs to detect a medium range of the signal by eliminating both the fastest and the slowest components. One of the possible ways to solve this problem is by using Band Pass filtering based on the sequential sliding averaging.

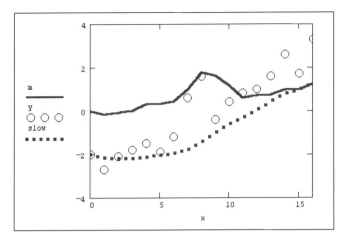

FIGURE 14.23. The result of Band Pass filtering (Listing 14.19)

The algorithm of Band Pass filtering is implemented in Listing 14.19, and the result of its usage is shown in Figure 14.23 by a solid curve. This algorithm implements the following sequence of operations:

1. A reduction of the data array y by subtracting the mean value from each element of the y array (the third and fourth lines of the listing).
2. Eliminating the high frequency component from the y signal in order to obtain a smoothed signal named middle. This can be achieved, for example, by using the sliding averaging with the small window w (in Listing 14.19 w = 3).
3. Separating the low frequency component slow from the middle signal. This can be done, for example, by using the sliding averaging with the large window w (for example, in Listing 14.19 w = 7), or by means of trend elimination *(see Section 14.3.3)*.
4. Subtract the slow trend from the middle signal (the last line of the listing), thus separating the medium range component of the initial signal y.

LISTING 14.19. Band Pass Filtering

```
x := ( 0   1   2   3   4   5   6   7   8   9   10   11   12   13   14   15   16 )ᵀ

y := ( 4.1 2.4 3 4.3 3.6 5.2 5.9 5 4.7 4 3.5 3.9 3 2.7 3.7 4.8 5.4 )ᵀ

meanY := mean ( y )

y := y − meanY

N := rows ( y )                    N = 17
```

$$i := 0 .. \quad N - 1$$

$$w := 3$$

$$middle_i := if\left(i < w, \frac{\sum\limits_{j\,=\,0}^{i} y_j}{i + 1}, \frac{\sum\limits_{j\,=\,i-w+1}^{i} y_j}{w} \right)$$

$$w := 7$$

$$slow_i := if\left(i < w, \frac{\sum\limits_{j\,=\,0}^{i} middle_j}{i + 1}, \frac{\sum\limits_{j\,=\,i-w+1}^{i} middle_j}{w} \right)$$

$$m := middle - slow$$

14.4. INTEGRAL TRANSFORMS

Integral transforms of the signal array $y(x)$ map the whole data set $y(x)$ to some function of another coordinate $F(v)$. Let us consider built-in functions provided by Mathcad for calculating integral transforms.

14.4.1. The Fourier Transform

The mathematical idea of the Fourier transform is found in the representation of the signal $y(x)$ as an infinite sum of sine curves of the following form: $F(v) \cdot \sin(v \cdot x)$. The $F(v)$ function is known as the *Fourier transform* or the *Fourier integral* or the *Fourier spectrum* of the signal. Its argument v has the physical sense of the frequency of an appropriate signal component. An inverse Fourier transform converts the spectrum $F(v)$ to an initial signal $y(x)$. According to its definition,

$$F(v) = \int\limits_{-\infty}^{\infty} y(x) \cdot \exp(-ivx)\,dx.$$

As you can see, the Fourier transform is a complex variable, even if the signal is real number.

Fourier Transform of the Real Data

The Fourier transform is of tremendous importance for various math applications. There is a very efficient algorithm for it, known as the Fast Fourier Transform (FFT). This algorithm is implemented in several built-in Mathcad functions.

- ▣ fft (y) — The vector of a direct Fourier transform
- ▣ FFT (y) — The vector of a direct Fourier transform in alternate normalizing
- ▣ ifft (v) — The vector of an inverse Fourier transform
- ▣ IFFT (v) — The vector of an inverse Fourier transform in alternate normalizing

Here:

- ● y — A vector of real data, taken with constant intervals of the argument values
- ● v — A vector of real data comprising the Fourier spectrum taken with constant intervals of the frequency values

Argument of the direct Fourier transform (i.e., the vector y) must have exactly 2^n elements (n is integer). The result is a vector of $1 + 2^{n-1}$ elements. On the other hand, the argument of an inverse Fourier transform must have $1 + 2^{n-1}$ elements, and the result of this transform will represent a vector comprising 2^n elements. If the number of data points doesn't coincide with the power of 2, if must be complemented with missing elements whose values are set to zero.

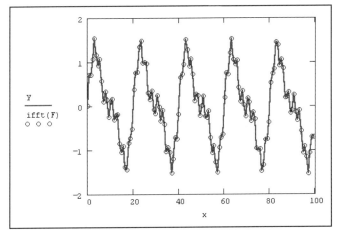

FIGURE 14.24. The initial data and an inverse Fourier transform (Listing 14.20)

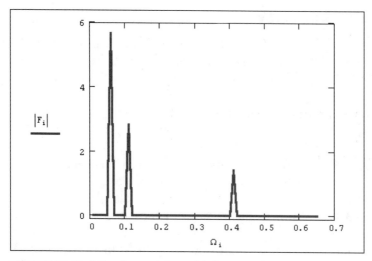

FIGURE 14.25. A Fourier transform (Listing 14.20)

Listing 14.20 provides an example of the calculation of the Fourier spectrum for the sum of three sinusoidal signals of different amplitude (shown in Figure 14.24 as a solid curve). The calculation was performed for N = 128 points, and it was assumed that the data-sampling interval was equal to Δ. In the next to last line of the listing, the built-in function ifft was used, while the last line correctly determines respective values of the frequency Ω_i. Notice that the calculation results are presented in the form of the modulus of Fourier spectrum (Figure 14.25) since the spectrum itself is complex. It is very useful to compare the obtained amplitudes and location of the spectrum peaks to the sine curves definition in Listing 14.20.

LISTING 14.20. A Fast Fourier Transform

$$N := 128$$

$$xMAX := 100 \qquad\qquad \Delta := \frac{xMAX}{N}$$

$$i := 0 .. \ N - 1$$

$$x_i := i \cdot \Delta$$

$$y_i := \sin(2 \cdot \pi \cdot 0.05 \cdot x_i) + 0.5 \cdot \sin(2 \cdot \pi \cdot 0.1 \cdot x_i) + 0.25 \cdot \sin(2 \cdot \pi \cdot 0.4 \cdot x_i)$$

$$F := fft \ (y)$$

$$\Omega_i := (i + 1) \cdot \frac{1}{xMAX}$$

The result of an inverse Fourier transform is shown as a circle in Figure 14.24, along with the initial data. It can be clearly seen that in the case under consideration, the y(x) signal is restored with high precision, which is characteristic for smoothly changing signals.

A Fourier Transform of the Complex Data

The algorithm of the Fast Fourier Transform for complex data is implemented by respective functions, whose names include the "c" character.

- ▪ cfft(y) — The vector of a direct complex Fourier transform
- ▪ CFFT(y) — The vector of a direct complex Fourier transform in alternate normalizing
- ▪ icfft(y) — The vector of an inverse complex Fourier transform
- ▪ ICFFT(v) — The vector of an inverse complex Fourier transform in alternate normalizing

Here:

- ● y — A vector of the data taken at equal intervals of the argument values
- ● v — A vector of the data forming the Fourier spectrum taken at equal frequency intervals

Functions performing a real Fourier transform make use of the fact that in the case of real data, the spectrum will be symmetric in relation to zero, and because

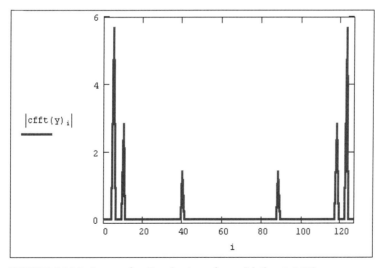

FIGURE 14.26. A complex Fourier transform (Listing 14.20)

of this, they return only half of the spectrum *(see the Section "Fourier Transform of the Real Data" earlier in this chapter)*. Thus, for 128 real data points, the function returned only 65 points of the Fourier spectrum. If the function implementing a complex Fourier transform is applied to the same data (Figure 14.26), the resulting vector will comprise 128 elements. Comparing Figure 14.25 and Figure 14.26, you'll notice the difference between the results of real and complex Fourier transforms.

Two-Dimensional Fourier Transform

Mathcad allows one to apply built-in functions for a complex Fourier transform to both one- and two-dimensional arrays (i.e., matrices). Listing 14.21 provides an appropriate example, which is illustrated by Fig 14.21 showing level curves of the initial data along with the calculated Fourier spectrum.

LISTING 14.21. Two-Dimensional Fourier Transform

$$N := 64$$

$$i := 0 .. \ N - 1 \qquad j := 0 .. \ N - 1$$

$$y_{i,j} := \sin\left(\frac{i+j}{10}\right) + \sin\left(\frac{i-j}{10}\right)$$

$$F := CFFT(y)$$

$$F := submatrix(F, 7, N - 7, 7, N - 7)$$

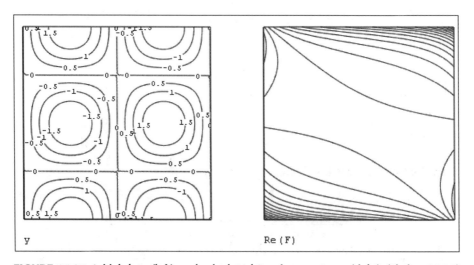

FIGURE 14.27. Initial data (left) and calculated Fourier spectrum (right) (Listing 14.21)

14.4.2. A Wavelet Transform

Recently, considerable attention has been drawn to other integral transforms, such as the wavelet transform (or the discrete wave transform). Mainly, it is used for analysis of non-stationary signals, and for most types of such problems, it proves to be more efficient than the Fourier transform. The main difference between wavelet transforms and Fourier transforms is the fact that in this case, the data is expanded by other functions known as wavelet generating functions rather than by sine curves (as in the case of the Fourier transform). Wavelet generating functions, in contrast to infinitely oscillating sine curves, are localized in some finite area of the function's argument. As the distance from this area grows, wavelet functions tend to zero or have infinitely small values. Figure 14.28 shows an example of such a function, which is also known as " the Mexican hat."

Due to its math sense, a wavelet spectrum has two arguments rather than one. Apart from the frequency, it accepts the second argument b — the location of the wavelet function. Hence, b has the same dimension as x.

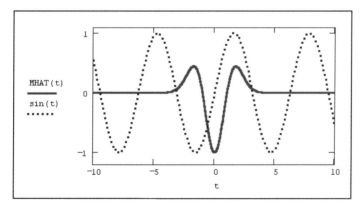

FIGURE 14.28. A wavelet function versus a sine curve

The Built-In Function for Wavelet Transform

Mathcad has one built-in function for calculating wavelet transform based on the Daubechies wavelet function.

- ▪ wave (y) — The vector of a direct wavelet transform
- ▪ iwave (v) — The vector of an inverse wavelet transform

Here:

- y — The vector of argument data at equal intervals
- v — The vector of a wavelet spectrum

Similar to the Fourier transform, an argument of the wavelet transform function (i.e., the y vector) must have exactly 2^n elements (where n is an integer). The result of the wave function is a vector composed of several coefficients of the two-parametric wavelet spectrum. Listing 14.22 illustrates the usage of the wave function in the example of the analysis of the sum of two sine curves. Figure 14.29 shows three families of the coefficients of the calculated wavelet spectrum.

LISTING 14.22. Searching the Daubechies Wavelet Spectrum

$$f(t) := \sin\left(2 \cdot \pi \cdot \frac{t}{50}\right) + 0.3 \cdot \sin\left(2 \cdot \pi \cdot \frac{t}{10}\right)$$

$$Nmax := 256 \qquad\qquad i := 0 .. \ Nmax - 1$$

$$y_i := f(i)$$

$$W := wave(y)$$

$$Nlevels := \frac{\ln(Nmax)}{\ln(2)} - 1$$

$$Nlevels = 7$$

$$k := 1, 2 .. \ Nlevels$$

$$coeffs(level) := submatrix\left(W, 2^{level}, 2^{level+1} - 1, 0, 0\right)$$

$$C_{i,k} := coeffs(k)_{\displaystyle floor\left[\left(\frac{i}{\frac{Nmax}{2^k}}\right)\right]}$$

Programming Other Wavelet Transforms

Besides using the built-in wave function, you can also program your own custom algorithms for calculating wavelet spectrums. Programming is reduced to accurate calculation of appropriate families of integrals. Listing 14.23 provides an example illustrating implementation of such a program and the results obtained using this program are shown in Figure 14.30. This example illustrates the analysis of the

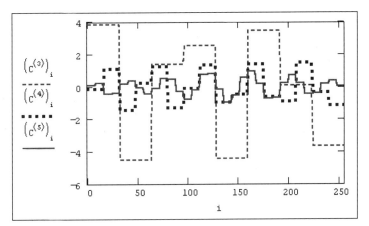

FIGURE 14.29. The wavelet spectrum based on the Daubechies function (Listing 14.22)

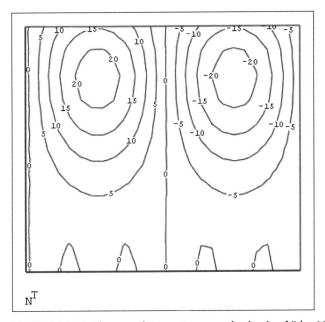

FIGURE 14.30. The wavelet spectrum on the basis of "the Mexican hat" (Listing 14.23)

same function as the one used in the previous listing (the function composed on the basis of the sum of two sine curves). The graph of the two-parameter spectrum c(a,b) is presented in the form of level curves on the plane (a,b).

LISTING 14.23. Searching the Wavelet Spectrum on the Basis of " The Mexican Hat"

$$f(t) := \sin\left(2 \cdot \pi \cdot \frac{t}{50}\right) + 0.3 \cdot \sin\left(2 \cdot \pi \cdot \frac{t}{10}\right)$$

$$\text{MHAT}(t) := \frac{d^2}{dt^2} \exp\left(\frac{-t^2}{2}\right)$$

$$\text{Nmax} := 256$$

$$\psi(a, b, t) := \text{MHAT}\left(\frac{t - b}{a}\right)$$

$$W(a, b) := \int_{-\text{Nmax}}^{\text{Nmax}} \psi(a, b, t) \cdot f(t) \ dt$$

$$i := 0 \, .. \, 10 \qquad\qquad b := 0, 1 \, .. \, \frac{\text{Nmax}}{10}$$

$$a_i := \frac{(i + 15)^4}{2 \times 10^4}$$

$$N_{i, b} := W\left[(a_i), 2 \cdot b - \frac{\text{Nmax}}{10}\right]$$

The program implemented in this listing is very simple. However, in terms of performance, it is far from perfect. Each integral is evaluated independently, without the kind of performance optimization used in an FFT algorithm.

15 ⋮ Data Input/Output

In this chapter we'll consider various aspects of data input into Mathcad documents and output of the calculation results. In the starting sections of this chapter, we'll provide a brief reference to numeric input and output, and we'll list data types that are applicable in the Mathcad environment *(see Section 15.1)* when defining variables and functions.

However, the main focus will be on graphs, since they are the most powerful tools used to represent the results of calculations in Mathcad. Therefore, the majority of this chapter is dedicated to efficient usage of the graphs *(see Sections 15.2–15.4).*

The mechanism of creating animation files *(see Section 15.5)*, a feature that makes the results of working with Mathcad especially efficient and illustrative, is based on the dynamic change of the graphs. Moreover, Mathcad provides a wide range of functional capabilities of data input and output to external text or graphic files *(see Section 15.6).*

15.1. NUMERIC INPUT AND OUTPUT

The simplest and the most widely used method of data input and output is implemented in Mathcad by assignment and output (either numeric or symbolic) directly within a document. Actually, a Mathcad document simultaneously represents a program code and the result returned by that program. Numeric input and output was covered in detail in *Chapter 4*, which, actually, is nearly entirely dedicated to this topic (details on data input are provided in *Section 4.1, "Data Types," Section 4.2, "Dimensional Variables,"* and *Section 4.3, "Arrays"*; data output was discussed in *Section 4.4, "Numeric Data Output Format"*). As a result, here we'll limit ourselves to providing a couple of examples illustrating this important component of the Mathcad system (Listings 15.1 and 15.2).

LISTING 15.1. Numeric Data Input

$i := 0 .. \ 4$

$x := 1.5257285$

$y := 1234.567890$

$A := \begin{pmatrix} 1 & 2 \\ 3 & 4 \end{pmatrix}$

$f(x) := x^2$

LISTING 15.2. Numeric Data Output *(Listing 15.1, Continued)*

$x = 1.526$ $f(x) = 2.328$

$y = 1234.568$ $y = 1.235 \times 10^3$

$A = \begin{pmatrix} 1 & 2 \\ 3 & 4 \end{pmatrix}$ $f(A) = \begin{pmatrix} 7 & 10 \\ 15 & 22 \end{pmatrix}$

$A_{1,0} = 3$

$i =$

0
1
2
3
4

$f(i) =$

0
1
4
9
16

15.2. CREATING GRAPHS

Mathcad 2001i provides several types of graphs (see the illustrations for this chapter) that can be classified into the following two large groups.

- Two-dimensional graphs:
  - An XY (Cartesian) graph (XY plot)
  - A Polar graph (Polar Plot)
- 3D graphs:
  - A 3D surface graph (Surface Plot)
  - A level curves graph (Contour Plot)
  - A 3D bar chart (3D Bar Plot)
  - 3D set of points (3D Scatter Plot)
  - A vector field (Vector Field Plot)

This classification of graphs is a matter of convention, since by managing the settings of various parameters, it is possible to create combinations of graph types. Furthermore, you can even create new types of graphs (for example, a 2D bar chart of the distribution can be thought of as a variation of a simple XY plot).

All graphs are created in the same way, using the **Graph** toolbar. Any differences are due to the difference of the data being plotted.

If the data to be plotted are defined incorrectly, an error message will appear instead of the graph.

To create a graph (for example, a simple XY plot), proceed as follows:

1. Position the cursor at the point where you need to insert a graph within your document.
2. If the **Graph** toolbar is not present on the screen, display it by clicking the button with the graph icon on the **Math** toolbar.
3. When the **Graph** toolbar appears, click the X-Y Plot button to create a new XY plot (Figure 15.1). If you need to create a graph of another type, click the appropriate button on the **Graph** toolbar.
4. As a result, a blank graph area will appear in the selected location of your document, containing one or more placeholders (Figure 15.1, left). Fill in the placeholders with the names of variables or functions that you want to plot on the graph. In our case (a simple XY plot), these will be two placeholders that must be filled in with the names of the data that are to be plotted along x- and y-axes.

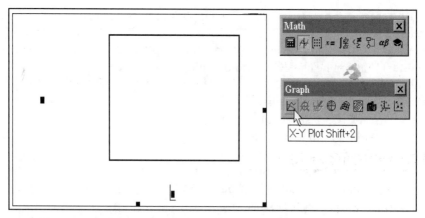

FIGURE 15.1. Creating an XY plot

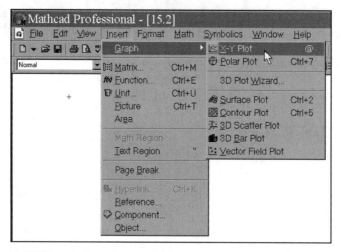

FIGURE 15.2. Creating a graph using menu commands

If data names are entered correctly, the required graph will appear on the screen. You can change the existing graph by changing the data, formatting the graph, or adding extra legend elements.

Several sections of this chapter are entirely dedicated to correct data specification and graph formatting.

The **Graph** toolbar provides the simplest and the most illustrative method of creating a graph. However, you can achieve the same result by selecting an appropriate item from the **Insert | Graph** submenu shown in Figure 15.2, or by pressing an appropriate keyboard shortcut.

To delete the existing graph, click it with the mouse and then select the **Cut** or **Delete** commands from the **Edit** menu.

15.3. TWO-DIMENSIONAL GRAPHS

The 2D graph group includes graphs in Cartesian and polar coordinate systems. In contrast to 3D graphs, the existing 2D graph can't be converted to a graph of another type. To create an XY plot, you'll need to define two sets of data that will be plotted along the x- and y-axes.

15.3.1. XY Graph of Two Vectors

The simplest and the most illustrative way to construct a Cartesian graph is by creating two vectors of data that will be plotted along the x- and y-axes. The sequence of procedures necessary to create a graph based on the two vectors x and y is shown in Figure 15.3. In this case, the placeholders near the axes must be filled in with the names of the appropriate vectors. Besides this, you can also plot vector elements along the axes, i.e., fill in the placeholders near the axes with x_i and y_i respectively (Figure 15.4). As a result of this experiment, you'll get a graph consisting of points corresponding to the pairs of vector elements connected with sections of straight lines. The resulting broken line is known as a *data series* or *trace*.

Notice that Mathcad automatically defines graph boundaries based on the range of values of vector elements.

It should be noticed that you can use the same method with little difficulty to create an XY plot of the matrix rows or columns. To do so, use the column extraction operator and plot appropriate calculations along the graph axes (a large variety of such examples can be found in illustrations for Chapters 11 and 12).

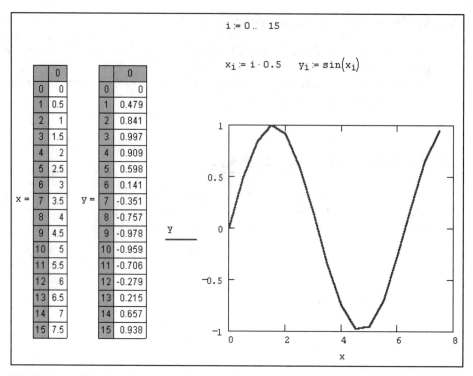

FIGURE 15.3. XY plot of two vectors

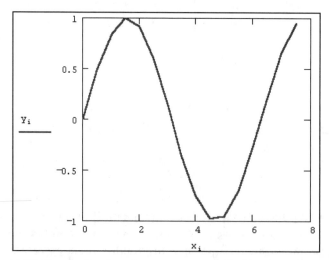

FIGURE 15.4. XY plot of two vectors specified by their elements

15.3.2. XY Plot of a Vector and Range Variable

You can use a range variable as a variable plotted along one of the axes (Figure 15.5). In this case, the data plotted along another axis must represent either an expression that explicitly contains this range variable or an element of the vector indexed by this range variable (but not the vector itself).

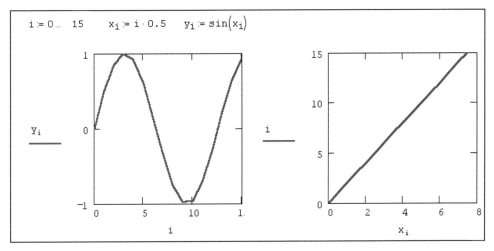

FIGURE 15.5. XY plots of a vector and range variable

15.3.3. XY Graph of a Function

You can plot a graph of any scalar function $f(x)$ using two methods. The first method supposes that you digitize the function value, assign these discrete values to a vector and plot the vector graph. Actually, this method was the one used to plot the sine curves shown in Figures 15.3–15.5. Another, simpler method, known as fast plotting, requires that you enter the function into one of the placeholders (for example, the one near the y-axis), and the placeholder next to another axis must be filled in with the argument name (Figure 15.6). As a result, Mathcad will automatically create a graph of the function within the limits of the argument values (which, by default, comprise the range from –10 to 10). Naturally, later you can change the range of argument values, and the graph will automatically change to fit this range.

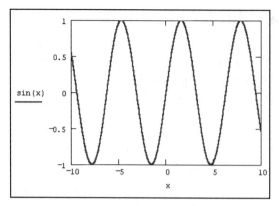

FIGURE 15.6. Fast plotting of the function graph

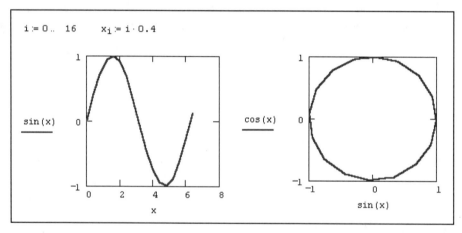

FIGURE 15.7. Graphs of functions accepting a vector argument

Notice that if the argument variable was assigned some value before plotting a graph within your document, the function dependence that takes this value into account will be drawn instead of fast plotting. Figure 15.7 shows examples of two such graphs.

15.3.4. A Polar Graph

To create a polar graph, click the **Polar Plot** button on the **Graph** toolbar (Figure 15.8) and fill in the placeholders with the names of the variables and the function that will be plotted in the polar coordinate system: the angle (lower placeholder) and radius vector (left placeholder). Similar to the Cartesian graph

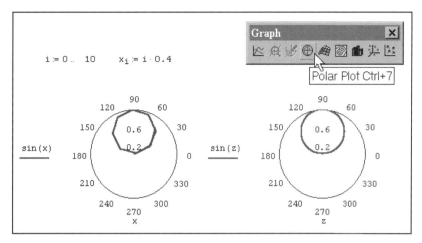

FIGURE 15.8. Polar plots

(see Sections 15.3.1–15.3.3), you can plot the following components along the axes: two vectors (Figure 15.8, left), vector elements, and range variables in various combinations, and, finally, you can also perform quick plotting of the function graph (Figure 15.8, right).

Formatting of the polar graphs is practically identical to formatting of XY plots, and thus, all the materials on formatting two-dimensional graphs illustrated by the examples of XY plots also relate to polar plots.

15.3.5. Creating Several Data Dependencies

You can plot up to 16 various dependencies within a single graph. To add another dependency curve to the existing graph, it is necessary to proceed as follows:

1. Position the insertion lines so that they completely enclose the expression in the y-axis label (Figure 15.9).
2. Press the <,> key.
3. An empty placeholder will appear, which you'll need to fill in with the expression for the second curve.
4. Click anywhere outside this expression (within the graph or outside it).
5. The second curve will be plotted on the graph. For example, Figure 15.9 shows two data dependencies, and pressing the <,> key will result in the appearance of a third placeholder, with which it is possible to define the third dependence.

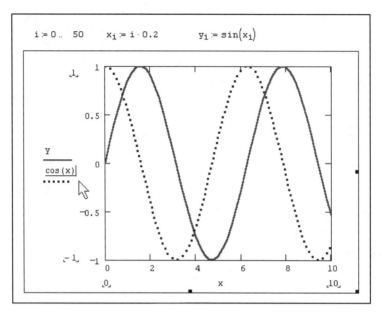

FIGURE 15.9. Plotting more than one expression within a single graph

To remove one or more data dependencies from the graph, use the <BackSpace> or <Del> key to delete their respective expression on the coordinate axes.

Using the method just described, you'll be able to create several dependencies related to the same argument. For example, Figure 15.9 shows the graphs for the pairs of points $y(x_i)$ and $\cos(x_i)$ of the same argument — elements of the vector x_i. This fact is evident in the presence of a single label x near the abscissa axis. On the other hand, Mathcad provides the powerful capability of displaying dependencies of different arguments within the same graph. To use this feature, sequentially enter all labels of all dependencies near both axes.

For example, if you need to replace the second graph shown in Figure 15.9 (the $\cos(x_i)$ graph shown with the dashed line) with the graph of the parametric dependence $\cos(\sin(x_i))$, simply type the comma key once more to enter another label near the x-axis and then enter the required expression – $\sin(x)$. The result of these operations is shown in Figure 15.10.

When plotting several dependencies of different arguments within the same graph, you need to make sure that the data types are matching for each separate pair of data points. For example, you can display a graph of the range variable function and the graph of function created in the quickplot mode (Figure 15.11).

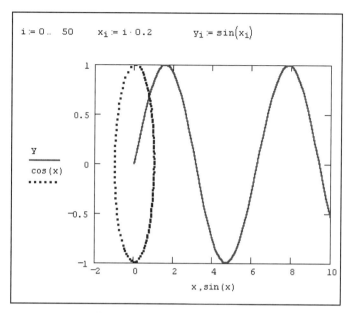

FIGURE 15.10. Plotting several dependencies from different arguments

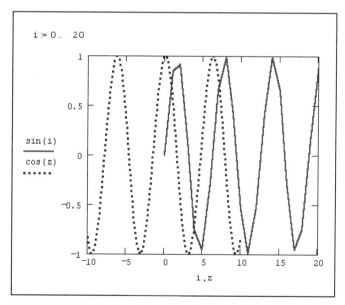

FIGURE 15.11. Plotting dependencies from different arguments
with different data types

15.3.6. Formatting Axes

Formatting capabilities for coordinate axes of the graphs include the capabilities of managing their range, scale, numbering, and display of several labels.

Changing the Range of the Axes

For newly created graphs, Mathcad selects the range for both coordinate axes automatically. To change this range, do the following:

1. Click anywhere within the graph to start editing it.
2. The graph will be highlighted, and next to each of its axes two fields with numeric values will appear. These numbers designate the boundaries of the graph range. To edit these values, click the required field with the mouse (for example, the upper boundary of the x-axis range, as shown in Figure 15.12).
3. Using the arrow keys along with <BackSpace> and <Del> keys, remove the whole contents of the field.
4. Enter the new value of the range boundary (20, for example).
5. Click anywhere outside the field, and the graph will automatically be redrawn within the new limits.

Figure 15.13 shows the results of changing the x-axis range to $(0,20)$ and y-axis range to $(-2,2)$.

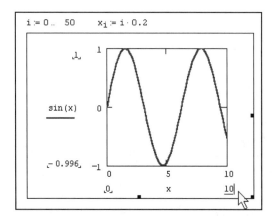

FIGURE 15.12. Changing the range of the x-axis

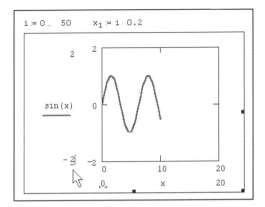

FIGURE 15.13. The result of changing the axes ranges

To return the default settings for a specific range, simply delete the value from the appropriate numeric field and then click anywhere outside that field. Mathcad will automatically set the range limit based on the data values that will be plotted on the graph.

Formatting the Scale

To change the current scale of the coordinate axis, open the Formatting Currently Selected X-Y Plot window and go to the X-Y Axes tab (Figure 15.14). To open this window, double-click anywhere within the graph and select the **Format I Graph I X-Y Plot** commands from the main menu or right-click the graph and select the Format command from the context menu.

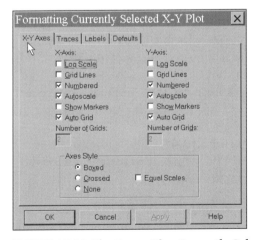

FIGURE 15.14. The **Formatting Currently Selected X-Y Plot** window

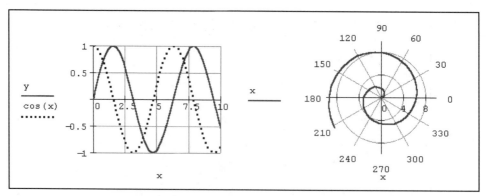

FIGURE 15.15. Grid lines on Cartesian and polar plots (Crossed axes style)

Using checkboxes and radio buttons presented on the **X-Y Axes** tab of the **Formatting Currently Selected X-Y Plot** window, you can easily change the formatting of each axis. Listed below are the available options with brief descriptions:

- **Log Scale** — The graph along this axis will be plotted on a logarithmic scale. This is rather useful if the data differ by several orders.
- **Grid Lines** — When this option is set, the grid lines will be displayed (see Figure 15.15).
- **Numbered** — Show scale numbering. If you clear this checkbox, the scale numbering will disappear (see Figure 15.16).
- **Autoscale** — Mathcad automatically performs axis scaling and range selection.
- **Show Markers** — Lets us add one or two horizontal or vertical lines marking specific values at the axes *(see the "Markers" section later in this chapter).*
- **AutoGrid** — When this option is set, Mathcad automatically determines the number of tick marks along the scale. If this checkbox is cleared, you'll need to enter the desired number of tick marks into the **Number of Grids** field directly below the **AutoGrid** checkbox.
- **Equal Scales** — When this checkbox is set, the x and y axes have equal scales.
- **Axes Style** — You can select one of the following three styles for the coordinate system:
  - **Boxed** — As shown in Figures 15.10–15.13.
  - **Crossed** — Coordinate axes in the form of two crossing lines (Figure 15.15).
  - **None** — Coordinate axes are not displayed.

For the polar plot there are other different axes styles: Perimeter, Crossed, and None. The coordinate system of the first style is shown in Figure 15.8, and the second style is illustrated in Figure 15.15.

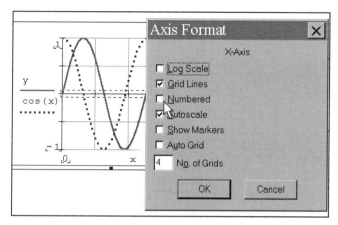

FIGURE 15.16. The **Axis Format** dialog

You can also change the parameters just described using the Axis Format dialog that opens if you double-click the required axis (Figure 15.16).

Markers

Markers are used to mark specific values at coordinate axes. The marker is the line perpendicular to the axis and labeled with a number or variable. To create a marker, proceed as follows:

1. Double-click the graph.
2. When the **Formatting Currently Selected X-Y Plot** window opens, go to the X-Y Axes tab (Figure 15.14) and set the **Show Markers** checkbox.
3. Click **OK**.
4. The placeholder will appear. Fill it in with the desired number or name of the variable that you want to emphasize on the marker (Figure 15.17, left).
5. Click anywhere outside the marker.

Markers are shown in Figure 15.17, on the right-hand side. For each axis, you can set up to two markers. If you define only one marker, the second one won't be displayed.

NOTE

When creating markers, graph tracing might prove to be rather useful, since it allows you to define the marker value with a high level of precision (see Section 15.3.10).

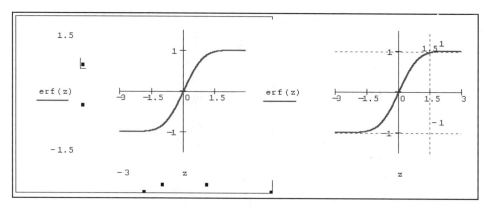

FIGURE 15.17. Creating markers (left) and defined markers (right)

15.3.7. Formatting Data Traces

You can easily specify a combination of parameters of the line and points for each of the data sequences (traces) represented on the graph. To do so, open the Formatting Currently Selected X-Y Plot and go to the Traces tab (Figure 15.18). To specify parameters for a specific trace, select the required trace from the list (the trace position within this list corresponds to the position of its dependency label at the y-axis) and then select the required option from the list to the right of the Legend Label field.

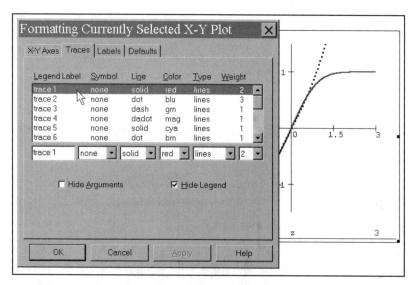

FIGURE 15.18. The **Traces** tab of the **Formatting Currently Selected X-Y Plot** window

The **Traces** tab provides the following options:

- ▦ **Legend Label** — The text of the legend label describing the specific data trace (in Figures 15.19–15.22 legend labels describe the meaning of different parameters)
- ▦ **Symbol** — The symbol designating specific data points (Figure 15.21)
- ▦ **Line** — Style of the line (Figure 15.19):
  - • **solid** (a solid line)
  - • **dot** (a dotted line)
  - • **dash** (a dashed line)
  - • **dadot** (a dash-and-dot line)
- ▦ **Color** — Specifies the color for the line and data points
- ▦ **Weight** — Specifies the thickness of the line (Figure 15.20) and data points
- ▦ **Type** — Type of data trace representation:
  - • lines
  - • points
  - • error
  - • bar
  - • step
  - • draw
  - • stem
  - • solid bar

Specific parameters might be unavailable for some types of graphs. For example, it is impossible to specify the draw character for step curve.

NOTE

Style, Color, and Thickness of the Line

You can improve the graph readability and make it more illustrative by changing line parameters for different data traces presented within a single graph (Figures 15.19 and 15.20).

Formatting Data Points

To create a graph using data points only, open the Formatted Currently Selected X-Y Plot window, go to the Traces tab (see Figure 15.18), select a trace, and then select the **points** option from the **Type** list. To display a data trace as a curve, select another option from this list (for example, lines).

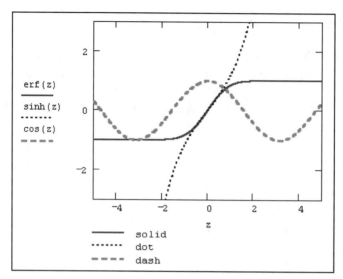

FIGURE 15.19. Lines of different styles

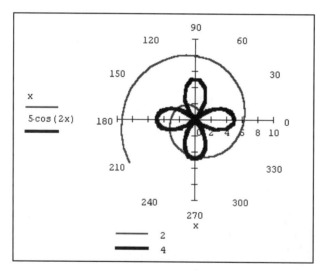

FIGURE 15.20. Lines of different thickness at the polar plot

The Symbol list specifies the symbol that will be used to represent the data point, while the **Weight** list enables you to specify the size of these symbols. The examples are shown in Figure 15.21.

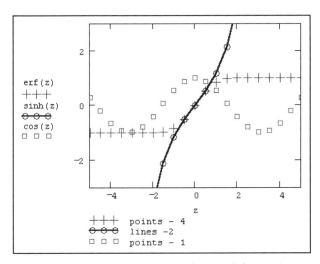

FIGURE 15.21. Different styles and sizes of data points

Types of Data Traces

Several types of data traces are represented in Figure 15.22.

Any of the types listed above can also be used for polar plots (see the example shown later in Figure 15.25).

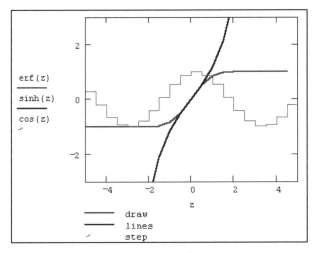

FIGURE 15.22. Different types of data traces

Bar Charts (Histograms)

Mathcad 2001i provides several types of bar charts suitable for constructing histograms (several notes on their usage are provided in *Chapter 13*). The illustration presented in Figure 15.23 shows three various types of bar charts.

Error Bar Graphs

The error bar graph differs significantly from other graph types since it requires three series of data rather than two. Besides pairs of Cartesian (XY) or polar coordinates, it is necessary to specify two more data series, representing appropriate error values for each pair of data points (Figure 15.24).

An error bar graph requires you to specify the error bar type for two sequential data series.

Figure 15.24 shows three plotted data series: y (the data itself), errorU (the upper error limit), and errorD (the lower error limit). For the last two data series the error type is specified.

Saving the Default Settings

The Defaults tab of the Formatting Currently Selected X-Y Plot window contains the following two controls:

- The Change to Defaults button — Switches all settings of the currently selected graphs to the default settings specified for the current document.
- The Use for Defaults checkbox — By setting this checkbox, you can specify the settings of the currently selected graph as the default settings for the current document.

15.3.8. Creating the Graph Title

To create the graph title, proceed as follows:

1. Double-click the graph.
2. Open the Formatting Currently Selected X-Y Plot window and go to the Labels tab.
3. Enter the text of the graph title into the Title field.

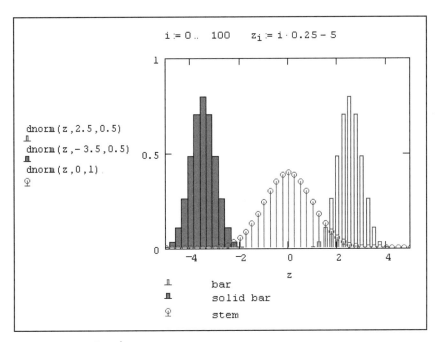

FIGURE 15.23. Bar chart types

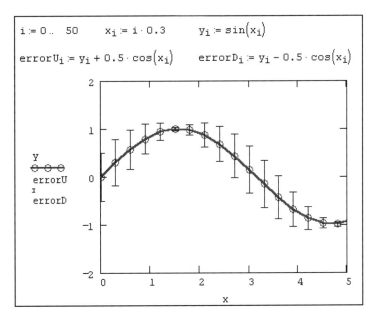

FIGURE 15.24. Creating an error bar graph

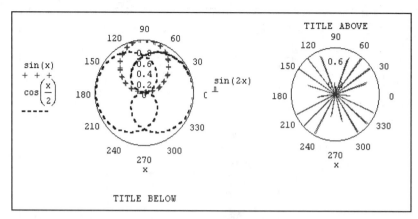

FIGURE 15.25. Selecting the position for the graph title

4. Set the Show Title checkbox.
5. Select the title position by setting the Above or Below radio buttons as shown in Figure 15.25.
6. Click OK.

15.3.9. Changing the Position and Size of the Graph

Before you move the graph to change its size, select the required graph by clicking it with the mouse. You can also change a graph's position within a document by simply dragging it to the required position. To resize the graph, move the cursor to the handle at the bottom or at the right border of the graphs, click the mouse button, and drag to stretch the graph's region in the direction you choose.

15.3.10. Tracing and Enlarging the Graphs

Tracing enables you to study graph behavior in detail. To enable the tracing mode, right-click the graph and select the **Trace** command from the right-click menu. The tracing window will appear (Figure 15.26), and you'll see two crossing dashed lines in the graph field.

You can move the crossing point of the tracing lines by moving the mouse cursor within the graph area. The coordinates of the point are displayed with high precision in the appropriate fields of the tracing window (the X-Value field displays the x coordinate, and Y-Value field displays the y coordinate). By clicking the Copy X or Copy Y buttons, you can copy the contents of the respective field

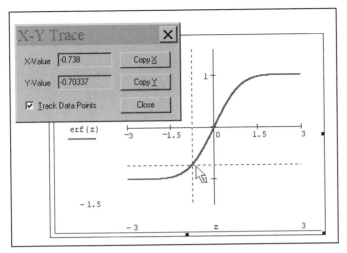

FIGURE 15.26. Tracing the graph

onto the clipboard. Later, you can insert the clipboard contents into any location within your document or onto the marker. To accomplish this, simply press the <Ctrl>+<V> keyboard combination.

If the **Track Data Points** checkbox is set (as shown in Figure 15.26), the tracing lines will follow exactly along the graph curve. If this checkbox is cleared, tracing lines can move anywhere within the graph area.

Besides tracing, Mathcad provides yet another convenient capability: zooming the graph. To open the **Zoom** dialog, choose the **Zoom** command from the right-click menu or select the **Format | Graph** commands from the main menu and then select the **Zoom** command from the appropriate submenu. Select the rectangular area of your graph with the mouse (Figure 15.27), then press the **Zoom** button. As a result, the selected part of your graph will be redrawn and zoomed (Figure 15.28). You can continue to re-scale the graph or return to the previous view by clicking the **Full View** button. To close the **Zoom** dialog and redraw the graph in an enlarged mode, click **OK**.

*You'll probably find that it is more convenient to open the tracing and scaling windows using the **Graph** toolbar (Figure 15.29). These toolbar buttons are available only when a two-dimensional graph is selected.*

NOTE

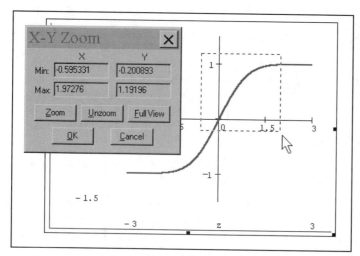

FIGURE 15.27. Zooming the graph

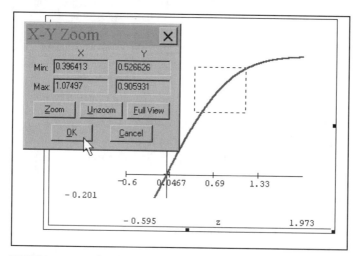

FIGURE 15.28. Viewing an enlarged graph

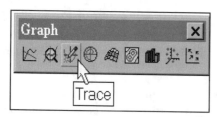

FIGURE 15.29. The tracing button on the Graph toolbar

15.4. 3D GRAPHS

The collection of 3D graphs is a remarkable feature that will be appreciated by all Mathcad users. In a matter of seconds, you can create an impressive presentation of the results of your calculations.

15.4.1. Creating 3D Graphs

To create a 3D graph, click any button labeled with the icon depicting any type of 3D graph on the **Graph** toolbar *(see Section 15.2)*. As a result, a blank graph area with three axes will appear (Figure 15.30) containing a single placeholder. Fill in this placeholder with the name of the function $z(x,y)$ for quick plotting of the 3D graph or with the name of matrix variable z, which will specify the data distribution $z_{x,y}$ for the XY plane.

Let us consider the procedure for creating 3D graphs with the simple example of the function $z(x,y)$ and matrix z (see Listings 15.3 and 15.4, respectively). 3D graphs of various types are created by clicking the appropriate button on the **Graph** toolbar. Notice that in order to construct a graph, the only text you need to enter is the appropriate functions' names or matrices that must be typed into the placeholders.

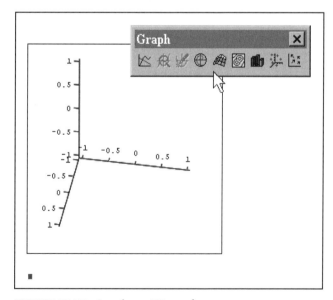

FIGURE 15.30. Creating a 3D graph

For the graphs specified by matrices, the scale of the xy plane must be specified manually. Mathcad simply draws the surface and points in 3D space or level curves based on the 2D structure of this matrix. When working in quickplot mode, you have the capability of constructing graphs in various ranges of arguments, similar to 2D graphs.

LISTING 15.3. Function for Quick Plotting of 3D Graphs

$$z(x,y) := x^2 + y^2$$

LISTING 15.4. Matrix for Displaying 3D Graphs

$$Z := \begin{pmatrix} 1 & 1 & 0 & 1.1 & 1.2 \\ 1 & 2 & 3 & 2.1 & 1.5 \\ 1.3 & 3.3 & 5 & 3.7 & 2 \\ 1.3 & 3 & 5.7 & 4.1 & 2.9 \\ 1.5 & 2 & 6.5 & 4.8 & 4 \end{pmatrix}$$

Surface Plot – 3D Surface Graph

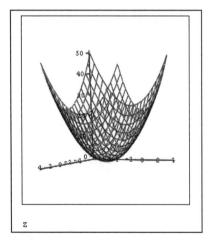

FIGURE 15.31. Quick plotting of the surface plot (Listing 15.3)

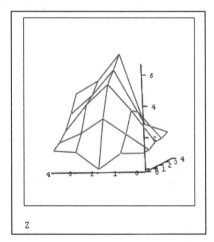

FIGURE 15.32. Surface graph specified by the matrix (Listing 15.4)

Contour Plot – Level Curves Graph

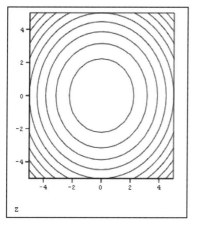

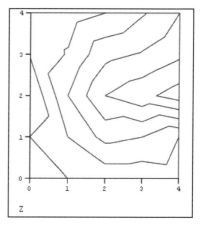

FIGURE 15.33. Quick plotting of the level curves (Listing 15.3)

FIGURE 15.34. Level curves specified by the matrix (Listing 15.4)

3D Bar Plot – 3D Histogram

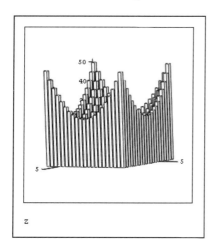

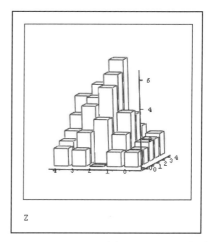

FIGURE 15.35. Quick plotting of the 3D histogram (Listing 15.3)

FIGURE 15.36. 3D histogram specified by the matrix (Listing 15.4)

3D Scatter Plot – Graph Represented as a Set of Points

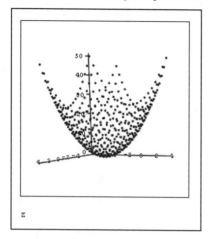

FIGURE 15.37. Quick plotting of the 3D scatter plot (Listing 15.3)

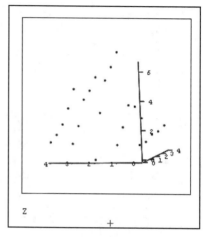

FIGURE 15.38. 3D scatter plot specified by the matrix (Listing 15.4)

The Vector Field Plot – Graph of the Vector Field

The vector field plot is slightly different from other types of 3D graphs. Its primary purpose lies in constructing a specific vector at each point of the XY plane. To specify a two-dimensional vector, it is necessary to specify two scalar numbers. As a result, a vector field in Mathcad is defined by a matrix of complex numbers. Real parts of each elements of this matrix specify the projection of the vector to the x-axis, while imaginary parts serve to specify vector projection to the y-axis.

The illustrations provided thus far are only the first step towards the creation of illustrative, colorful graphs. The following sections describe the rules of correct graph formatting. Using these rules, you'll be able to create illustrative graphs that, mathematically speaking, have an optimal look.

TIP

Applying interpolation to initial data often improves the 3D representation of the graph (see Section 14.1.5, "Multidimensional Interpolation," in Chapter 14).

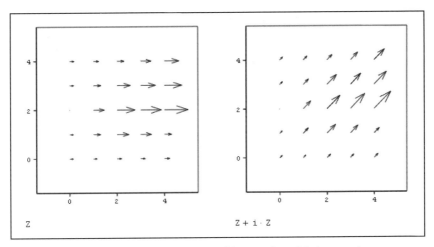

FIGURE 15.39. Vector field plots specified by matrices (Listing 15.4)

15.4.2. Formatting 3D Graphs

To format a 3D graph, open the **3-D Plot Format** window by double-clicking the graph area (Figure 15.40). All parameters of 3D graphs of all types are set using the options available in this dialog.

The **3-D Plot Format** dialog provides a large number of parameters, which, if changed, can significantly influence the graph's look. These options are grouped by their working principles on several tabs. Let us consider 3D graph formatting options in brief, explaining them with illustrative examples.

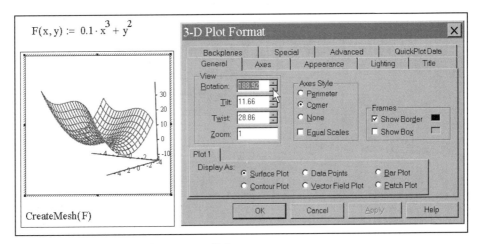

FIGURE 15.40. The **3-D Plot Format** dialog

Changing the Graph Type

To change the type of the existing graph (for example, to create a level curves graph instead of a 3D surface graph, etc.), simply set the appropriate radio button on the bottom part of the General tab and then click OK. The graph will be re-drawn.

Rotating the Graph

The easiest way to position the coordinate system with the graph in 3D space is to simply drag it with the mouse. Click and hold on the left mouse button, and try to drag the mouse pointer within the graph area. You'll see how the graph rotates.

NOTE

Certainly, only graphs in 3D space can be rotated. The graphs of three-dimensional data that by definition are plotted on the plane (such as vector fields and level curves) can't be rotated.

There is an alternative method used in changing the graph orientation, which calls for the usage of the Rotation, Tilt, and Twist fields on the General tab of the **3-D Plot Format** window (see Figure 15.40). Numeric values entered into these fields specify appropriate angles (in degrees) and thus set the direction of all three coordinate axes in 3D space. To better understand the influence of these angles on the graph orientation, compare the examples shown in Figures 15.40–15.42.

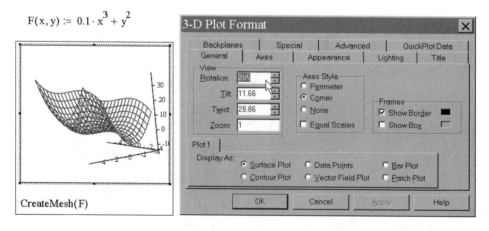

FIGURE 15.41. Changing the Rotation parameter (compare to Figure 15.40)

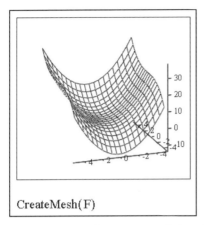

CreateMesh(F)

FIGURE 15.42. The same graph (see Figure 15.41) redrawn for Rotation = 0, Tilt = 20 and Twist = 200

Axes Style

Using the Axes Style option group, you can set one of the following styles of the coordinate axes:

- Perimeter — See the example shown in Figure 15.43.
- Corner — See the example shown in Figures 15.41 and 15.42.
- None — No axes are displayed.

If you set the Show Box checkbox, the coordinate space will be displayed as the box (Figure 15.44).

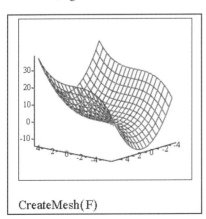

CreateMesh(F)

FIGURE 15.43. Positioning coordinate axes along the perimeter

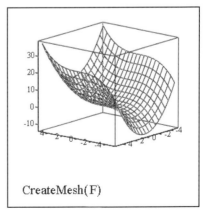

CreateMesh(F)

FIGURE 15.44. The effect of enabling the Show Box option

Zooming the Graph

The Zoom field on the General tab (see Figure 15.40) allows you to set the numeric value of the graph scale (Figure 15.45).

Formatting Axes

The Axes tab contains three nested tabs, where you can specify parameters for each of the three coordinate axes. Particularly, you can enable or disable the grid lines display, and enable or disable numbering and specify the range for each axis (Figure 15.46). The main purpose of these operations is much like that of similar operations for two-dimensional graphs. Another tab — Backplanes — specifies the display mode for the projections of the coordinate grid to the three hidden planes of the 3D graph (an example of formatting the xy plane is shown in Figure 15.47).

Style of Lines and Filling

Figures 15.48 and 15.49 illustrate the influence of the filling and line style on the view of the graph. These options are specified using the Appearance tab of the 3-Plot Format window. When the Fill Surface radio button in the Fill Options group (Figure 15.48) is set, you have access to the color options available in the Color Options group. If you select the Solid Color option, the surface will be filled with a solid color, as shown in Figure 15.49. If you select the Colormap option, the surface or contour graphs will be filled by various colors and shades (Figure 15.48). Notice that you can select the color map on the Advanced tab (Figure 15.50).

We advise you to experiment with different color maps, fillings, and line styles specified by the Line Options field (Figure 15.48) in order to get acquainted with the rich capabilities provided by Mathcad in this area. Some parameters influencing the graph contours display are available on the Special tab (Figure 15.51). There is such a large variety of combinations of different color maps, fillings, and other parameters, that it is much better to let the reader try everything on his own, to apply various combinations and select the most preferable ones.

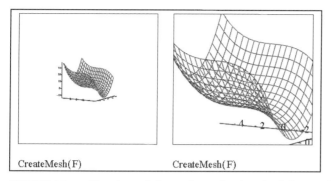

FIGURE 15.45. The view of the surface plot (on the left — scale set to 0.5, on the right — scale set to 1.5)

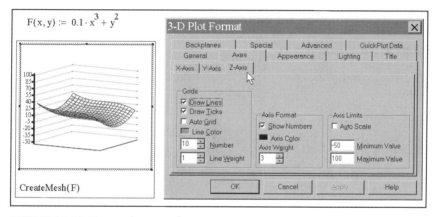

FIGURE 15.46. Formatting coordinate axes

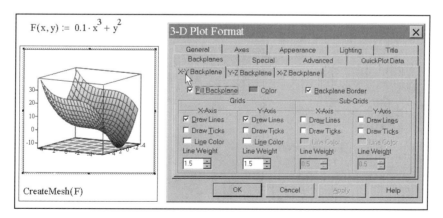

FIGURE 15.47. Formatting hidden planes of the graph

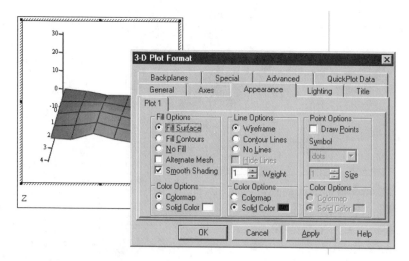

FIGURE 15.48. Surface filling

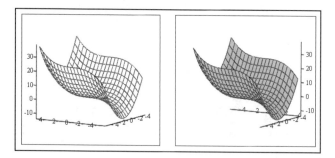

FIGURE 15.49. Filling the surface with white (left) and dark gray (right)

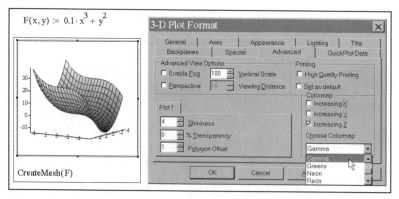

FIGURE 15.50. Selection of the color scheme

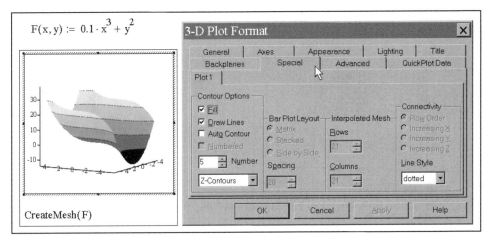

Figure 15.51. The **Special** tab

Special Effects

The Advanced tab (Figure 15.50) provides access to management of several special effects of graph representation, helping to make the graphs look more impressive. Let us list these effects:

- Shininess — Lets you regulate the shininess from 0 to 128
- Fog — Lets you regulate the effect of fog (Figure 15.52)
- Transparency — Specifies the percentage of the graph's transparency (Figure 15.53)
- Perspective — Displays perspective with the definition of the visibility distance (Figure 15.52)

There is another special effect, namely, lighting, which is specified on the **Lighting** tab (Figure 15.54). Notice, that you can use built-in lighting schemes (in Figure 15.54 the mouse cursor points to them), as well as specify the lighting color and direction on your own.

Graph Title

The graph title is defined on the **Title** tab and can be located both at the top and bottom of the graph (Figure 15.55).

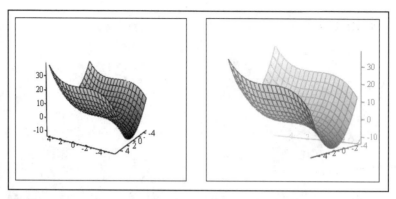

FIGURE 15.52. Effects of perspective (left) and fog (right)

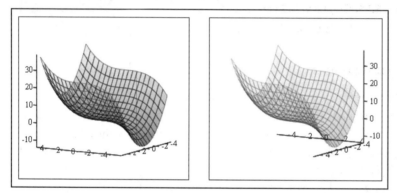

FIGURE 15.53. Transparency effect: left illustration at 40% and right illustration at 80%

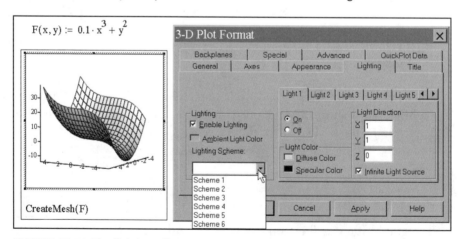

FIGURE 15.54. The lighting effect

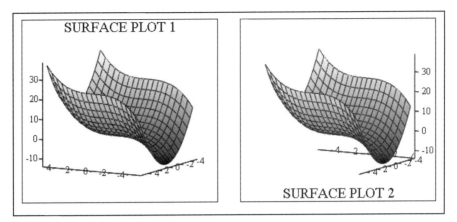

FIGURE 15.55. Graphs with titles

Editing Data Points

Most types of graphs allow you to display data points. The format parameters of the data points (including the symbol type and size), the connection of data points by the line, and so on, are specified on the Appearance tab (Figure 15.56). Options of point formatting are the same as those used for two-dimensional graphs *(see Section 15.3.7)*.

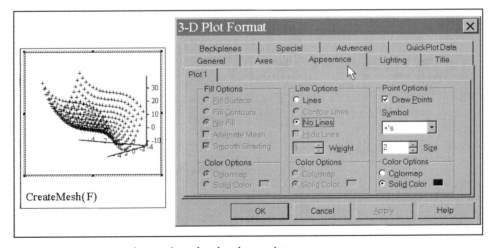

FIGURE 15.56. Formatting options for the data points

Quick Change of the Graph Format

The right-click menu (Figure 15.57) provides a quick and convenient method of formatting some types of 3D graphs. To apply this method, right-click the graph with the mouse and select the required formatting option from the right-click menu. The graph will be redrawn immediately.

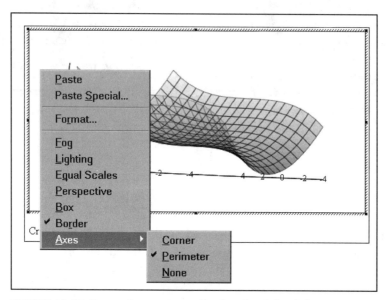

FIGURE 15.57. Formatting a 3D graph using the right-click menu

15.5. CREATING ANIMATION

In most cases, animation is the most illustrative method of presenting the results of math calculations. Mathcad enables you to create animated movies and save them in video files.

The basic principle of animation in Mathcad is sequential frame animation. The movie is simply a sequence of frames representing a specific part of the document selected by the user. The calculations are performed independently for each frame, and all formulae and graphs contained within the frame must represent functions of the frame number. The frame number is specified by the system variable FRAME, which can take only natural values. By default, if the animation preparation mode is not enabled, FRAME = 0.

Let us consider in brief the sequence of actions required to create an animated clip (for example, one demonstrating the propagation of the harmonic wave). Each moment of time will be specified by a specific value of the FRAME variable.

1. Enter all required expressions and graphs that include the FRAME variable into your document. Prepare the part of your document that you want to animate so as to place it on the screen within your range of view. For our example, we need to define the `f(x,t) := sin(x - t)` function and create its Cartesian `y(x, FRAME)`.
2. Perform the **View** I **Animate** commands.
3. The **Animate** dialog will appear. In this dialog, specify the number of the first frame in the **From** field, then specify the number of the last frame in the **To** field and the animation speed (in frames per second) in the **At** field (Figure 15.58).
4. Drag with the mouse to select the part of your document that will become the animated clip.
5. In the **Animate** window click the **Animate** button. After that, the preview field of the **Animate** dialog will display the results of the selected document area's calculations accompanied by a display of the current value of the FRAME variable. After the process is complete, the animation player window will appear on the screen (Figure 15.59).
6. Start the animation playback by clicking the playback button (in Figure 15.59 this is the one that the mouse cursor points to).

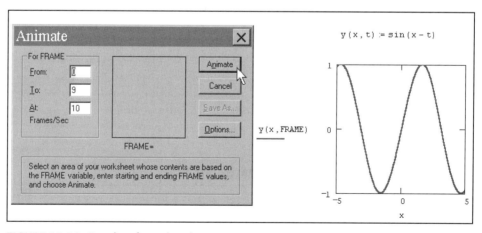

FIGURE 15.58. Creating the animation

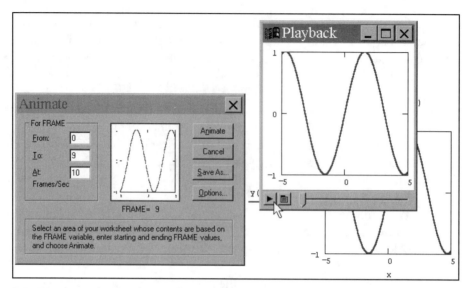

FIGURE 15.59. Viewing the newly created animation

7. If you are satisfied with the results of the animation, save it as a video file by clicking the Save As button in the **Animate** dialog. The Save Animation window will appear, where you can specify the file name and location in the standard way you do for most Windows applications.

8. Close the **Animate** dialog by clicking the Cancel button.

9. The resulting video file can be used independently from Mathcad. If you double-click this file when working with Windows Explorer, it will probably be loaded into the Windows Media Player, and you'll be able to play the animation using this application. Thus, by starting video files in a normal way, you'll be able to produce an impressive presentation of the results of your work, either on your own or on someone else's computer.

When creating animated files, it is possible to use the video compression program. To do so, click the Options button in the Animate dialog.

15.6. INPUT/OUTPUT TO EXTERNAL FILES

Data input and output to external files is an important component of any input/output system. Input of external data to Mathcad documents is used more frequently than data output to external documents, because Mathcad provides much better capabilities of representing the results of the calculations than most other application programs. Mathcad 2001i provides several methods of interacting with external files, the simplest being the family of built-in Mathcad functions.

15.6.1. Text Files

Listed below are Mathcad built-in functions intended for working with text files:

- READPRN("file") — Reading data from text file into the matrix
- WRITEPRN("file") — Writing data to a text file
- APPENDPRN("file") — Appending data to the end of the existing text file

Here:

- file — File pathname

You can specify both a fully qualified pathname, such as C:\Documents and Settings\<Username>\My Documents\, and a relative pathname (bear in mind that the relative pathname will be interpreted starting from the folder where the current Mathcad document is).

Examples illustrating the usage of Mathcad built-in functions are illustrated in Listings 15.5–15.7. You will be able to understand the results of Listings 15.5 and 15.7 better after viewing the resulting text files using any text editor (Notepad.exe, for example — see Figures 15.60 and 15.61, respectively).

LISTING 15.5. Writing the Matrix I to a Text File

$$I := \text{identity}(5)$$

$$I_{3,4} := 99$$

LISTING 15.6. Reading Data from a Text File into the C Matrix

$$\text{WRITEPRN}(\text{"datafile.prn"}) := I$$

$$C := \text{READPRN}(\text{"datafile.prn"})$$

$$C = \begin{pmatrix} 1 & 0 & 0 & 0 & 0 \\ 0 & 1 & 0 & 0 & 0 \\ 0 & 0 & 1 & 0 & 0 \\ 0 & 0 & 0 & 1 & 99 \\ 0 & 0 & 0 & 0 & 1 \end{pmatrix}$$

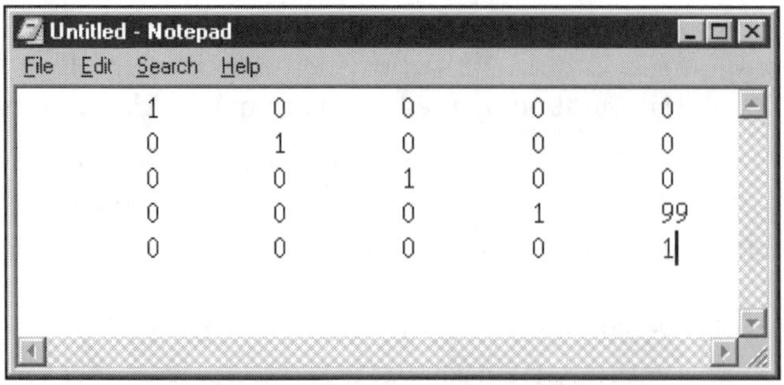

FIGURE 15.60. A file created by Listing 15.5

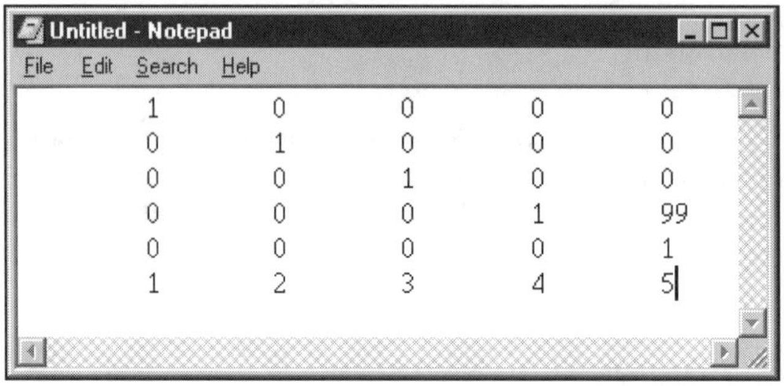

FIGURE 15.61. A file created by Listings 15.5 and 15.7

LISTING 15.7. Appending the k Vector to the Existing Text File

```
k := ( 1   2   3   4   5 )

APPENDPRN ( "datafile.prn" ) := k
```

15.6.2. Graph Files

Similar to input and output to text files, you can organize input and output to graph files of different formats.

When performing graphical input and output, the data is identified by intensity of a specific color in the image pixel. The main functions are listed below:

- READRGB ("file") — Reads the color image
- READBMP ("file") — Reads the grayscale image
- WRITERGB ("file") — Writes the color image
- WRITEBMP ("file") — Writes the grayscale image

Here:

- `file` — File pathname

There is a large variety of special functions providing special access to graph files, for example, those which read only one of the main colors and so on. These special functions are covered in detail in the Mathcad on-line Help system.

Working principles of graph access functions are illustrated by Listings 15.8–15.10. The image created by Listing 15.8, is shown in Figure 15.62.

LISTING 15.8. Writing the `I` Matrix to a Graph File

```
I := identity (100) · 100

I₃,₉ := 500

WRITEBMP ("data.bmp") := I
```

LISTING 15.9. Reading Data from a Graph File

```
C := READBMP ("data.bmp")
```

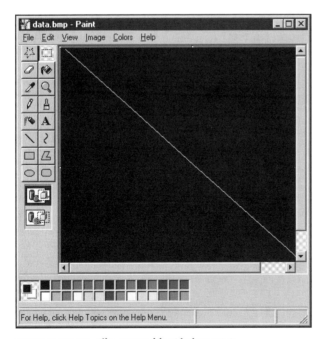

FIGURE 15.62. File created by Listing 15.8

LISTING 15.10. Writing to a Color Graph File

```
R := identity (100) · 100

G := identity (100)

B := identity (100)

WRITERGB ( "color.bmp" ) := augment (R , G , B)
```

15.6.3. Sound Files

Mathcad 2001i is capable of writing and reading the amplitude of acoustic signals written with the .wav filename extension:

- ■ READWAV("file") — Reading the sound file into the matrix
- ■ WRITEWAV("file",s,b) — Writing data to a sound file
- ■ GETWAVINFO("file") — Creates a vector comprising four elements holding information on the sound file

Here:

- ● file — File pathname
- ● s — Sample frequency specified by the matrix
- ● b — Sound resolution (in bits)

16 Formatting Documents

I n this chapter we'll consider some techniques for formatting the results of your work in Mathcad documents. Besides the fact that Mathcad is a powerful mathematical editor that enables you to perform rather complex symbolic and numeric calculations, it also provides a rich array of capabilities for formatting the results of your calculations. If you use the instruments provided in Mathcad for document formatting correctly, you'll be able to produce rather impressive and illustrative documents that present results in a mathematically accurate and intuitive style.

The main capabilities of the Mathcad editor were considered in the first part of this book *(see Chapter 2)*. In the initial sections of this chapter, we'll list the main formatting elements, both built-in and external, that are allowed for usage in Mathcad documents for selecting regions *(see Section 16.1)*; formatting fonts and formulae *(see Section 16.2)*; and specifying page layout, headers, and footers *(see Section 16.3)*. Hyperlinks are rather important elements of the document, enabling you to organize efficient navigation within Mathcad documents *(see Section 16.4)*.

16.1. ELEMENTS OF DOCUMENT FORMATTING

There are different ways of presenting the calculation results in Mathcad, including the following:

- Hard copy — Printed documents
- Web pages — Documents that can be published on the Internet and viewed using browsers
- Mathcad documents — Documents that can be presented to your audience using the Mathcad application itself
- E-books — Electronic books, representing interactive Mathcad documents that are specially formatted and constructed according to the principle that can be viewed with an example from the Resource Center
- Fragments of documents exported and formatted in other applications (for example, Microsoft Word documents or Microsoft PowerPoint presentations)

16.1.1. Formatting Elements

Let us consider formatting elements that are allowed for usage in Mathcad for both math calculations and for purely decorative purposes (Figure 16.1, from top to bottom):

- Text areas
- Math regions or math areas
- Graph regions
- Components of other applications
- Embedded objects

Outside the region boundaries of the document is a blank area. Besides the elements listed above, it often proves advantageous to use the following add-on formatting elements:

- A locked and highlighted area
- Headers and Footers
- Document layout elements, such as page breaks, styles, and margins
- References
- Hyperlinks

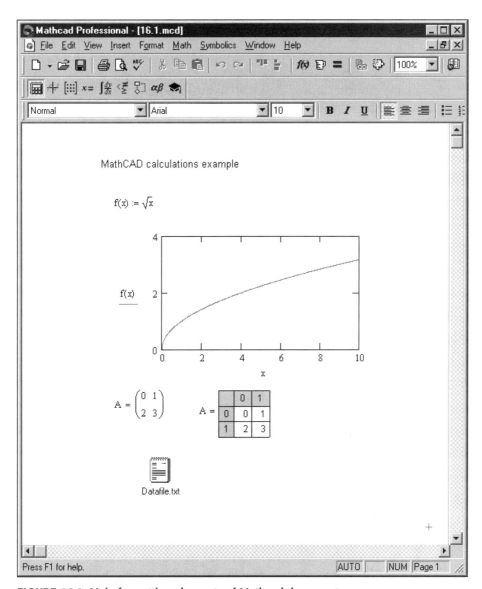

FIGURE 16.1. Main formatting elements of Mathcad documents

16.1.2. Placement of Formatting Elements in Your Documents

Correct and intuitive positioning of the objects within Mathcad documents is an important component of the presentation of calculation results.

Inserting a New Region

Before you insert a specific element into your document, it is first necessary to select the position where the new elements will be inserted. This is accomplished using the insertion cursor. To position the insertion cursor, click the desired location with the mouse. Then you can use the required item from the Insert menu, one of the available toolbars, or by simply typing in the characters from the keyboard.

Examples illustrating the insertion of various regions were considered earlier in this book (for example, formulae in Chapter 2, graphs in Chapter 15, and so on).

Remember, to insert a component, it is necessary to use the Insert | Component menu commands; on the other hand, an embedded object can be inserted by copying it from another application to the clipboard, and by then pressing the <Ctrl>+<V> keyboard combination in Mathcad.

Moving Regions within a Document

To move a specific region to another location within a Mathcad document, proceed as follows:

1. Click the desired region with the mouse to select it. After this, the region will be highlighted, and as long as the mouse cursor is within the limits of that region, it will turn into the insertion lines. The selection of various elements is shown in Figure 16.2.
2. Without clicking the mouse buttons, point the mouse pointer to the region boundary. After that, the pointer will change to the shape of a hand.
3. Click and hold on the left mouse button, and drag the object to the desired location.

Remember, the order of formulae and graphs within your document influences the results of your calculations. The calculation of formulae is performed in the following order: from left to right and from top to bottom.

To create a copy of a specific region in another location of your document, start dragging it in the way you normally would, then press the <Ctrl> key and hold it until you release the mouse button at the desired location.

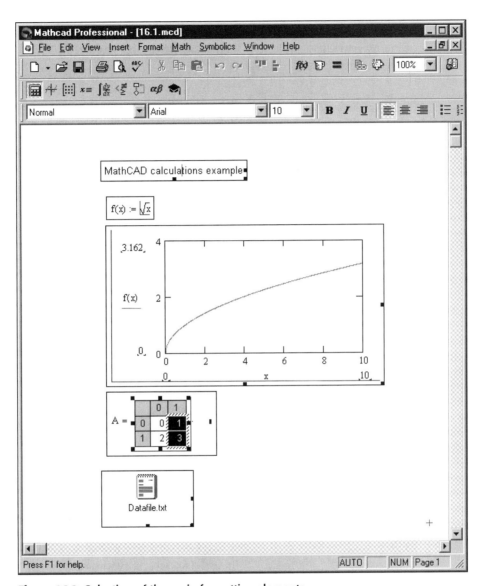

Figure 16.2. Selection of the main formatting elements

Changing the Region Size

If you look carefully at Figure 16.2, you will notice that all regions, with the exception of the math regions, have small black rectangles (known as markers or handles) at their boundaries when selected. Using these handles, you can stretch the regions in the direction of the mouse cursor's movement. If you point the mouse cursor to the handle, it'll turn into the double arrow, which, if dragged, will let you resize the selected region.

In contrast to the method described above, the size of the formula can't be changed in the same way. To format both formulae and text regions (including font type and size), use the **Formatting** toolbar *(see Section 16.2)*.

Remember that changing the scale via the View I Zoom menu commands influences only the screen representation of the document when viewing it with Mathcad. It will not be saved in the document and doesn't influence a printed copy when sending the document to the printer.

Separating the Regions

When working with Mathcad, as you drag the regions from one location to another, you will notice that they start overlapping and may even become invisible if the neighboring region covers them completely. To separate the regions, Mathcad developers have provided a very convenient functionality.

1. Drag with the mouse to select a group of overlapping regions (Figure 16.3).
2. Select the **Format | Separate Regions** commands from the main menu.

As a result, the selected regions will be separated both vertically and horizontally.

Deleting Regions

Mathcad provides a universal method for deleting the whole region. To use this method, select the region, and then choose the **Edit I Delete** commands from the main menu. Various regions can also be deleted by pressing repeatedly <Del> or <BackSpace> keys on the keyboard (notice, however, that these keys are mainly used for deleting the contents within the regions rather than for deleting the whole region with all its contents).

The Mathcad editor, due to its math-specific features, is different from most text processors (which are intuitive and don't cause any difficulties for most users). If you encounter problems when working with the Mathcad editor (which wouldn't be surprising), revisit the first part of this book (see Chapter 2).

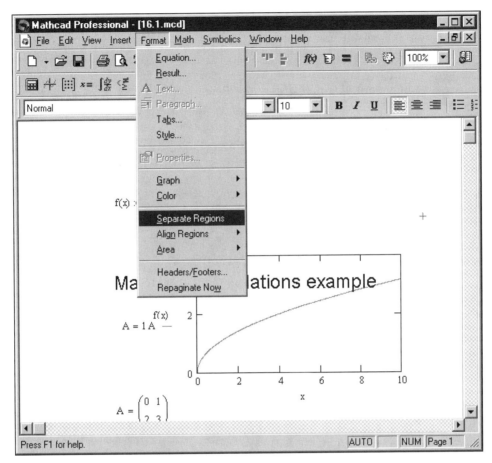

FIGURE 16.3. Separating regions

16.1.3. Highlighting Regions

In Mathcad documents, you can highlight specific regions with color or borders.

Highlighting Regions by Color

To select a specific region by its color, right-click that region and select the **Prop-erties** command from the right-click menu. Alternately, you can select the region and then select the **Properties** command from the **Format** menu. The **Properties** dialog will open, where you need to set the **Highlight Region** checkbox and click OK (Figure 16.4). The region will be highlighted by a color (yellow, by default).

After you set the **Highlight Region** checkbox, the **Choose Color** button becomes available, allowing you to select any other highlighting color from those available on the color palette.

Most formulae provided in the Resource Center are highlighted by some color. After you copy formulae to your document, you can remove the highlighting by clearing the checkbox in the Properties dialog.

To specify a background color for the whole document, select the **Format |
Color | Background** commands from the main menu and select the desired background color from the color palette.

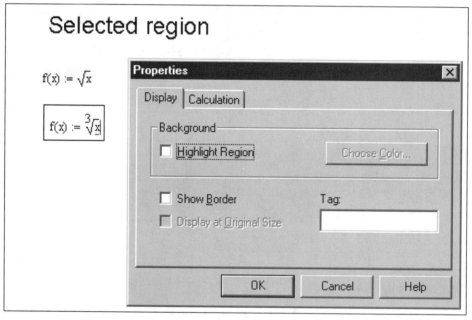

FIGURE 16.4. Highlighting regions with a color

Highlighting Regions with Borders

If necessary, you can highlight the region not only with a color, but also with borders (Figure 16.5). To enable bordering, set the **Show Border** checkbox in the **Properties** dialog. Notice that you can highlight the region both with color and borders.

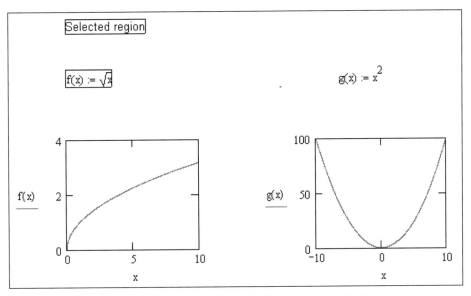

FIGURE 16.5. Regions with a border (left) and without a border (right)

NOTE

*Notice that by using the **Calculations** tab of the **Properties** dialog, you can include or exclude specific formulae from the calculation process (see Chapter 3). Such formulae are designated by the rectangular dot at the top right corner of the formulae text box (Figure 16.6).*

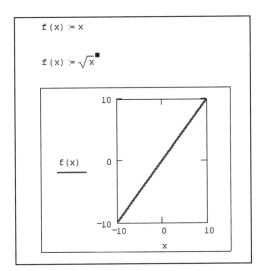

FIGURE 16.6. The central formula has been excluded from the calculation process

16.1.4. Working with Areas

Fragments of Mathcad documents can be placed into specific areas. Generally speaking, areas can include different regions. These areas might be required for the following purposes:

- Delimiting document fragments in terms of their meaning
- Temporarily hiding specific fragments of the document
- Locking specific fragments of the document in such a way as to prevent unauthorized access by any users other than the ones explicitly given such access by the developers of the document

Areas can be created when developing Mathcad-based tutorials in order to prevent unauthorized modifications.

TIP

Creating a New Area

To create a new area in your document, simply position the insertion cursor into the required location, and then select the Insert I **Area** commands from the main menu. As a result, a pair of horizontal lines will appear at the selected locations, marked by a sign in the form of a black triangle (Figure 16.7). The fragment of the document that falls between these lines makes up an area.

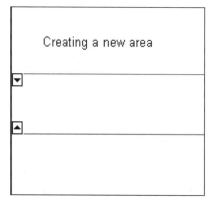

FIGURE 16.7. Creating an area

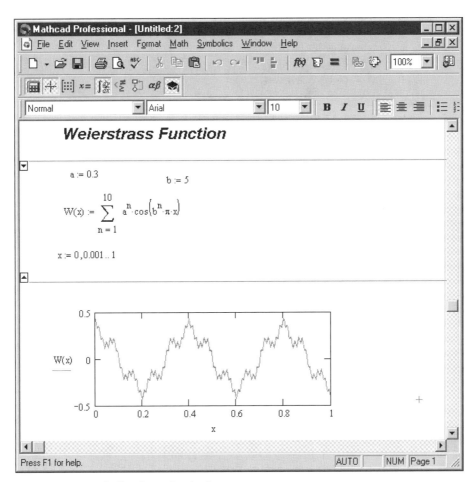

FIGURE 16.8. Including formulae in the area

You can change the size of the area any time by clicking any of the horizontal lines with the mouse (it'll become selected as a result) and moving it to the desired location. To delete the area from your document, select any of the horizontal lines and press the <Del> key.

To insert a formula inside the area, simply drag it there with the mouse (Figure 16.8).

Hiding the Area

To hide the area, double-click any of the horizontal lines that designate it, or, alternately, place the cursor within the area and then select the Format I Area I

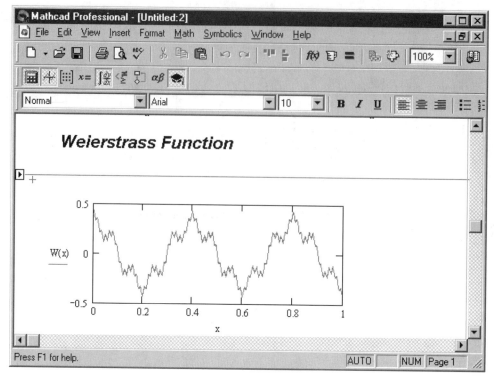

FIGURE 16.9. Hiding the area

Collapse commands from the main menu. Immediately after that, all the contents of the area will be hidden from the screen, but they will still be taken into account when calculations are performed (Figure 16.9).

To return the area to the screen, proceed in the same way: double-click the line designating the presence of a hidden area (Figure 16.9), or select the **Format | Area | Expand** commands from the main menu.

The usage of hidden areas is efficient in large documents.

TIP

Locking the Area

Besides hiding and expanding areas, you can also lock them, i.e., prevent them from any type of editing. To lock an area, select the **Format | Area | Lock** commands from the main menu. When the **Lock Area** dialog appears (Figure 16.10),

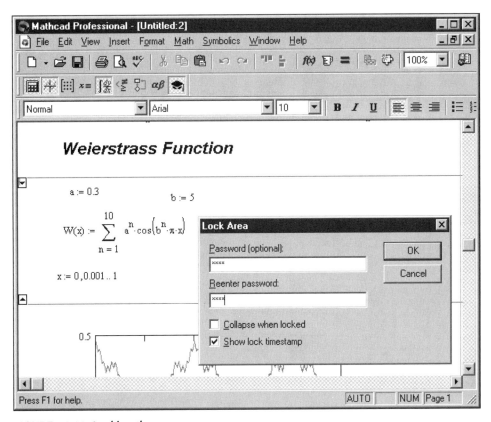

FIGURE 16.10. Locking the area

click OK. If you need to prevent unauthorized editing of the area by another user, you can protect it with a password by entering the password in the Password text field, retyping it in the field below, and then clicking OK.

To unlock the locked area, select the Format I Area I Unlock commands from the main menu. If the area was locked with password protection, you'll be prompted to enter the password when unlocking the area (Figure 16.11). The locked area is designated by the symbols of the locks at the left edges of the horizontal lines framing the locked area.

Figure 16.11 displays the locked area with information on the time when it was locked. This information is displayed since the Show lock timestamp checkbox was set in the Lock Area dialog. Notice that if you set the Collapse when locked checkbox in the Lock Area dialog, the locked area won't be displayed on the screen.

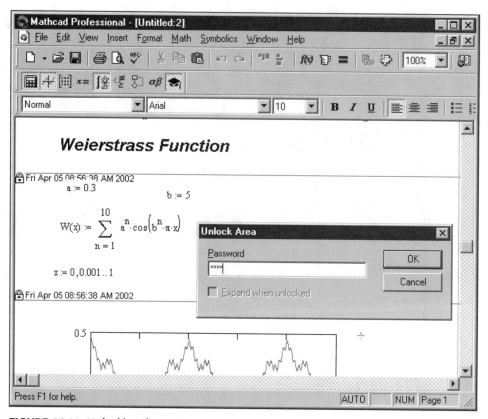

FIGURE 16.11. Unlocking the area

16.2. FORMATTING TEXT AND FORMULAE

To format text and formulae, use the **Formatting** toolbar. It provides two ways of formatting text regions:

- Applying text styles, which influences the format of the whole text region *(see Section 16.2.1)*
- Formatting specific text elements

To apply a style to the whole text region or formula, use the drop-down list of available styles in the **Formatting** toolbar (Figure 16.12). All controls available on this toolbar to the right of the style list are intended for formatting separate text

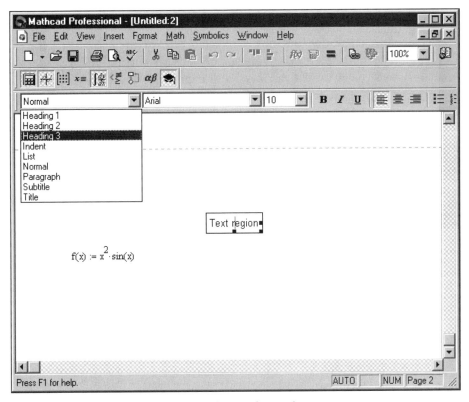

FIGURE 16.12. A list of available styles for text formatting

fragments. One specific feature of applying these formatting capabilities to formulae is the fact that when you change the font of a specific formula, the respective parameters of the math style in the whole document change.

Formatting of the text (and, sometimes, even formula formatting) in Mathcad has many of the same formatting capabilities as most other text editors. Therefore, let us consider in brief its formatting functionality, including fonts, paragraph settings, and page layout parameters. To better understand Mathcad's capabilities, let us consider the second method first, which directly relates to text formatting.

16.2.1. Formatting Text

Text formatting implies management of the following two main formatting components:

■ Font format
■ Paragraph format

Font

You can change the font of the selected text fragment by using the **Formatting** toolbar (Figure 16.13), or by means of the **Text Format** dialog (Figure 16.14). To open this dialog, select the **Text** command from the **Format** menu. Alternately, you can select the text fragment, right-click it with the mouse, and select the **Font** command from the context menu. Available font parameters and appropriate controls available on the **Formatting** toolbar and in the **Text Format** dialog are listed below:

- Font — Specifies font type
- Size — Specifies font size (in Figure 16.13 this field is pointed to by the cursor)
- Font Style — The field in the **Text Format** dialog. On the **Formatting** toolbar, the following buttons **B** *I* U correspond to this field:
  - Bold
  - Italic
  - Underlined
- Strikeout — The checkbox in the **Text Format** dialog
- Color — The field of the **Text Format** dialog, containing a list of available colors
- Position — Checkboxes in the **Text Format** dialog:
  - Superscript
  - Subscript

Examples illustrating text formatting are shown in Figure 16.13.

Paragraph

The following controls are used to specify the parameters:

- Paragraph indent — Three markers on the ruler at the top of the screen specify the left boundary of the first line of the paragraph (the top-left marker), its remaining lines (bottom-left marker), and the right boundary of the paragraph (the right marker).
- Numbered and bulleted lists — Two rightmost buttons at the **Formatting** toolbar.
- Alignment — Specified by the following buttons ▤ ▤ ▤ on the **Formatting** toolbar (Figure 16.15). The following alignment options are available:
  - Left
  - Centered
  - Right

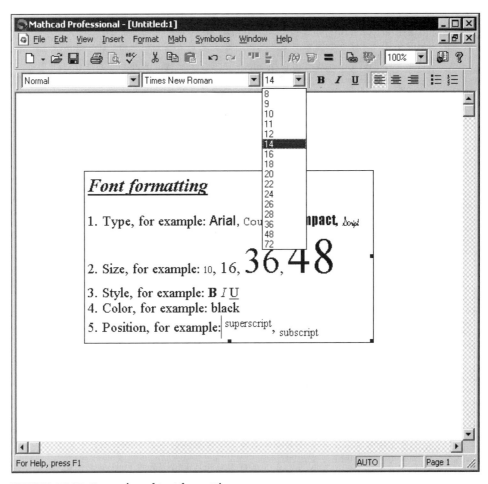

FIGURE 16.13. Examples of text formatting

All paragraph formatting parameters can also be changed in the **Paragraph Format** window that can be opened by selecting the **Paragraph** command from the **Format** menu or by selecting the command with the same name from the right-click menu (Figure 16.16).

There is yet another additional formatting capability — specification of the tab settings. Instructions on specifying the tab settings are available in Mathcad's on-line Help system.

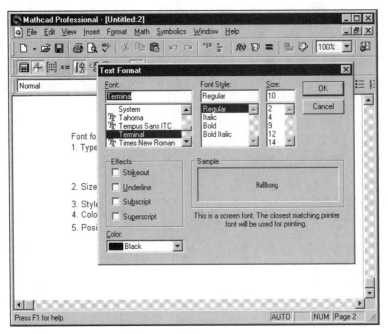

FIGURE 16.14. The **Text Format** dialog

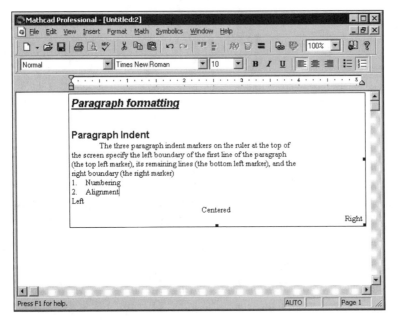

FIGURE 16.15. Examples of paragraph formatting

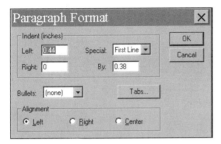

FIGURE 16.16. The **Paragraph Format** dialog

16.2.2. Text and Formula Styles

When you start entering the text into the text region or a formula into the math region, the font and paragraph format is selected according to the default styles saved in the document template.

The text style contains information on all font and paragraph settings of the text region. The math style contains all information on the font setting of all formula elements (notice that the math style doesn't contain paragraph settings since each formula by default can't consume more than one line).

Formatting Formulae

When working with math regions, you can apply all above-considered methods of font formatting using the **Formatting** toolbar. The principal feature of formula formatting allows for the fact that font changes applied to a separate parameter of a specific formula immediately change the same parameter in all formulae existing in the document (where that parameter is applied) (Figures 16.17 and 16.18). It should be noticed that formulae contain elements to which different math styles are applied. Consider, for example, Figure 16.17, which illustrates the change in the formula elements to which the **Variables** style has been applied. Figure 16.18 illustrates the change in the formula elements to which the **Constants** style has been applied. The name of the style applied to the formula fragment being edited is displayed on the left boundary of the **Formatting** toolbar.

Applying Style to a Formula or Text

You can change formula or text paragraph formatting by applying a specific style to it. To perform this task, select a formula or text paragraph, and then select the style that you want to apply from the list of available styles (see Figure 16.12).

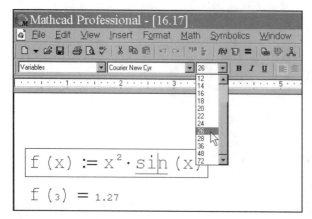

FIGURE 16.17. Changing the font of the formula elements to which the variables style has been applied

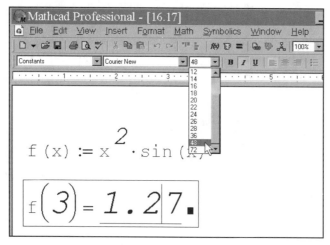

FIGURE 16.18. Changing the font of the formula elements to which the constants style has been applied

Changing the Style

To change the settings of the text or math styles or in order to create a new custom style, select the **Style** or **Equation** commands from the **Format** menu (Figure 16.19). In the **Text Style** dialog, click the **Modify** button. When the **Define Style** window opens, edit the font and paragraph parameters that you want to specify for the current style. The **Description** field at the bottom of the **Define Style** dialog lists the currently selected parameters for the style.

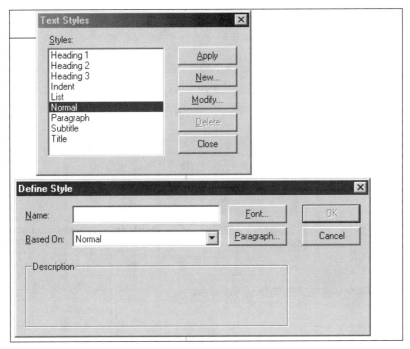

FIGURE 16.19. Changing the text style

Applying a Style to the Formula

Sometimes it might be necessary to apply a font style to one of the variables or to a number to make it look different from the other ones. For this purpose, you can change the math style of the variable by doing the following (Figure 16.20):

1. Click the required variable name or specific number.
2. From the main menu, select the **Format | Equation** commands.
3. In the **Equation Format** dialog select the formula style from the **Style Name** list.
4. If you need to change font settings, click the **Modify** button.
5. Click OK.

As a result, the variable font will be formatted according to the selected style.

Variables with the same names but formatted using different styles are different variables! If you want to change the math style of a variable, change it anywhere where this variable is encountered in the document. Figure 16.21 gives an appropriate example in which Mathcad distinguishes between x variables formatted in different styles.

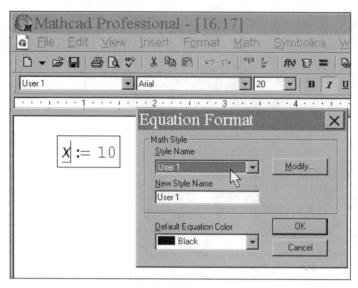

FIGURE 16.20. Changing the style of the variable

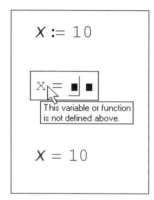

FIGURE 16.21. In Mathcad, the style influences variable identification

If you don't need to change the style, you can apply it to the formula in an easier way, simply by selecting it from the list of available styles on the **Formatting** toolbar.

16.3. FORMATTING PAGES

Mathcad has a set of page formatting tools, which can be further classified as page layout management and creating headers and footers.

16.3.1. Page Layout

Page layout of the printed document can be specified in the Page Setup dialog (Figure 16.22), which can be opened by selecting the command with the same name from the File menu. This dialog provides the following settings:

- Size (page size)
- Source (source of the paper)
- Orientation (paper orientation):
  - Portrait (vertical paper orientation)
  - Landscape (horizontal paper orientation)
- Margins (paper margins):
  - Left (left margin)
  - Right (right margin)
  - Top (top margin)
  - Bottom (bottom margin)

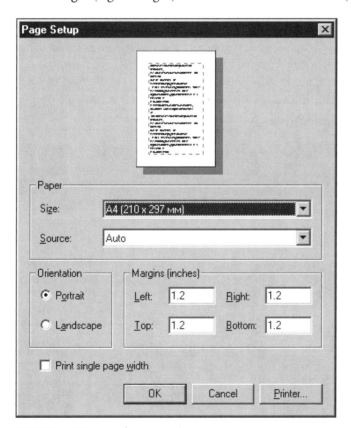

FIGURE 16.22. Page layout settings

The preview area at the top of this window allows you to evaluate the influence of a specific setting on the page layout. In addition, setting the margins in the Page Setup dialog influences the position of the boundary lines that you see in the working area of your Mathcad document.

16.3.2. Headers and Footers

Headers and footers are formatting elements that appear in a unified form on each page for the printed copy of the document. To insert headers and footers into your document, proceed as follows:

1. Select the Headers/Footers command from the Format menu.
2. The Header/Footer dialog will open (Figure 16.23). Depending on the type of the element that you want to insert, go to the Header or Footer tab.
3. Depending on the position where you need to insert a header or footer, click one of the text fields labeled Left, Center, or Right.
4. Insert the header or footer text, combining it with information that can be inserted automatically (for example, current data, page number, filename, and so on.). To insert this information, simply click one of the buttons labeled with an appropriate icon located at the left bottom part of this dialog.
5. Repeat steps 3–4 as necessary.

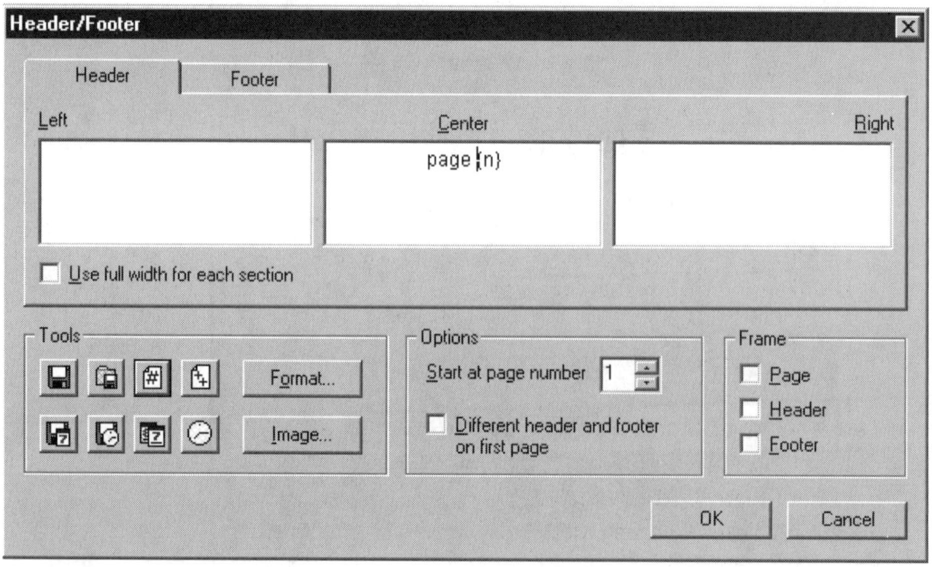

FIGURE 16.23. Inserting headers and footers

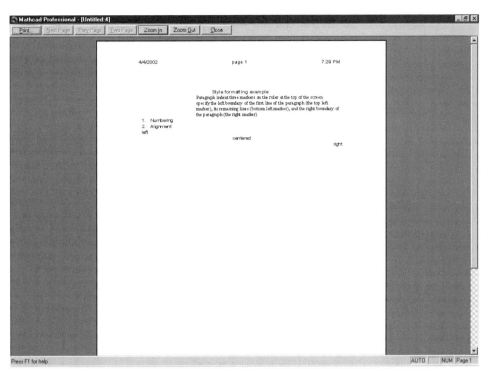

FIGURE 16.24. Viewing headers and footers in the Print Preview mode

Headers and footers influence only printed copies of the document. To view headers and footers as they will look in the printed copy, select the File | Print Preview commands from the main menu (Figure 16.24).

Let us provide a brief listing of the buttons that automatically insert information into the header or footer. The symbol in braces defines appropriate information in the header or footer field.

- File Name {f}
- File Path {p}
- Page Number {n}
- Number of Pages {nn}
- Date Last Saved {fd}
- Time Last Saved {ft}
- Date {d}
- Time {t}

To specify the starting page number, open the Headers/Footers dialog and insert the required number into the Start at page number field.

16.3.3. Document Settings

The main parameters of the document are saved in the document settings. They are automatically saved, along with the document contents in the Mathcad file, and can be used as default settings when creating new documents on the basis of document templates *(see Chapter 2)*.

Main document settings are listed below:

- Default text properties
- Definitions of all text and math styles
- Headers and footers
- Margins for printed copies of the document
- Numeric format of the calculation results
- Values of the built-in variables
- Main dimensions of all variables
- Default number system
- Default calculation mode

16.4. REFERENCES AND HYPERLINKS

Hyperlinks are active areas in Mathcad documents that direct you to another location within the currently active document, another document created in Mathcad, another software application, or to a specific Internet site. Hyperlinks are efficient in large documents, tutorial systems, and presentations created using Mathcad. If you decide to develop electronic books, the skills of hyperlink usage will become especially important.

16.4.1. Inserting the Tag

Before defining the hyperlink, it is necessary to specify the position to which this hyperlink must direct the cursor, which in Mathcad is known as the tag. To set the tag, do the following:

1. Right-click the location where you want to insert the tab.
2. Select the **Properties** command from the right-click menu.

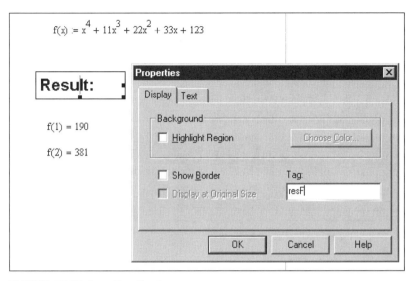

$$f(x) := x^4 + 11x^3 + 22x^2 + 33x + 123$$

Result:

$f(1) = 190$

$f(2) = 381$

FIGURE 16.25. Inserting the tag

3. In the Properties dialog, go to the Display tab.
4. Go to the **Tag** field and insert the tag name that will identify this location in the document (Figure 16.25).
5. Click OK.

16.4.2. Inserting the Hyperlink

Having defined the tag, you can start creating a hyperlink that will direct the cursor to the location of this tag. The hyperlink may reside in any location of any document. To insert the hyperlink, do the following:

1. Click the text or formula region that you need to make a hyperlink.
2. Select the **Hyperlink** command from the **Insert** menu.
3. The **Edit Hyperlink** window will appear. Go to the **Link to file or URL** text field and define the path to the document that contains the tag, along with the name of the tag itself, in the following format `filename#tagname` (Figure 16.26).
4. If you desire, you can go to the text field at the bottom of this window and specify the text that will be displayed in the status bar when the user points the cursor to the hyperlink.
5. Click OK.

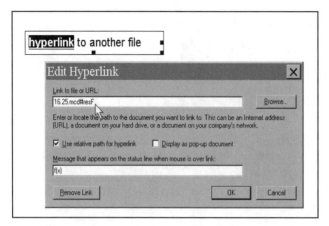

FIGURE 16.26. Inserting the hyperlink

The file name must be specified even if the tag is located in the same document as the hyperlink.

If you have done all the operations correctly, then double-clicking the hyperlink will redirect you to the location where the tag resides, (in our example you'll be redirected to the 16.25.mcd document). To edit the hyperlink, it is sufficient to select the Insert I Hyperlink commands from the main menu when insertion lines are in the hyperlink's area. The Edit Hyperlink dialog will appear in which you can edit its parameters (see Figure 16.26). To delete the hyperlink, click the **Remove Link** button.

Besides hyperlinks that refer to Mathcad documents, you can create hyperlinks that redirect you to other files (such as video files or HTML files, including those that can be found on the Internet). To refer to the files on the Internet, simply specify an appropriate URL in the top text field of the Edit Hyperlink dialog.

16.4.3. References

Besides hyperlinks, sometimes it makes sense to use similar objects known as *references*. The reference to document A, inserted into a specific position of document B, results in the recalculation of the entire document A within document B. Thus, references enable you to store nested calculations in different files.

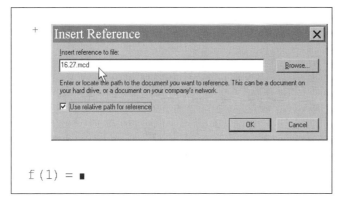

FIGURE 16.27. Creating a reference

References might be needed if a group of developers works on a single large problem. In this case, after distributing the tasks and developing naming conventions for local and global variables, each participant of the project can develop his or her own file with calculations.

To set a reference, select the Insert | Reference commands from the main menu. The **Insert Reference** window will appear (Figure 16.27) where you'll need to specify the path to the referred file. In the example shown in Figure 16.27, the description of the f(x) function is missing. However, this description is present in the referred file. Therefore, after you click **OK**, information on the referred file will appear at the place of the insertion cursor, and the result returned by a function (f(1)) will be calculated according to the formulae present in that file.

16.5. PICTURES

Mathcad 2001i provides new powerful tools for formatting documents, which enable you to enter and edit pictures created in various graphical formats. These tools provide Mathcad 2001i with the basic functionality of a graphical editor.

To insert a picture into your document, proceed as follows:

1. Save the picture in a graphical file and place it into the same folder as a Mathcad document.
2. Display the **Matrix** toolbar if it is not present on the screen.

3. Click the **Picture** button on the **Matrix** toolbar (Figure 16.28).
4. Fill in the placeholder with the quoted filename.

In the example shown in Figure 16.28, the picture was saved in the file named rocket1.gif. As a result of your actions, the contents of the graphical file will be displayed in the picture area.

If you fill in the placeholder with the name of the matrix defined earlier in the document, the picture that will be created will reflect the structure of that matrix. This option is very efficient for rendering the structure of large matrices, especially the sparse ones (see the example in Section 15.6.2, "Graph Files," in Chapter 15).

When the user selects the picture by clicking it with the mouse, Picture Toolbar will appear automatically (Figure 16.29). This toolbar enables you to edit the picture using advanced graphical tools that perform such tasks as mirroring the picture, zooming in/out on its fragment, and so on. The functions of most buttons at Picture Toolbar are similar to the ones present in the most popular graphical editors. Insertion of the picture's regions enables you to perform efficient document formatting.

FIGURE 16.28. Inserting a picture region

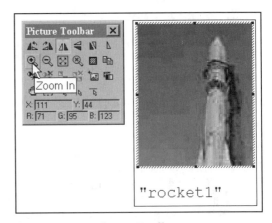

FIGURE 16.29. Picture Toolbar

17 Resource Center

This chapter is dedicated to e-books — interactive collections of Mathcad documents, which you can use to compose your calculations and reference information. Most of this chapter is dedicated to the Resource Center, an electronic book supplied with Mathcad that represents a large storage of calculation examples and various math and engineering information. In this chapter we'll discuss both efficient methods of working with the Resource Center *(see Section 17.1)*, and its contents *(see Section 17.2)*. To conclude, we'll provide some additional information on other e-books and add-on packages for Mathcad, which you can purchase from their respective vendors. Those features that allow you to create your own custom e-books are also covered *(see Section 17.3)*.

17.1. GENERAL INFORMATION

The Resource Center is an electronic book (e-book) supplied with Mathcad 2001i. It provides all the capabilities of e-books that can be used as Mathcad plug-ins.

Remember that the Resource Center is not the only e-book. Besides the Resource Center, Mathcad developers provide a wide range of e-books and add-on software. Moreover, the user can also supply the results of his calculations in the form of a custom e-book (see Section 17.3).

The Resource Center provides a set of examples illustrating solutions of various math, physical, and engineering problems using Mathcad 2001i, as well as its combinations with other software applications, such as Axum™, Excel, MATLAB, and so on. The Resource Center provides the following functionality:

- Access to reference information on Mathcad capabilities
- Full-featured interactive help and a tutorial environment on mathematics, combined with examples of real-world calculations
- The capability of inserting fragments from the Resource Center into Mathcad documents in order to simplify the process of inserting math formulae and avoiding input errors

Let us consider the main techniques involved in working with the Mathcad Resource Center.

17.1.1. Starting the Resource Center

To open the Resource Center, click the button with the icon depicting an open book on the standard toolbar in the main Mathcad window (Figure 17.1) or use the Resource Center command from the Help menu. You'll see the new Resource Center window with its home page loaded.

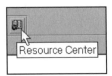

FIGURE 17.1. Starting the Resource Center

17.1.2. The Contents of the Resource Center Windows

The Resource Center home page contains the following components (from top to bottom):

- A window header — Contains the header of the current Resource Center page.

- A menu bar — Much like the Mathcad menu bar because e-books such as the Resource Center perform all calculations in the same way as they are performed in Mathcad.

- A navigation toolbar — A toolbar similar to the navigation bar present in most browsers. Located directly under the menu bar (Figure 17.2), this toolbar significantly simplifies the navigation between the Resource Center's pages.

- An address bar — This toolbar simplifies navigation in the Internet (it is not shown in Figure 17.2; to display it on the screen, select the **Address Bar** command from the View menu).

- The contents of the page itself — Hyperlinks to the Resource Center sections:
  - Overview and Tutorial — Basic information on Mathcad capabilities and the main techniques of working with this system
  - QuickSheets and Reference Tables — A set of templates and examples of calculations for solving various math problems
  - Extending Mathcad — A collection of examples dedicated to the usage of the components of other applications in Mathcad

- Hyperlinks for working on the Internet (left part of the window):
  - Collaboratory — The link to the Internet forum of Mathcad users created for exchanging ideas and Mathcad documents
  - Web Library — A collection of e-books and files containing calculations
  - Mathcad.com — A link to the Mathcad Web page on the server of its developer, MathSoft
  - Support — A support line for users to help them solve problems that might arise when working with Mathcad
  - Web Store — MathSoft's e-shop where you can get information on new software (including Mathcad extensions) and purchase it

- Status bar — Contains help strings for the user informing him of the actions that he can perform.

17.1.3. Resource Center Navigation

Navigating through the contents of the Resource Center is as simple as viewing documents in Web browsers. When pointing to the hyperlink, the mouse cursor changes its shape to that of an icon depicting a hand. By clicking specific hyperlinks, you can go to their respective pages of the Resource Center. Special arrow buttons on the navigation panel are intended for a quick transition to recently viewed Resource Center pages, previous and proceeding chapters, the home page, and for searching the Resource Center.

When viewing the Resource Center's pages, it is necessary to remember that besides viewing text and graphics, it performs calculations similar to those of a Mathcad calculation. Therefore, some time will be spent on these calculations (this fact is especially important if your computer is not particularly powerful). On the other hand, interactive features compensate for the extra time required, since it is possible to change any numeric parameters right on the Resource Center page and to immediately view the changes to the calculation results. In addition, you can select a fragment of the required Resource Center page, copy it to the clipboard by pressing <Ctrl>+<C>, and then, after switching to the Mathcad window, insert it into your document by pressing <Ctrl>+<V> or by clicking an appropriate button on the standard toolbar.

When you start to perform new calculations, take time to find an appropriate example in the Resource Center. It is highly probable that the small amount of time you spend finding the required page will be compensated by a boost in your productivity that results from faster formula input (you can just copy and paste formulae via the clipboard). Furthermore, since the Resource Center contains a large number of debugged and tested calculations, you will be able to avoid possible errors.

17.2. THE RESOURCE CENTER GUIDE

The Resource Center contains a large amount of information that is constantly and steadily increasing from one Mathcad version to another. The Resource Center contains reference information on practically any area of math and any method of solving a particular problem. In this section, we'll briefly describe some chapters included in the Resource Center. The structure of this section completely follows that of the Resource Center itself, which, we hope, will help the reader to better navigate the material.

17.2.1. Overview and Tutorial

The first section of the Resource Center contains examples of Mathcad usage for math calculations, information on basic system capabilities, and techniques for working with the Mathcad editor, including presentation techniques for the results of your calculations in the form of e-books.

An Overview of Mathcad 2001i

Here the user will find the most important information on Mathcad 2001i:

- New Features — Advanced Mathcad users who have experience with its previous versions will be interested in getting acquainted with real-world examples of using the new capabilities implemented in the version that was released in 2001.
- What is Mathcad? — Introductory information on the problems that can be solved using Mathcad and a description of Mathcad's main capabilities.

The Getting Started Tutorial

This is a very useful interactive tutorial, which will demonstrate step-by-step all of Mathcad capabilities, without which it would be impossible to perform any calculations. For example, the first article titled A Quick Tour of Mathcad Features provides a brief overview outlining the main objective of Mathcad; the second one — Mathcad Toolbars — describes the usage of Mathcad's toolbars; and the article titled Creating and Using Graphs (Figure 17.2) will teach the reader how to construct various graphs and so on.

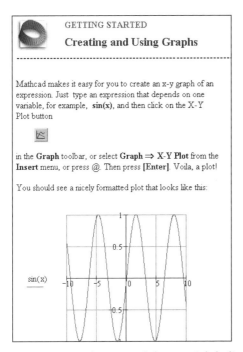

FIGURE 17.2. The page of the tutorial dedicated to graphs

The next articles are intended for those users who have already mastered the basics of using Mathcad and need to get acquainted with the methods of solving specific math problems using Mathcad tools.

Creating 3D Graphs

The most impressive procedures in Mathcad are those involved in preparing and constructing 3D graphs. The section dedicated to the creation and formatting of such graphs includes several articles containing rather useful information and examples (Figures 17.3 and 17.4). It's highly recommended that you view this part of the Resource Center, especially if you encounter such problems. The examples provided there contain a large number of useful tips on the usage of built-in functions intended for fast and optimal construction of 3D graphs.

Chapter 15 of this book can be used as auxiliary material intended to help you understand the contents of this Resource Center material.

TIP

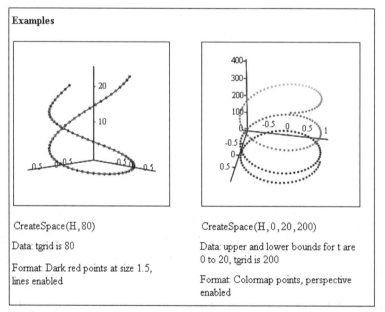

FIGURE 17.3. A fragment of the article describing 3D graph construction

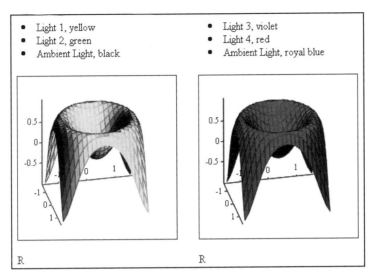

- Light 1, yellow
- Light 2, green
- Ambient Light, black

- Light 3, violet
- Light 4, red
- Ambient Light, royal blue

FIGURE 17.4. A fragment of the article on the techniques of the graph formatting

Analyzing Your Data

This section describes some methods of solving various data analysis problems. It provides a wide range of practical examples illustrating data regression, interpolation, and smoothing *(see Chapter 14)*.

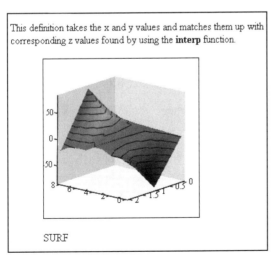

This definition takes the x and y values and matches them up with corresponding z values found by using the **interp** function.

SURF

FIGURE 17.5. A fragment of the article illustrating two-dimensional spline interpolation

- Data Management — Techniques of correct data preparation for performing specific operations, and working with vectors, matrices, or range variables *(see Chapters 4 and 9)*.
- Regression — A large number of practical examples illustrating the usage of various forms of regression.
- Interpolation — This article provides an illustrative example of two-dimensional spline interpolation (Figure 17.5).
- Smoothing — This section considers examples of the usage of built-in functions such as `ksmooth`, `supsmooth`, and `medsmooth`.

Solving Differential Equations

This section provides a complete interactive reference on solving differential equations, starting with basic terms and definitions and concluding with illustrations of real physical applications, such as spring oscillations models, the Lorenz attractor, etc. Naturally, this section provides detailed coverage of the specific features of solving Cauchy or boundary value problems using the solve block and Mathcad built-in functions.

You'll probably be interested in acquainting yourself with the basics of numeric algorithms in the Mathcad developers' interpretation. These descriptions are provided in the article titled **Numerical Differential Equations Solvers**. The author hopes that the information provided in this book *(see Chapters 11 and 12)* will spark the reader's interest in the Resource Center pages dedicated to differential equations.

Documenting and Publishing Your Work

This section mainly concentrates on presenting the results of your work *(see Chapter 16)* and is intended to demonstrate the techniques of documenting your work and publishing Mathcad documents.

17.2.2. *QuickSheets* and *Reference* Tables

The most useful parts of the Resource Center are concentrated within this section. Before solving a specific problem it's often useful to view the solutions of similar problems provided by professionals. Therefore, it is highly probable that QuickSheets contain an example very similar to yours. If so, you'll be able to part of the Resource Center page to your document and change a couple of parameters.

Listed below are the contents of the QuickSheets accompanied by the references to the related chapters of this book. Some sections of QuickSheets contain dozens of practical examples illustrating the usage of most Mathcad built-in functions and other features (Figure 17.6).

■ Reference Tables — Reference information from various scientific and technical areas, such as math, physics, chemistry, and so on (see Figure 17.7).

FIGURE 17.6. A fragment of the contents of one of the QuickSheet sections

CALCULUS REFERENCE FORMULAS

Table of Derivative Formulas

Constant Multiple Rule
$$\frac{d}{dx} c \cdot f(x) = c \cdot \frac{d}{dx} f(x)$$

Power Rule
$$\frac{d}{dx} x^n = n \cdot x^{n-1}$$

Product Rule
$$\frac{d}{dx}(f(x) \cdot g(x)) = f(x) \cdot \frac{d}{dx} g(x) + \frac{d}{dx} f(x) \cdot g(x)$$

Sum Rule
$$\frac{d}{dx}(f(x) + g(x)) = \frac{d}{dx} f(x) + \frac{d}{dx} g(x)$$

Quotient Rule
$$\frac{d}{dx} \frac{f(x)}{g(x)} = \frac{g(x) \cdot \frac{d}{dx} f(x) - f(x) \cdot \frac{d}{dx} g(x)}{g(x)^2}$$

Chain Rule
$$y = f(u) \; u = g(x) \quad \frac{d}{dx} f(u) = \frac{d}{du} f(u) \cdot \frac{d}{dx} g(x)$$

Derivatives of Trigonometric Functions

FIGURE 17.7. One of the tables explaining differentiation rules

- About QuickSheets — Reference information on QuickSheets usage.
- Personal QuickSheets — Here you can combine a set of the most frequently used Resource Center articles.
- Arithmetic and Algebra — *See Chapters 3, 4, and 10.*
- Business and Finance— *See Section 10.11, "Financial Functions," in Chapter 10.*
- Units — *see Section 4.2, "Dimensional Variables," in Chapter 4.*
- Vectors and Matrices — *See Section 4.3, "Arrays," and Chapter 9.*
- Solving Equations — *See Chapters 8, and 9.*
- Graphing and Visualization — *See Chapter 15.*
- Calculus and Differential Equations — *See Chapters 7, 11, and 12.*
- Data Analysis — *See Chapter 14.*
- Statistics — *See Chapter 13.*
- Components in Mathcad — Using other applications in combination with Mathcad (*see Section 17.2.3*).
- Special Functions — *See Chapter 10.*

■ Programming — *See Chapter 6.*
■ Animations — *See Section 15.5, "Creating Animation," in Chapter 15.*
■ Extra Math Symbols — *See Section 3.2.6, "Creating User-Defined Operators," in Chapter 3.*

17.2.3. Extending Mathcad

Sometimes the problem of combined usage of Mathcad and other math software becomes a matter of importance. First, it's necessary to performed combined calculations when part of your work has already been done using another system, such as Excel. This external system may simply be more common but less powerful than Mathcad, and therefore unsuitable for complex calculations. Secondly, the inverse situation might arise, in which other applications prove to be more efficient than Mathcad when solving specific math problems, and therefore, some part of the problem might be solved by such applications more efficiently. For example, the MATLAB system provides many more capabilities of complex resource-consuming matrix calculations.

This section of the Resource Center provides a set of examples illustrating the combined use of other application components in Mathcad calculations along with appropriate information on the specific features of using a specific component. Some of these components are supplied with Mathcad 2001i, while others must be purchased separately.

Acquisition Data: Waveform Input

This Resource Center article is dedicated to accessing an external device that might be installed on your computer from Mathcad. For example, various automated engineering systems are controlled by the PC with installed sensors and other external peripheral devices. Quite often, the data obtained from such devices require serious mathematical and statistical processing, which can be easily organized using Mathcad.

Mathcad provides a special ActiveX control intended for managing external data acquisition (Figure 17.8). This control enables you to easily acquire input data from some external device, accumulate the data, and send output data to it. The example provided in the Resource Center is dedicated to data acquisition from a digital voltmeter switched to the PC.

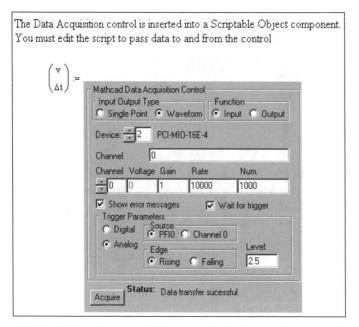

FIGURE 17.8. A fragment of the page dedicated to data acquisition and management control

VisSim Add-in: Simulations with Mathcad

The VisSim application is a built-in service interface module intended for use with Mathcad 2001i. This package combines Drag-and-Drop technology with powerful calculation and modeling tools for fast and accurate calculations. VisSim includes a set of interconnected units such as: signal sources, encrypting and de-crypting units, modulators and demodulators, filters, channels, radio frequency components, and so on. Combined use of the visual capabilities of the VisSim units with powerful calculation tools provided by Mathcad simplifies the proc-esses of designing, developing, constructing, and using complex communication systems.

MathSoft Custom Controls: Buttons, Boxes, and Sliders

This article provides some examples of creating GUI applications based on Math-cad 2001i. One can create such applications by using MathSoft custom controls, such as buttons, lists, text fields, and so on (Figure 17.9).

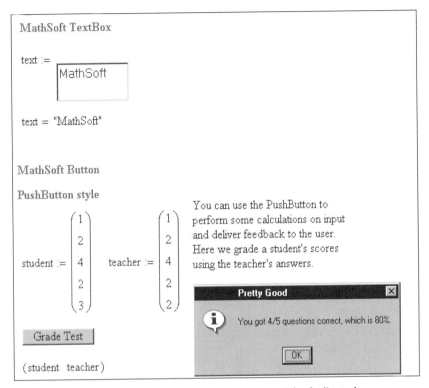

FIGURE 17.9. A fragment of the Resource Center article dedicated to MathSoft custom controls

Excel: Calculating the Periodicity of Sunspots

This document illustrates the usage of the Excel component.

The example illustrating combined usage of Excel and Mathcad relates to the calculation of sunspot periodicity. The article describes how, in 1848, Rudolph Wolf invented a daily method of evaluating the solar activity by calculating individual spots and groups of spots on the solar surface. The number of sunspots increases and decreases approximately every 11 years. This cycle is asymmetric; the article shows that the number of sunspots rises from the minimum level to the maximum during the first 4 years and during the subsequent 7 years it drops from the maximum back to the minimum (Figure 17.10).

By viewing this page, you'll become familiar with the procedures of working with Excel components in Mathcad, and at the same time, you'll get a broader vision of science by studying problems of solar and Earth physics.

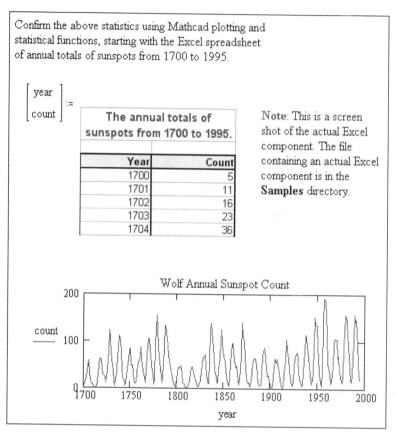

FIGURE 17.10. A fragment of the Resource Center article describing Wolf's numbers

Visio Add-in: Analysis of a Circuit

The article is dedicated to combined use of Mathcad and Visio. The built-in Visio component enables you to insert a Mathcad object into the Visio application.

Axum: Poisson Distribution

This document illustrates Mathcad usage with the components of the Axum application, which represents a powerful graphical system.

The example illustrating combined usage of Mathcad and Axum relates to Poisson's distribution. Poisson's distribution is used in many problems, such as modeling the number of atoms emitted by a radioactive source that struck a specific target during a specified time period.

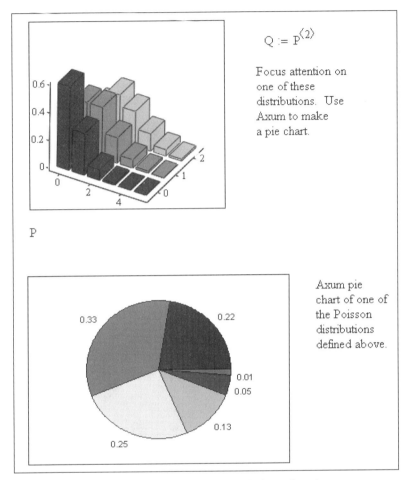

FIGURE 17.11. A Resource Center fragment dedicated to the usage of Axum graphical capabilities

MATLAB: Drawing a Buckyball

This document illustrates the usage of MATLAB components.

New Jersey Pick-It Lottery

This document illustrates the combined usage of Axum and Excel components in Mathcad.

Visio: RC Circuit Solver

Yet another example is dedicated to the Visio component. It describes the model of an electric circuit containing a resistor and capacitor (RC circuit). This example draws the circuit using the Visio application based on the user-defined circuit parameters (Figure 17.12), and it then uses the same parameters to model the circuit in Mathcad.

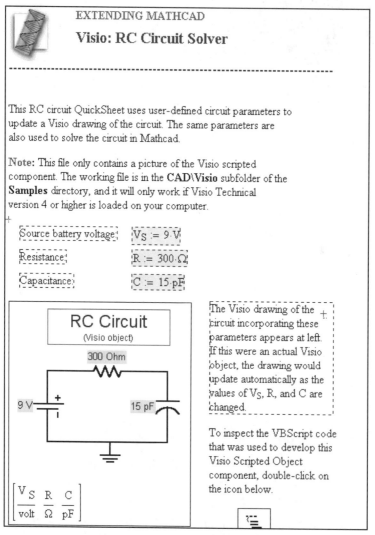

EXTENDING MATHCAD

Visio: RC Circuit Solver

This RC circuit QuickSheet uses user-defined circuit parameters to update a Visio drawing of the circuit. The same parameters are also used to solve the circuit in Mathcad.

Note: This file only contains a picture of the Visio scripted component. The working file is in the **CAD\Visio** subfolder of the **Samples** directory, and it will only work if Visio Technical version 4 or higher is loaded on your computer.

Source battery voltage: $V_S := 9 \cdot V$

Resistance: $R := 300 \cdot \Omega$

Capacitance: $C := 15 \cdot pF$

RC Circuit
(Visio object)

300 Ohm

9 V   15 pF

The Visio drawing of the circuit incorporating these parameters appears at left. If this were an actual Visio object, the drawing would update automatically as the values of V_S, R, and C are changed.

To inspect the VBScript code that was used to develop this Visio Scripted Object component, double-click on the icon below.

$$\begin{bmatrix} V_S & R & C \\ \hline \text{volt} & \Omega & pF \end{bmatrix}$$

FIGURE 17.12. A fragment of the page dedicated to the usage of Visio capabilities

Visio: Mathcad-Driven CAD Drawing

This example illustrates the usage of the CAD drawing capabilities directly from Mathcad.

Lotus® 1-2-3®: Inserting a Scripted Component

This example demonstrates how to use the object component of the Mathcad script to develop the component of Lotus 1-2-3, which is functionally similar to the built-in Excel component supplied with Mathcad.

SmartSketch: Mathcad-Driven CAD Drawing

This example (Figure 17.13) and the two following are dedicated to the combined usage of Mathcad and the SmartSketch system.

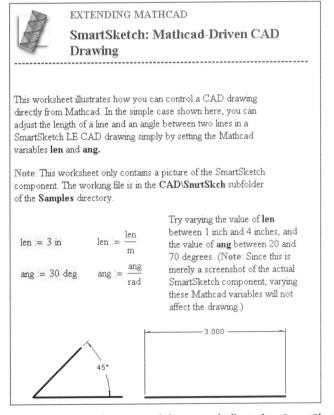

FIGURE 17.13. A fragment of the page dedicated to SmartSketch usage

SmartSketch: Mathcad-Driven Backhoe Drawing

This example uses Mathcad calculations for drawing the position of a backhoe arm in SmartSketch.

SmartSketch: Belt and Pulley Example

This example illustrates the calculation of the minimal force required to prevent belt slippage in a two-pulley belt system. The two-pulley belt system is drawn in SmartSketch, while the calculations are performed in Mathcad.

SPLUS: Binomial Random Variables

This article illustrates how to work with the script and graphical components of the SPLUS application.

SPLUS: Generating Random Samples

This is simply another example of the SPLUS usage (Figure 17.14).

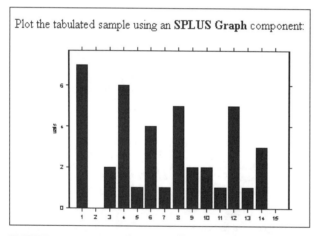

FIGURE 17.14. A page fragment illustrating SPLUS graphical capabilities

17.3. ELECTRONIC BOOKS

Electronic books are collections of calculations supplied with hyperlinks and interactive examples of Mathcad calculations. An electronic book is characterized by the following properties that distinguish it from normal Mathcad documents:

- Each page of the e-book is a full-featured Mathcad document with real calculations and capabilities of changing numeric values and parameters.
- The user of an electronic book can copy its fragments into his or her documents.
- An electronic book is supplied with a table of contents, an index, and an advanced navigation system implemented by hyperlinks.
- Pop-up windows provide the user with an additional source of information.
- Navigation from page to page is implemented using a special-purpose navigation panel.
- The user is able to include and save the notes on the pages of the electronic book.

17.3.1. Add-on Electronic Books and Extension Packages

In the previous sections of this chapter we got acquainted with the example of the e-book Resource Center. Certainly, you have noticed that the Resource Center has all the specific features of an electronic book. Besides the Resource Center, the MathSoft company provides a whole collection of e-books dedicated to various areas of math and technical applications. You can get information on the available e-books and how to purchase them on the Mathcad site *http://www.Mathcad.com*.

To open an e-book for viewing, select the **Open Book** command from the **Help** menu and specify the path to the file containing the book. As a result, a new window similar to that of the Resource Center will appear on the screen displaying the loaded e-book and providing the navigation toolbar.

Besides e-books, MathSoft distributes various Mathcad extension packages related to various special branches of mathematical science. For example, there is a special add-on package for solving equations and a package for wavelet analysis. Besides the properties of the e-books listed above, they are able to add a new built-in function to the standard set of such functions provided by Mathcad. If you are interested, you can get the required information on available extension packages from the Mathcad Web site.

17.3.2. Creating Custom E-Books

If needed, you can present the results of your calculations saved in several Mathcad documents in the form of an electronic book. All the steps required to create an e-book are listed below, supplied with a brief description. Notice that this process is not all that simple, and, probably won't be necessary for all users.

1. Create a set of normal Mathcad documents that will become the pages of your e-book, and save them in appropriate files *(see materials from Part I)*.
2. Insert tags and hyperlinks in such a way as to provide the user of your book with the capabilities of full-featured navigation between the book's pages *(see Section 16.4)*.
3. Create a new Mathcad document that will contain the book's table of contents and save it in the file named TOC.MCD. In the TOC, insert all required hyperlinks for the appropriate pages.
4. Create a specially formatted file, known as *HBK-file* containing the description of the e-book to be created. To create this file, follow the instructions provided in the Help system for authors. To view this information, select the **Author's Reference** command from the **Help** menu.
5. If desired, create an index (also following the instructions provided in the **Author's Reference**.
6. Debug your e-book by testing the accuracy of the HBK-file, hyperlinks, and index. To do so, close the Mathcad window, and then start Mathcad again in the HBK-mode *(see Author's Reference)*.

After performing all above-described actions, you'll be able to view your e-book in the same way as any other e-book provided by MathSoft.

Appendixes

Appendix A
Built-in Operators and Functions

TABLE A.1. Arithmetic Operators

Operator	Keys	Scalar	Vector	Matrix
:=	<:>	Definition		
≡	<~>	Global definition		
=	<=>	Numeric output		
→	<Ctrl>+<=>	Symbolic output		
+	<+>	Addition		
-	<->	Subtraction or negation (unary operator)		
·	<*>	Multiplication	Matrix multiplication, Multiplication by scalar	
			Dot product	
×	<Ctrl>+<8>		Cross product	
/ or ÷	</> or <Ctrl>+</>		Division	

TABLE A.1. Arithmetic Operators (*Continued*)

Operator	Keys	Scalar	Vector	Matrix
!	<!>	Factorial		
$\bar{}$	<">	Complex conjugation		
$\sqrt{}$	<\>	Square root		
$\sqrt[n]{}$	<Ctrl>+<\>	N-th root		
(■)	<>	Parentheses (priority change)		
■.$_{.}$	<[>		Subscript	
■^{T}	<Ctrl>+<1>		Transpose	
|■|	<Shift>+<\>	Absolute value	Absolute value of the vector	Determinant
Σ■	<Ctrl>+<4>		Sum or elements	
■$^{-1}$		Inverse value		Inverse matrix
■^{n}	<^>+n	Raise to power n		Raise matrix to power n
→	<Ctrl>+<->		Vectorize	
■$^{<■>}$	<Ctrl>+<6>		Matrix column	

Unless otherwise specified, scalar operations on vectors and matrices are performed independently with each element, like on a scalar value.

NOTE

TABLE A.2. Calculus Operators

Operator	Keys	Description	Reference
$\int_{■}^{■} ■ \, d■$	<Shift>+<7>	Definite integral	7.1

TABLE A.2. Calculus Operators (*Contunued*)

Operator	Keys	Description	Reference
$\int \blacksquare \, d\blacksquare$	\<Ctrl>+\<I>	Indefinite integral	7.1.3
$\dfrac{d}{d\blacksquare}\blacksquare$	\<?>	Differentiation	7.2
$\dfrac{d^n}{d\blacksquare^n}\blacksquare$	\<Ctrl>+\<?>	Calculating n-th derivative	7.2
$\displaystyle\sum_{\blacksquare=\blacksquare}^{\blacksquare} \blacksquare$	\<Ctrl>+\<Shift >+\<4>	Summation	3.2.2
$\displaystyle\sum_{\blacksquare} \blacksquare$	\<Ctrl>+\<4>	Range variable summation	3.2.2
$\displaystyle\prod_{\blacksquare=\blacksquare}^{\blacksquare} \blacksquare$	\<Ctrl>+\<Shift >+\<3>	Iterated product	3.2.2
$\displaystyle\prod_{\blacksquare} \blacksquare$	\<Ctrl>+\<3>	Range variable iterated product	3.2.2
$\lim_{\blacksquare \to \blacksquare} \blacksquare$	\<Ctrl>+\<L>	Two-sided limit	3.2.2
$\lim_{\blacksquare \to \blacksquare^-} \blacksquare$	\<Ctrl>+\<A>	Left-sided limit	3.2.2
$\lim_{\blacksquare \to \blacksquare^+} \blacksquare$	\<Ctrl>+\<B>	Right-sided limit	3.2.2

TABLE A.3. Alphabetic List of the Built-in Functions

Function	Arguments	Description	Reference
a*(z)	z — Argument	Inverse trigonometric or hyperbolic function *	10.4, 10.5

TABLE A.3. Alphabetic List of the Built-in Functions (*Contunued*)

Function	Arguments	Description	Reference
Ai(x)	x — Argument	Airy function of the first kind	10.1.3
angle(x,y)	x,y — Coordinates of the point	Angle formed by the positive OX axis and the straight line containing coordinate origin (0, 0) and the point (x, y)	10.4
APPENDPRN (file)	file — String representation of the path and file name	Appends data to the end of the existing text file	15.6.1
arg(z)	z — Function argument	Argument of a complex number	10.2
atan2(x,y)	x,y — Coordinates of the point	Returns the angle (in radians) from the x-axis to a line containing the origin (0,0) and the point (x,y)	10.4
augment (A,B,C,...)	A,B,C,... — Vectors or matrices	Returns an array resulting from joining vectors or matrices from left to right	9.2.2
bei(n,x) ber(n,x)	n — Order x — Argument	Imaginary and real parts of the Bessel Kelvin function	10.1.4
Bi(x)	x — Argument	Airy function of the second kind	10.1.3
bspline (x,y,u,n)	x,y — Data vectors u — Vector of knots n — Order of polynomials	Vector of B-spline coefficients	14.1.3
Bulstoer (y0,t0,t1,M,D)	See rkfixed	Returns the matrix containing the solution of a Cauchy problem for a system or ordinary differential equations solved using the Bulirsch-Stoer method	11.3

TABLE A.3. Alphabetic List of the Built-in Functions (*Contunued*)

Function	Arguments	Description	Reference
bulstoer (y0,t0,t1, acc,D,k,s)	See rkadapt	Returns the matrix containing the solution of a Cauchy problem for a system or ordinary differential equations solved using the Bulirsch-Stoer method (only for the last point of the interval)	11.3.2
bvalfit (z1,z2,x0,x1, xf,D,load1, load2,score)	z1,z2 — Vector of initial values for missing left and right boundary conditions x0 — Left boundary x1 — Right boundary xf — Internal point D(x,y) — Vector function specifying a system of ordinary differential equations load1(x0,z), load2(x1,z) — Vector functions specifying left and right boundary condition score(xf,y) — Vector function specifying the knot of solutions at xf	Returns the vector of missing boundary conditions for a boundary value problem with an additional condition at the intermediate point	12.1.4
ceil(x)	x — Argument	Returns the smallest integer not less than x	10.8
cfft(y) CFFT(y)	y — A data vector	Vector of a direct complex Fourier transform (cfft(y) and CFFT(y) are practically identical except for different normalizing factors and sign conventions)	14.4.1

TABLE A.3. Alphabetic List of the Built-in Functions (*Contunued*)

Function	Arguments	Description	Reference
cholesky(A)	A — A square definite matrix	Cholesky decomposition	9.5.1
cols(A)	A — A matrix or vector	Returns number of columns	9.2.3
concat (S1,S2,...)	S1, S2, ... — Strings	Concatenation of string variables	10.7
cond1(A) cond2(A) conde(A) condi(A)	A — Square matrix	Condition numbers in different norms (L1, L2, Euclidean, ∞)	9.2.6
cos(z)	z — Argument	Cosine	10.4
cosh(z)	z — Argument	Hyperbolic cosine	10.5
cot(z)	z — Argument	Cotangent	10.4
coth(z)	z — Argument	Hyperbolic cotangent	10.5
csort(A,i)	A — Matrix i — Column index	Sorts the matrix with the elements of the i-th column	9.2.4
CreateMesh (F,s0,s1, t0,t1,sgr, tgr,fmap)	F(s,t) — Vector function of three elements t0,t1 — Limits of t s0,s1 — Limits of s tgr,sgr — Number of grid points by t and s fmap — Coordinate mapping function	Creates the nested array representing x-, y- and z-coordinates of the parametric surface specified by the function F	9.2.1
CreateSpace (F[,t0,t1, tgr,fmap])	F(t) — Vector function of three elements t0,t1 — Limits of t tgr — Number of grid points by t fmap — coordinate mapping function	Creates nested array representing x-, y- and z-coordinates of the 3D parametric curve specified by the function F	9.2.1

TABLE A.3. Alphabetic List of the Built-in Functions (*Contunued*)

Function	Arguments	Description	Reference
csc(z)	z —Argument	Cosecant	10.4
csch(z)	z — Argument	Hyperbolic cosecant	10.5
csgn(z)	z — Argument	Complex sign of a number	10.2
cspline(x,y)	x,y — Data vectors	Vector of coefficients of the cubic spline	14.1.2
cyl2xyz(r,θ,z)	r,θ,z — Cylindrical coordinates	Transforms cylindrical coordinates to Cartesian	10.10
D* (x,par)	x — Random value par — A list of the distribution parameters *	Probability density with statistic distribution*	13.1.4
diag(v)	v — Vector	Diagonal matrix containing the specified vector on its diagonal elements	9.2.1
eigenvals(A)	A — Square matrix	Eigenvalues of the matrix	9.4
eigenvec(A,λ)	A — Square matrix λ — Eigenvalue	Eigenvector of the matrix corresponding to the specified eighenvalue	9.4
eigenvecs(A)	A — Square matrix	Eigenvectors of the matrix	9.4
erf(x)	x — Argument	Error function	13.1.1
erfc(x)	x — Argument	Inverse error function	13.1.1
error(S)	S — String	Returns the string S as error message	10.7
exp(z)	z — Argument	Exponential function (the number e raised to power z)	10.3
expfit(x,y,g)	x,y — Data vectors g — Vector of initial values a,b,c	Exponential regression $a \cdot e^{bx} + c$	14.2.3

TABLE A.3. Alphabetic List of the Built-in Functions (*Contunued*)

Function	Arguments	Description	Reference
fft(y) FFT(y)	y — Data vector	Vector of direct Fourier transform (in different norms)	14.4.1
fhyper(a,b,c,x)	a,b,c — Parameters x — Argument, $-1<x<1$	Gaussian hypergeometric function	10.6
Find (x1,x2,...)	x1,x2,... — Variables	Returns the root of an algebraic equation (scalar) or a system of algebraic equations (vector) defined in the block with Given	8.3–8.4
floor(x)	x — Argument	The largest integer less than or equal to x	10.8
Gamma(x) Gamma(a,x)	x — Argument	Euler's gamma function or incomplete gamma function with parameter a	10.6
genfit(x,y,g,G)	x,y — Data vectors g — Vector of initial values of the regression parameters G(x,C) — Vector function composed of the custom functions and their partial derivatives by each parameter	Vector of regression coefficients by the generic custom function	14.2.4
geninv(A)	A — Matrix	Creates an inverse matrix	9.2.1
genvals(A,B)	A,B — Square matrices	Generalized eigenvalues	9.4
genvecs(A,B)	A,B — Square matrices	Generalized eigenvectors	9.4
Given		The keyword used for insertion of the systems of equations, inequalities, and so on	8.3

TABLE A.3. Alphabetic List of the Built-in Functions (*Contunued*)

Function	Arguments	Description	Reference
heaviside step(x)	x — Argument	Heaviside function	10.9
Her(n,x)	x — Argument n — Order	Hermite polynomial	10.6
I0(x) I1(x) In(m,x)	x — Argument	Modified Bessel function of the first kind, of 0-th, first, and m-th order	10.1.2
ibeta(a,x,y)	x,y — Argument a — Parameter	Incomplete beta function	10.6
identity(N)	N — Matrix dimension	Creates an identity matrix	9.2.1
Icfft(v) ICFFT(v)	v — Vector of partial derivatives of the Fourier spectrum	Vector of an inverse complex Fourier transform (in different norms)	14.4.1
if(cond,x,y)	cond — Logical condition x,y — Values returned if the condition is true (false)	Conditional function	10.9
ifft(v) IFFT(v)	v — Vector of partial derivatives of the Fourier spectrum	Vector of an inverse Fourier transform (in different norms)	14.4.1
IsString(x)	x — Argument	Returns 1 if x is a string, otherwise returns 0	10.7
iwave(v)	v — Vector of the frequency data of the wavelet spectrum	Vector of an inverse wavelet transform	14.4.2
Im(z)	z — Argument	Imaginary part of the complex number	10.2

TABLE A.3. Alphabetic List of the Built-in Functions (*Contunued*)

Function	Arguments	Description	Reference
interp(s,x,y,t)	s — Vector of second derivatives x,y — Data vectors t — Argument	Spline interpolation	14.1.2
intercept(x,y)	x,y — Data vectors	Coefficient b of the linear regression $b + a \cdot x$	14.2.1
J0(x) J1(x) Jn(m,x)	x — Argument	Bessel functions of the first kind of the 0-th, first, and m-th order	10.1.1
Jac(n,a,b,x)	x — Argument a,b — Parameter n — Order	Jacobi polynomial	10.6
js(n,x)	n — Order x — Argument	Spherical Bessel function of the first kind	10.1.5
K0(x) K1(x) Kn(m,x)	x — Argument	Modified Bessel function of the second kind of the 0-th, first, and m-th order	10.1.2
Kronecker delta(x,y)	x,y — Argument	Kronecker delta-function	10.9
ksmooth(x,y,b)	x,y — Data vectors b — Width of the smoothing window	Smoothing with the Gauss function	14.3.1
Lag(n,x)	x — Argument n — Order	Laguerre polynomial	10.6
last(v)	v — Vector	Index of the last element of the specified vector	9.2.3
Leg(n,x)	x — Argument n — Order	Legendre polynomial	10.6

TABLE A.3. Alphabetic List of the Built-in Functions (*Contunued*)

Function	Arguments	Description	Reference
length(v)	v — Vector	Number of elements of the specified vector	9.2.3
line(x,y)	x,y — Data vectors	Vector composed of the coefficients of the linear regression $b + a \cdot x$	14.2.1
linfit(x,y,F)	x,y — Data vectors F(x) — User-defined function	Vector composed of the regression coefficients of the custom function	14.2.4
linterp(x,y,t)	x,y — Data vectors t — Argument	Partially-linear interpolation	14.1.1
lgsfit(x,y,g)	x,y — Data vectors g — Vector of initial values a,b,c	Regression by logistic function $a / (1 + b \cdot e^{-cx})$	14.2.3
ln(z)	z — Argument	Natural logarithm	10.3
lnfit(x,y)	x,y — Data vectors	Regression by logarithmic function $a \cdot \ln(x) + b$	14.2.3
loess(x,y,span)	x,y — Data vectors span — Parameter defining the size of the neighborhood	Vector of coefficients for regression by polynomial fractions (used in combination with interp)	14.2.2
log(z)	z — Argument	Decimal logarithm	10.3
log(z, b)	z — Argument	Base b logarithm of z	10.3
logfit(x,y,g)	x,y — Data vectors g — Vector of initial values a,b,c	Regression by a logarithmic function $a \cdot \ln(x + b) + c$	14.2.3
lsolve(A,b)	A — Matrix of a system of linear algebraic equations b — Vector of right parts	Solution of a system of linear algebraic equations	9.3

TABLE A.3. Alphabetic List of the Built-in Functions (*Contunued*)

Function	Arguments	Description	Reference
lspline(x,y)	x,y — Data vectors	Vector of the coefficients of a linear spline	14.1.2
lu(A)	A — Square matrix	LU-decomposition	9.5.3
matrix(M,N,f)	M — Number of rows N — Number of columns f(i,j) — Function	Creates matrix composed of elements f(i,j)	9.2.1
Maximize (f,x1,...)	f(x1,...) — Function x1,... — Arguments by which maximization is done	Vector of argument values, which makes the function f take its maximum value (additional constraints might be specified in the Given block)	8.6
mhyper(a,b,x)	x — Argument a,b — Parameters	Confluent hypergeometric function	10.6
Minerr (x1,x2,...)	x1,x2,... — Variables	Returns the vector of the approximate values that come close to satisfying the system of equations and inequalities defined in the Given block	8.5
Minimize (f,x1,...)	f(x1,...) — Function x1,... — Arguments by which minimization is done	Vector of argument values, at which the function f reaches its minimum (additional constraints might be set in the Given block)	8.6
medsmooth(y,b)	y — Data vector b — Smoothing window width	Smoothing using the "running medians" method	14.3.1
multigrid (F,ncycle) —	F — Matrix of the right part of Poisson's equation ncycle — Number of cycles at each level of the multigrid iteration (usually set to 2)	Solution matrix of Poisson's equation for the square area with zero boundary conditions	12.4.1

TABLE A.3. Alphabetic List of the Built-in Functions (*Contunued*)

Function	*Arguments*	*Description*	*Reference*
norm1(A) norm2(A) norme(A) normi(A)	A — Square matrix	Norms of matrices (L1, L2, Euclidean, ∞)	9.2.5
num2str(z)	z — Number	Returns the string whose characters correspond to the decimal value of the specified number z	10.7
Odesolve (t,t1[,step])	t — Integration variable t1 — End point of the integration interval step — Number of integration steps	Returns the matrix representing the solution of the Cauchy problem for a single ordinary differential equation defined in the Given block with initial conditions at the point t0	11.1.1, 11.2
p* (x,par)	x — Random value par — List of the distribution parameters*	Cumulative probability distribution*	13.1.4
pol2xy(r,θ)	r,θ — Polar coordinates	Converts polar coordinates to Cartesian coordinates	10.10
polyroots (v)	v — Vectors composed of the polynomial coefficients	Returns the vector composed of all polynomial roots	8.2
predict(y,m,n)	y — Initial vector m — Number of elements y by which extrapolation is performed n — Number of elements to be predicted	Prediction function extrapolating the vector	14.1.4
pspline(x,y)	x,y — Data vectors	Vector of the coefficients of a quadratic spline	14.1.2

TABLE A.3. Alphabetic List of the Built-in Functions (*Contunued*)

Function	Arguments	Description	Reference
pwrfit(x,y,g)	x,y — Data vectors g — Vector of initial values a,b,c	Regression by power function $a \cdot xb + c$	14.2.3
q* (p,par)	p — Probability value par — List of distribution parameters*	Quantile (function inverse to the cumulative probability distribution)*	13.1.4
qr(A)	A — Vector or matrix	QR-decomposition	9.5.2
r* (M,par)	M — Dimension of the vector par — List of distribution parameters *	Vector of random values with statistics*	13.1.4
rank(A)	A — Matrix	Rank of the matrix	9.2.7
Re(z)	z — Argument	Real part of the complex number	10.2
READPRN(file)	file — String representation of the pathname	Returns an array formed by the file contents	15.6
regress(x,y,k)	x,y — Data vectors k — Polynomial order	Vector of coefficients for polynomial regression (used in combination with interp)	14.2.2
relax (a,b,c,d,e,F, v,rjac)	a,b,c,d,e — Matrices of difference scheme coefficients F — Matrix of the right part of the equation v — Matrix of boundary conditions rjac — Parameter of the algorithm (0...1)	Matrix of the solution of a partial differential equation obtained using the grid method for a square area	12.4.1, 12.4.3
reverse(v)	v — vector	Reverses vector elements	9.2.4

TABLE A.3. Alphabetic List of the Built-in Functions (*Contunued*)

Function	Arguments	Description	Reference
rkadapt (y0,t0,t1, acc,D,k,s)	y0 — Vector of initial values (t0,t1) — Integration interval acc — Calculation error D(t,y) — Vector function specifying a system of ordinary differential equations k — Maximum number of integration steps s — Minimal integration steps	Returns the matrix with the solution of a Cauchy problem for a system of ordinary differential equations using the Runge-Kutta method with a variable step and specified precision (for defining only the last point of the interval)	11.3
Rkadapt (y0,t0,t1,M,D)	*See* rkfixed	Returns the matrix with the solution of a Cauchy problem for a system of ordinary differential equations using the Runge-Kutta method with a variable step	11.3
rkfixed (y0,t0,t1,M,D)	y0 — Vector of initial conditions (t0,t1) — Integration interval M — Number of integration steps D(t,y) — Vector function specifying a system of ordinary differential equations	Returns the matrix with the solution of a Cauchy problem for a system of ordinary differential equations using the Runge-Kutta method with a fixed step	11.1.2, 11.3
root (f(x,...), x[a,b])	f(x,...) — Function x — Variable (a,b) — Interval of the search	Returns the root of the function	8.1

TABLE A.3. Alphabetic List of the Built-in Functions (*Contunued*)

Function	Arguments	Description	Reference
`round(x, n)`	`x` — Argument	Rounding	10.8
	`n` — Number of significant digits after the decimal point		
`rows(A)`	`A` — Matrix or vector	Number of rows	9.2.3
`rref(A)`	`A` — Matrix or vector	Converts the `A` matrix to the row-reduced echelon form	9.2.1
`rsort(A,i)`	`A` — Matrix	Sorts the matrix with the elements of the `i`-th row	9.2.4
	`i` — Row index		
`sbval (z,x0,x1,D, load,score)`	`z` — Vector of initial guess values for the missing initial conditions	Returns the vector of the missing initial values for the two-point boundary value problem for a system of ordinary differential equations	12.1.3
	`x0` — Left boundary		
	`x1` — Right boundary		
	`D(x,y)` — Vector function specifying a system of ordinary differential equations		
	`load(x0,z)` — Vector function specifying initial conditions		
	`score(x1,y)` — Vector function specifying right boundary conditions		
`search (S,Subs,m)`	`S` — String	Finds the starting position of the `Sub` substring in the specified string `S`, beginning from the position `m`	10.7
	`Sub` — Substring		
	`m` — Starting position for the search		
`sec(z)`	`z` — Argument	Secant	10.4

TABLE A.3. Alphabetic List of the Built-in Functions (*Contunued*)

Function	Arguments	Description	Reference
sech(z)	z — Argument	Hyperbolic secant	10.5
sign(x)	x — Argument	Sign of the number	10.9
signum(z)	z — Argument	Complex sign of the number $z/\|z\|$	10.2
sin(z)	z — Argument	Sine	10.4
sinh(z)	z — Argument	Hyperbolic sine	10.5
sinfit(x,y,g)	x,y — Data vectors g — Vector of initial values a,b,c	Regression by sine curve $f(x)=a\cdot\sin(x+b)+c$	14.2.3
slope(x,y)	x,y — Data vectors	Coefficient a of linear regression $b+a\cdot x$	14.2.1
sort(v)	v — Vector	Sorts vector elements	9.2.4
sph2xyz(r,θ,φ)	r,θ,φ — Spherical coordinates	Converts spherical coordinates to Cartesian coordinates	10.10
stack (A,B,C,...)	A,B,C,... — Vectors or matrices	Joins A, B, C, ... matrices top to bottom	9.2.2
Stiffb (y0,t0,t1, M,D,J)	*See* rkfixed J(t,y) — Function returning a matrix whose first column contains derivatives and whose remaining rows and columns form the Jacobian matrix for D(t,y)	Returns a matrix representing a solution of the Cauchy problem for a stiff system of ordinary differential equations using the Bulirsch-Stoer method	11.5.2
stiffb (y0,t0,t1, acc,D,J,k,s)	*See* rkadapt J(t,y) — Function returning the matrix whose first column contains derivatives and whose remaining rows and columns form the Jacobian matrix for D(t,y)	Returns a matrix representing a solution of a Cauchy problem for a stiff system of ordinary differential equations using the Bulirsch-Stoer method (for defining only the last point of the interval)	11.5.2

TABLE A.3. Alphabetic List of the Built-in Functions (*Contunued*)

Function	Arguments	Description	Reference
Stiffr (y0,t0,t1, M,D,J)	See Stiffb	Returns a matrix containing a solution of a Cauchy problem for a stiff system of ordinary differential equations using the Rosenbrock method	11.5.2
stiffr (y0,t0,t1,acc, D,J,k,s)	See stiffb	Returns a matrix containing a solution of a Cauchy problem for a stiff system of ordinary differential equations using the Rosenbrock method (for defining only the last point of the interval)	11.5.2
str2num(S)	S — String	Converts the string representation to a real number	10.7
str2vec(S)	S — String	Converts the string representation to a vector of ASCII codes	10.7
strlen(S)	S — String	Number of characters in the specified string	10.7
submatrix (A,ir,jr, ic,jc) ,	A — Matrix ir,jr — Rows ic,jc — Columns	Returns part of the matrix between rows ir,jr and columns ic,jc	9.2.2
substr(S,m,n)	S — String	Substring of the string S obtained by selecting n characters starting from position m	10.7
supsmooth(x,y)	x,y —Data vectors	Smoothing using the adaptive algorithm	14.3.1
svd(A)	A — Real matrix	Singular decomposition	9.5.4
svds(A)	A — Real matrix	Returns a vector containing singular values of A	9.5.4

TABLE A.3. Alphabetic List of the Built-in Functions (*Contunued*)

Function	Arguments	Description	Reference
tan(z)	z — Argument	Tangent	10.4
tanh(z)	z — Argument	Hyperbolic tangent	10.5
Tcheb(n,x)	x — Argument n — Order	Chebyshev polynomial of the first kind	10.6
tr(A)	A — Square matrix	Trace of the matrix	9.1.8
trunc(x)	x — Argument	Returns integer part of x	10.8
Ucheb(n,x)	x — Argument n — Order	Chebyshev polynomial of the second kind	10.6
vec2str(v)	v — Vector of ASCII codes	String representation of the elements of vector v	10.7
wave(y)	y —Data vector	Vector of direct wavelet transform	14.4.2
WRITEPRN(file)	file — String representation of the file path	Writes data to the file	15.6
xy2pol(x,y)	x,y — 2D Cartesian coordinates	Converts Cartesian coordinates to polar coordinates	10.10
xyz2cyl(x,y,z)	x,y,z — Cartesian coordinates	Converts Cartesian coordinates to cylindrical coordinates	10.10
xyz2sph(x,y,z)	x,y,z — Cartesian coordinates	Converts Cartesian coordinates to spherical coordinates	10.10
Y0(x) Y1(x) Yn(m,x)	x — Argument, x>0	Bessel function of the second kind, of 0-th, first, and m-th order	10.1.1

TABLE A.3. Alphabetic List of the Built-in Functions (*Contunued*)

Function	Arguments	Description	Reference
ys(n,x)	n — Order x — Argument	Spherical Bessel function of the second kind	10.1.5

Financial functions are not listed here. Information on these functions is provided in Chapter 10.

　　Some functions that make up the families of typical functions are listed using shorthand notation, with the missing part of the name replaced by the wildcard character *. This form is used for presenting most statistical functions describing various distributions or file output functions. More detailed information on each of the listed functions is provided in the section specified in the Reference column of the table above.

Appendix B
Error Messages

TABLE B.1. Error messages

Error	Possible Cause	Troubleshooting Recommendations
Numeric Calculations Errors		
A "Find" or "Minerr" must be preceded by a matching "Given".	This error arises when `Find` or `Minerr` functions do not have matching `Given` keywords.	Each solve block that ends with `Find` or `Minerr` must start with the `Given` keyword.
All evaluations result in either an error or a complex result.	Mathcad can't plot specific points, since there are no real values to be plotted on the graph.	This error message may appear if there is an error or if all values are complex.
Arguments in function definitions must be names.	The selected function definition contains an invalid argument list.	The argument list must contain correctly named variables, or the names in the list must be comma-delimited.

TABLE B.1. Error messages (*Continued*)

Error	Possible Cause	Troubleshooting Recommendations
At least one limit must be infinity.	When you select the infinite limit algorithm for integration, at least one of the integral limits must be infinity.	To enter the infinity character (∞) press \<Ctrl>+\<Shift>+\<Z>. To change the algorithm, right-click the integral and select the required algorithm from the right-click menu.
Can only evaluate an nth order derivative when n = 0, 1, ..., 5.	Order of derivative must be one of the following numbers: 0, 1, 2, ..., 5.	If you need to calculate the derivative of higher order, you must do it using symbolic differentiation.
Can't evaluate this function when its argument is less than or equal to zero.	This error message might relate to XY- or polar graphs having logarithmic axes at which either limits or some values aren't positive.	Neither negative numbers nor zero can be plotted on logarithmic axes. Change the type of the graph axes or create the graph for another values.
Can't converge to a solution.	Numeric method is divergent (can't find a solution).	Make sure that you aren't applying the operation to the function in close proximity to the singularity point. Try to change parameters of the numeric method (for example, the initial guess). Try to increase the TOL constant (i.e., try to find the solution with inferior precision). Try to change the numeric algorithm, if possible. To do so, right-click the error message and select another algorithm from the right-click menu.
Can't define the same variable more than once in the same expression.	You are attempting to calculate the same variable twice within the same expression.	Example: If you create a new vector a := and use the same name in the right part of this definition, you'll get this error message.

TABLE B.1. Error messages (*Continued*)

Error	Possible Cause	Troubleshooting Recommendations
Can't determine what units the result of this operation should have.	You have raised the expression containing dimension units to the power that is a vector or a variable changing within specific limits. The dimension of the result can't be defined.	If the expression includes dimension units, you can only raise it to a real fixed power.
Can't divide by zero.	The divide by zero condition has occurred somewhere within your program or the numeric method.	Try to locate the exact position where the divide by error occurs, and eliminate it. Try to change parameters of the numeric method, the precision constants, or the numeric algorithm itself.
Could not find a solution.	Numeric method diverges (can't find a solution).	*See "Can't converge to a solution"*
Can't find the data file you're trying to use.	The file that you are trying to access can't be found.	Make sure that the file exists in a specified location.
Can't have anything with units or dimensions here.	This expression uses dimension units somewhere where they arent allowed.	Dimension units arent allowed in the following locations: • In arguments of most functions • In exponential functions • In subscripts and superscripts To use the dimensional expressions, first cancel the units by dividing the expression by UnitsOf(*expression*).
Can't have more than one array in a contour plot.	You have entered more than one array for the placeholder of the contour or surface plot.	You can enter only one array to this placeholder, since graphs can plot only one surface at a time.

TABLE B.1. Error messages (*Continued*)

Error	Possible Cause	Troubleshooting Recommendations
Can't perform this operation on the entire array at once. Try using "vectorize" to perform it element by element.	This error message can appear when the user is attempting to divide one vector by another vector.	To apply a function or operator to each element of the vector or matrix, apply the vectorize operator.
Can't plot this many points.	You attempt to plot more points than the allowed number.	Try to make the number of points less than 150,000.
Can't put a ":=" inside a solve block.	Within the solve block there mustn't be any definitions. It must contain only Boolean expressions.	Use the **Boolean** toolbar.
Can't raise an expression having units to a complex power.	This expression contains dimension units, and you try to raise it to a complex power.	The expression with dimension units can only be raised to a real power. If you need to raise this expression to a complex power, first cancel the dimension units by dividing this expression by UnitsOf(*expression*).
Can't solve a system having this many equations.	Mathcad is unable to solve this system.	*See the "Solve block" definition (Chapter 8).*
Can't understand something in this data file.	The file that you are trying to access using READ or READPRN functions doesn't have the required format.	• The file must contain ASCII text • All lines in the file must have the same number of values if you are using READPRN • If the file has the required format, but if this error message persists, try to remove any text from the file

TABLE B.1. Error messages (*Continued*)

Error	Possible Cause	Troubleshooting Recommendations
Can't understand the name of this function.	This error may appear if the expression is used as a function name, and this name doesn't satisfy the function naming requirements (for example, you have used a number as a function name, 6(x).	The expression must satisfy the Mathcad requirements for function names.
Can't understand the way this range variable is defined.	Range variable definition is invalid.	When specifying the range of the range variable, use one of the following forms: • `Rvar := n1 ... n2` • `Rvar := n1, n2 ... n3`
Can't understand this number.	This expression contains a symbol or decimal point in a position where it isn't allowed.	You'll encounter this error if you accidentally write the number with two decimal points, for example, as follows: `.452.`
Can't use a range variable in a solve block.	This error will appear if you use a range variable in an inappropriate location, such as in the solve block.	Implement an algorithm that avoids using range variables within a solve block.
Cannot evaluate this accurately at one or more of the values you specified.	This error will appear if you attempt to calculate the function for the argument that falls outside the accurate range of the function.	Check the function's domain of definition.
Cross product is defined only for vectors having exactly three elements.	Cross product is defined only for vectors that have exactly three elements.	*See Section 9.1.7, "Cross Product," in Chapter 9.*

TABLE B.1. Error messages (*Continued*)

Error	Possible Cause	Troubleshooting Recommendations
Can't evaluate this expression. It may have resulted in an overflow or an infinite loop.	This expression contains too many nested functions. If the function is defined recursively (i.e., in terms of itself), it might result in an infinite loop.	Check several iterations of the loop.
Degree of the polynomial must be between 1 and 99.	Vector passed to the `polyroot` function must contain at least 2 and no more than 99 elements.	Check the number of elements of the vector passed to the `polyroot` function.
Dimensions must be >4.	This matrix must have at least 4 rows and 4 columns.	Check the number of matrix rows and columns.

End of file.	You've tried to read more data values from a datafile than there are values in that file.	For example, if the file has 10 values, and you have written the following expression:
		
		`I := 1 ... 100`
		
		`xi: = READPRN(file)`
		
		this error message will appear.

End points cannot be the same.	This error appears when differential equations are solved incorrectly.	The endpoints of the interval on which the solution will be evaluated must be different.
Equation too large.	This expression is too complicated to evaluate.	Break the equation down into two or more smaller expressions.
Floating point error.	You are evaluating a function at a point that is not allowed.	Check if the point at which you are evaluating the function is legal: for example, the fifth argument to the `relax` function can't have zeros in the non-boundary positions of the coefficient matrix.

TABLE B.1. Error messages (*Continued*)

Error	Possible Cause	Troubleshooting Recommendations
Found a singularity while evaluating this expression. You may be dividing by zero.	You evaluated a function or performed an operation on an illegal value.	For example, this error will appear if you attempt to divide by zero or if you try to invert a singular matrix. If this happens, try to locate and eliminate the error.
Found a number with a magnitude greater than 10^307.	Because of the inherent limitations, numbers that are too large can't be represented. Sometimes, especially in complex calculations, intermediate results might be too large to be represented. If this happens, you'll encounter this error message.	Try to change the parameters of the numeric algorithm or the algorithm itself.
Illegal context. Press <F1> for help.	Most often, this error message is caused by syntax errors.	Check the syntax and order of formulae in your document.
Illegal dimensions.	This matrix does not have the rows or column you're referring to.	Type the matrix name and press the <=> key to verify the number of rows and columns the matrix contains.
Integer too large/ Integer too small.	This value is too large/small to work with.	If you are working with built-in functions, click the function name and then view on-line help information by pressing <F1>.
Invalid format.	Arguments of this function may be invalid.	If you are working with built-in functions, click the function name and then view on-line help information by pressing <F1>.
Live symbolics not available.	Live symbolic evaluation is not available in this particular version of this software.	*See Chapter 5.* To overcome the problem, consider updating your Mathcad version.

TABLE B.1. Error messages (*Continued*)

Error	Possible Cause	Troubleshooting Recommendations
Must be <= 10,000.	This value must be less than or equal to 10,000.	If you are working with built-in functions, click the function name and then view on-line help information by pressing <F1>.
Must be >= 10^–16.	This value must be greater than or equal to 10^{-16}.	If you are working with built-in functions, click the function name and then view on-line help information by pressing <F1>.
Must be string.	The function or operator requires a string.	Check the argument passed to the function or operator and make sure that it is a string. *See Chapter 10*
Must be between two locked regions.	The **Area | Lock** command from the **Format** menu is available only if the cursor is positioned within a locked area.	In order to use the **Area | Lock** command from the **Format** menu, you must click in an area of the worksheet that is in a locked area. *See Chapter 16*
Must be function.	This argument must be a function.	If you are working with built-in functions, click the function name and then view on-line help information by pressing <F1>.
Must be increasing.	The values in this vector must be in ascending order.	Type the name of the vector and press '=' to verify what values are in the vector.
Must be less than the number of data points.	This argument must be smaller than the number of data points you have.	If you are working with built-in functions, click the function name and then view on-line help information by pressing <F1>.
Must be positive.	This function can't be calculated when its values are less than or equal to zero.	This error message might relate to XY plots or polar plots that have a logarithmic axis in which either the limits, or some of the values are not positive. Negative numbers and zero cannot be placed anywhere on a logarithmic axis.

TABLE B.1. Error messages (*Continued*)

Error	Possible Cause	Troubleshooting Recommendations
Must be real.	This value must be real. Its imaginary part must be equal to zero.	Examples are subscripts and super-scripts, differential equation solvers, `mod`, and `angle`.
Must be real scalar.	This value cannot be complex or imaginary.	If you are working with built-in functions, click the function name and then view on-line help information by pressing <F1>.
Must be real vector.	This vector can't have complex or imaginary elements. Also, notice that it must be a column vector rather than a row vector.	If you are working with built-in functions, click the function name and then view on-line help information by pressing <F1>.
Must be square.	This error marks a non-square matrix in an operation or function that requires a square matrix.	For example, the matrix must be square when performing operations such as calculating determinants, obtaining an inverse matrix, raising to a power, or applying `eigenvals` and `eigenvec` functions.
No solution found.	Mathcad could not find a solution.	If you are working with built-in functions, click the function name and then view on-line help information by pressing <F1> to make sure that the function is used correctly. Notice, however, that the solution might be simply non-existent. *See also "Can't converge to a solution."*
Not enough memory for this operation.	There is not enough memory to complete this computation.	Try to free up some memory by making the array or matrix smaller (Mathcad uses approximately 8 bytes of memory per matrix element) or by deleting any large bitmaps, arrays, or matrices.

TABLE B.1. Error messages (*Continued*)

Error	Possible Cause	Troubleshooting Recommendations
Singular matrix.	This matrix must be neither singular nor nearly singular.	A matrix is singular if its determinant is equal to zero. A matrix is nearly singular if it has a high condition number (*see Chapter 9*).
The number of rows and/or columns in these arrays do not match.	This message flags a function or operator that requires two matrices having a certain number of rows and columns.	For example, addition of two matrices of different sizes is illegal. Multiplication of matrices requires that the number of columns of the first matrix match the number of rows of the second matrix (*see Section 9.1, "The Simplest Matrix Operations," in Chapter 9*).
The units in this expression do not match.	This message will appear if you are adding two elements of different dimensions, if you have created a matrix whose elements have different dimensions, or if you are trying to solve a system of equations with the unknowns that have different dimensions.	Check the usage of dimensional variables.
There is an extra comma in this expression.	An extra comma is present.	Commas can be used to separate the following: • Arguments of the function • The first two elements of a range when defining a range variable • Expressions in a plot • Elements of the input table • Subscripts of the matrix Any other use of comma results in this error message. For example, if you need to write the number "four thousand", then the notation 4,000 is incorrect. Simply type 4000 to enter this number.

TABLE B.1. Error messages (*Continued*)

Error	Possible Cause	Troubleshooting Recommendations
This expression is incomplete. You must fill in the placeholders.	You have not completed the expression by filling in all indicated placeholders.	Enter the required numbers or expressions into the indicated placeholders.
This expression is incomplete. You must provide an operator.	You have not filled in the operator placeholders or there is a blank space between two operands.	Check the correctness of the expression. You might have accidentally deleted an operator.
This function has too many arguments.	The selected expression contains the function with an invalid number of arguments.	Check the correctness of the function usage.
This function is undefined at one or more of the points you specified.	You have attempted to calculate a function or operator at an inappropriate value.	For example, the expression $-3!$ will produce this error, since the factorial is undefined for negative values, $\ln(0)$ will result in this error message since logarithmic function is undefined at 0.
This function needs more arguments.	The selected expression contains a function that requires more arguments than you have specified.	For built-in functions, click the function name and press <F1> to check the correctness of the number and type of arguments. For custom functions, check the function definition.
This operation can only be performed on a function.	This argument must be a function.	For built-in functions, click the function name and press <F1> to view additional information.
This operation can only be performed on an array. It can't be performed on a number.	You have attempted to perform an operation that requires an array over a scalar value.	For example, this error will appear if the superscripted variable is defined as scalar value. Since a superscripted variable represents a column of a matrix, you must define it as a vector.

TABLE B.1. Error messages (*Continued*)

Error	Possible Cause	Troubleshooting Recommendations
		For surface plots or contour plots, the array plotted must have at least two rows or two columns.
This operation can only be performed on a number or an array.	The function or operator you are using requires a single constant, a matrix, or a vector.	Check the argument passed to the function or operator that you are using and make sure that it is either a constant or an array (matrix or vector)
This operation can only be performed on a string.	The function or operator you are using requires a string argument. For example, string functions usually require at least one string argument.	Make sure that the argument that you pass to the function or operator is a string.
This subscript is too large.	You tried to use a subscript or superscript that exceeds the limit.	The limit for subscript or superscript values is 8 million. Make sure that the values that you are using don't exceed this limit.
This value must be a matrix.	You are trying to perform a matrix operation on something other than a matrix.	Make sure that the object on which you are trying to perform this operation is a matrix (an array comprising more than one column).
This value must be a vector. It can be neither a matrix nor a scalar.	This error marks a matrix or scalar in an operation that requires a vector (a one-column array).	Make sure that the value on which you are trying to perform the operation is of the vector type. For example, the vector sum operator is only applicable to vectors.
This value must be an integer greater than 1.	This value must be an integer ≥ 1.	If you are working with built-in functions, click the function name and then view on-line help information by pressing <F1>.

TABLE B.1. Error messages (*Continued*)

Error	Possible Cause	Troubleshooting Recommendations
This variable or function is not defined above.	The name of the undefined function will be highlighted by red.	Make sure that this function or variable is defined earlier in your document. This message will appear if the variable is incorrectly used in its global definition. This error often is the evidence or the fact that another equation earlier in the document is erroneous. If this is the case, all expressions using the erroneous expression will be marked with red.
Underflow.	Because of limitations inherent in representing numbers in a computer, numbers that are too small cannot be represented. When an expression involves such a number, you'll see this error message. Sometimes, especially in a complicated calculation, an intermediate result will be too small to represent.	This error is rather hard to diagnose and eliminate. Sometimes, especially in a complicated calculation, an intermediate result will be too small to represent. Try to simplify your calculation by dividing it into several simpler tasks.
Value of subscript or superscript is too big (or too small) for this array.	This expression uses a subscript or superscript that refers to a nonexistent array element.	The subscript or superscript should not be smaller than the ORIGIN. Type "ORIGIN=" to see what the ORIGIN is set to.
This is not a scalar. Press <F1> for help.	You have used the vector or expression with intervals or another type of expression where the scalar must be used.	Check that the value that you are using is of the scalar type.

TABLE B.1. Error messages (*Continued*)

Error	Possible Cause	Troubleshooting Recommendations
You have one solve block inside another. Every "Given" must have a matching "Find" or "Minerr".	You have specified two `Given` keywords, one following another without `Find` or `Minerr` between them. The solve block can't contain nested solve blocks.	As an alternative, you can define a function in terms of one solve block and use it within another solve block. In most cases this produces the same effect.
You interrupted calculation. To resume, click here and choose "Calculate" from the "Math" menu.	The user has interrupted the calculation process by pressing <Esc>.	To recalculate the selected equation, point at it with the cursor and select the **Math | Calculate** menu commands.

Error Messages in Symbolic Calculations

Error	Possible Cause	Troubleshooting Recommendations
Argument too large (Integer too large in context, Object too large).	Usually the result of evaluating an expression with a floating point value larger than about 10 billion.	Make sure that the limit of 10 billion isn't exceeded.
Discarding a large result.	The answer is too large to display as a formatted math region.	You can place the answer on the clipboard.
Expecting the array or list.	Operators in the expression you are simplifying or evaluating require vector or matrix operands.	Make sure that the operands that you are using are of the correct type.
Expression contains non-symbolic operators.	You're applying a symbolic operation to an expression containing an operator or variable placeholders.	Fill the placeholders before applying a symbolic operation to the expression.

TABLE B.1. Error messages (*Continued*)

Error	Possible Cause	Troubleshooting Recommendations
Floats not handled.	The `factor` command has been applied to an expression containing decimals.	The `factor` keyword is used to transform an expression into a product. It is applicable to polynomials, integers, and sums of rational expressions. It isn't valid for numbers with decimal points, so make sure that the expression to which you apply the command doesn't contain floating-point numbers.
Illegal function syntax.	This message appears when you make syntax errors when entering functions.	Check the syntax and make sure that there are not any syntax errors, such as `(f)(x)`
Invalid arguments.	The symbolic processor can't carry out the requested operation for the given arguments.	This message will appear, for example, if you apply a scalar function to the array without using the `vectorize` operator and then select the **Symbolics | Simplify** commands from the menu.
Invalid range.	In looking for a numerical solution to an equation, the symbolic processor tried to evaluate one of its built-in functions at a point outside its domain.	Pay special attention to the domain of the definition of the built-in functions.
No answer found; stack limit reached.	The symbolic processor reached the limit of its internal expression stack without completing the evaluation or simplification you requested.	Make sure that you have correctly formulated the problem and are using the correct method of solving it.

TABLE B.1. Error messages (*Continued*)

Error	Possible Cause	Troubleshooting Recommendations
No answer found.	The symbolic processor could not find an exact solution for the equation you were solving.	Make sure that an exact solution of the equation actually exists. If you are sure that it exists, check to make sure you are using a suitable method of solving the task.
No closed form found for …	The symbolic processor couldn't find an indefinite integral for the expression you were integrating or a closed form for the sum or product you were evaluating.	If you see this error message, check to make sure that the solution you are looking for actually exists.
Syntax error.	Usually the result of applying a symbolic operation to an inappropriate or incorrectly typed expression. Symbolic processing of expressions with units will also generate this error.	Check the syntax.

Index

Changing origin 118

Prime 300
Ctrl + F7

Prine 〔⁷〕

ScJscript Dot:"